Clinical and Biochemical Luminescence

CLINICAL AND BIOCHEMICAL ANALYSIS

A series of monographs and textbooks

EDITORS

1. Colorimetric and Fluorimetric Analysis of Organic Compounds and Drugs, *M. Pesez and J. Bartos*
2. Normal Values in Clinical Chemistry: Statistical Analysis of Laboratory Data, *Horace F. Martin, Benjamin J. Gudzinowicz, and Herbert Fanger*
3. Continuous Flow Analysis: Theory and Practice, *William B. Furman*
4. Handbook of Enzymatic Methods of Analysis, *George G. Guilbault*
5. Handbook of Radioimmunoassay, *edited by Guy E. Abraham*
6. The Hemoglobinopathies, *Titus H. J. Huisman and J. H. P. Jonxis*
7. Automated Immunoanalysis (in two parts), *edited by Robert F. Ritchie*
8. Computers in the Clinical Laboratory: An Introduction, *E. Clifford Toren, Jr. and Arthur A. Eggert*
9. The Chromatography of Hemoglobin, *Walter A. Schroeder and Titus H. J. Huisman*
10. Nonisotopic Alternatives to Radioimmunoassay: Principles and Applications, *edited by Lawrence A. Kaplan and Amadeo J. Pesce*
11. Biochemical Markers for Cancer, *edited by T. Ming Chu*
12. Clinical and Biochemical Luminescence, *edited by Larry J. Kricka and Timothy J. N. Carter*

ADDITIONAL VOLUMES IN PREPARATION

Clinical and Biochemical Luminescence

EDITED BY

Larry J. Kricka
University of Birmingham
Birmingham, England

Timothy J. N. Carter
Battelle
Centre de Recherche de Genève
Carouge, Genève, Switzerland

MARCEL DEKKER, INC. New York and Basel

This book is dedicated to
J.R.C. and T.D.M.K.

Library of Congress Cataloging in Publication Data
Main entry under title:

Clinical and biochemical luminescence.

(Clinical and biochemical analysis ; 12)
Includes bibliographies and indexes.
1. Fluorimetry. 2. Phosphorimetry.
3. Chemistry, Clinical--Technique. 4. Biological chemistry--Technique. I. Kricka, Larry J., [date]. II. Carter, Timothy J. N., [date].
III. Series. [DNLM: 1. Luminescence--Laboratory manuals. 2. Luminescent proteins--Laboratory manuals. W1 CL654 v.12 / QY 25 C641]
QP519.9.F58C58 1982 574.19'125 82-8976
ISBN 0-8247-1857-7 AACR2

Marcel Dekker, Inc.
270 Madison Avenue, New York, New York 10016

Current printing (last digit):
10 9 8 7 6 5 4 3 2 1

Printed in the United States of America

FOREWORD

The phenomenon of chemiluminescence has been known from earliest times, and one can find examples of its manifestations in both the animal and vegetable kingdoms (e.g., glowworms, St. Elmo's Fire, and will-o'-the-wisp). The first clearly recorded example of luminescence in minerals was not noted until 1602. This was the so-called Bolognian Stone prepared by the Italian cobbler Casciarolo.

Although Robert Boyle (the first great British analytical chemist, and the first to use the term "chemical analysis") examined various materials that produced luminescence, as far as is known it did not occur to him that chemiluminescence might have possibilities as an analytical technique.

The first systematic studies on the application of chemiluminescence for general analytical purposes were initiated some thirty years ago by the late Professor L. Erdey at the Technical University, Budapest, although one can find occasional applications throughout the preceding four or five decades. Professor Erdey and his co-worker, Mrs. I. Buzás, examined the substances luminol, lophine, and lucigenin as indicators in acid-base, redox, compleximetric, and precipitation systems. The experimental conditions they so carefully established should be of great value in developing new methods which use the more modern techniques of luminometric measurement.

Biochemists have more readily utilized these processes than analytical chemists, and there are many methods based on bioluminescent processes for analytes of clinical interest. In the field of general chemical analysis attention has been focused on reactions in the gas phase and on their use in commercial instruments used for environmental analysis. However, during the last few years, numerous papers have been published which discuss the utilization of chemiluminescence in solution for a variety of determinations. These reactions are selective and sensitive and seem to have great possibilities although there is clearly room for improvement.

The appearance of this book is timely. The editors, Dr. Larry J. Kricka and Dr. Timothy J. N. Carter, have extensive experience with chemiluminescent (and bioluminescent) processes, and they have been fortunate in obtaining contributors highly experienced in various applications. The book will help draw the attention of analytical chemists and biochemists to the extensive possible applications of this fascinating technique in analysis.

Ronald Belcher†
Department of Clinical Chemistry
University of Birmingham
Birmingham, England

†Dr. Belcher is deceased.

PREFACE

Considerable interest has been shown in the application of bio- and chemiluminescent analytical techniques in the clinical laboratory. In spite of such interest, witnessed by the explosive increase in the number of papers, articles, and reviews of luminescence appearing in the scientific press in the last five years, the actual number of routine applications remains very small.

It is our opinion that this situation is the result of two major problems which must be surmounted by any potential user before entering the field. First, there has been, until quite recently, a dearth of satisfactory equipment for performing luminescent analyses, and thus in many cases it has been necessary for experimenters to construct their own instrumentation. This is a course of action not commonly contemplated in the routine laboratory. Second, and we feel more important, there has been an insufficient understanding, by potential users, of the basic properties of the available luminescent systems, which has prevented their potential from being realized. Coupled with this has been the poor quality of commerical reagents, a situation which is only now being remedied.

In this volume, therefore, we have attempted to answer these two interrelated problems. The initial chapters stress the most important facets of the basic chemical properties of the luminescent systems which have applications in the clinical laboratory. Subsequent chapters deal with the practical applications of luminescence. Finally, there is a review of current instrumentation that compares and contrasts the features of commercially available luminometers.

Larry J. Kricka
Timothy J. N. Carter

CONTRIBUTORS

Anthony K. Campbell Department of Medical Biochemistry, Welsh National School of Medicine, Heath Park, Cardiff, Wales

Timothy J. N. Carter* Department of Clinical Chemistry, Wolfson Research Laboratories, Queen Elizabeth Medical Centre, Edgbaston, Birmingham, England

Marlene DeLuca Department of Chemistry, University of California at San Diego, La Jolla, California

Maurice B. Hallett Department of Medical Biochemistry, Welsh National School of Medicine, Heath Park, Cardiff, Wales

Michael J. Harber Department of Renal Medicine, K.R.U.F. Institute, Royal Infirmary, Welsh National School of Medicine, Cardiff, Wales

Edward G. Jablonski† Department of Chemistry, University of California at San Diego, La Jolla, California

Larry J. Kricka‡ Department of Clinical Chemistry, University of Birmingham, Birmingham, England

---

*Dr. Carter is currently with the Biomedical Group, Battelle, Centre de Recherche de Genève, Carouge, Genève, Switzerland.

†Dr. Jablonski is currently with the Department of Biology, The Johns Hopkins University, McCollum-Pratt Institute, Baltimore, Maryland.

‡Dr. Kricka is currently with the Department of Chemistry, University of California at San Diego, La Jolla, California, on a leave of absence from the University of Birmingham.

Peter H. Lloyd Coulter Electronics Ltd., Luton, Bedfordshire, England

Arne Lundin* Department of Biochemistry, Arrhenius Laboratory, University of Stockholm, Stockholm, Sweden

Philip E. Stanley† Department of Clinical Pharmacology, The Queen Elizabeth Hospital, Woodville, South Australia

Gary H. G. Thorpe Department of Clinical Chemistry, Wolfson Research Laboratories, Queen Elizabeth Medical Centre, Edgbaston, Birmingham, England

---

*Dr. Lundin is currently with the Bioluminescence Centre, LKB-Produkter AB, Bromma, Sweden.

†Dr. Stanley is currently a consultant scientist based in Cambridge, England.

CONTENTS

Foreword *Ronald Belcher* iii
Preface v
Contributors vii

1. INTRODUCTION 1

Timothy J. N. Carter and Larry J. Kricka

I. Luminescence: Terminology and Definitions 1
II. Historic Aspects 2
III. Role of Luminescence in the Clinical Laboratory 6
References 8

2. QUANTUM MECHANICAL ASPECTS OF CHEMILUMINESCENCE 11

Peter H. Lloyd

I. Introduction 11
II. Energetics of Luminescent Reactions 12
III. Summary 19
References 20

3. CHEMILUMINESCENCE 21

Gary H. G. Thorpe, Larry J. Kricka, and Timothy J. N. Carter

I. Introduction 21
II. Liquid Phase Chemiluminescence 23
III. Gas Phase Chemiluminescence 34

IV. Solid Phase Chemiluminescence 36
V. Conclusion 37
References 37

4. ANALYTICAL APPLICATIONS OF BIOLUMINESCENCE: THE FIREFLY SYSTEM 43

Arne Lundin

I. Introduction 44
II. Recent Improvements of the Kinetics of the Firefly Reaction 44
III. Technical Considerations 49
IV. Applications 54
V. Conclusion 66
References 66

5. ANALYTICAL APPLICATIONS OF BIOLUMINESCENCE: MARINE BACTERIAL SYSTEM 75

Edward G. Jablonski and Marlene DeLuca

I. Introduction and General Properties 76
II. Materials 78
III. Assays 80
IV. Conclusion 84
References 84

6. APPLICATIONS OF COELENTERATE LUMINESCENT PROTEINS 89

Maurice B. Hallett and Anthony K. Campbell

I. Coelenterate Luminescence 90
II. Possible Applications of Coelenterate Luminescence 97
III. Preparation of Ca^{2+}-Activated Photoproteins 100
IV. Introduction of Ca^{2+}-Activated Photoproteins into Cells 103
V. Applications of Membrane Vesicles 111
VI. Quantification of Cytoplasmic Photoprotein 113
VII. Relationship between Free Ca^{2+} and Light Emission 114
VIII. Quantification of Intracellular Ca^{2+} 116
IX. Ca^{2+} Distribution within the Cell; Image Intensification 119
X. Problems Associated with the Use of Photoproteins 121
XI. Conclusion 123
Note Added in Proof 124
References 124

7. ANALYTICAL APPLICATIONS OF CHEMILUMINESCENCE 135

Timothy J. N. Carter and Larry J. Kricka

I. Introduction 135
II. Activation and Inhibition Assays 136
III. Enzymic Production of Hydrogen Peroxide Monitored by Chemiluminescence 140
IV. Peroxidase-Mediated Chemiluminescence 143
V. Conclusion 148
References 148

8. LUMINESCENT IMMUNOASSAYS 153

Larry J. Kricka and Timothy J. N. Carter

I. Introduction 154
II. Classification of Luminescent Immunoassays 156
III. Advantages and Disadvantages of Luminescent Immunoassays 157
IV. Luminescent Immunoassay 159
V. Luminescent Enzyme Immunoassay 167
VI. Luminescent Enzyme Multiplied Immunoassay Technique 169
VII. Luminescent Cofactor Immunoassay 170
VIII. Conclusion 174
References 174

9. IMMOBILIZED LUMINESCENCE REAGENTS 179

Larry J. Kricka and Timothy J. N. Carter

I. Introduction 180
II. Immobilization Techniques 180
III. Solid Supports 182
IV. Choice of Immobilization Method 182
V. Immobilized Bacterial Luciferase 183
VI. Immobilized Firefly Luciferase 183
VII. Luciferase Coimmobilized with Other Enzymes 186
VIII. Conclusion 186
References 186

10. APPLICATIONS OF LUMINESCENCE IN MEDICAL MICROBIOLOGY AND HEMATOLOGY 189

Michael J. Harber

I. Introduction 190
II. Measurement of Bacterial Adherence 192
III. Screening Test for Bacteriuria 195
IV. Antimicrobial Susceptibility Testing 197

V. Antiobiotic Assay 199
VI. Vitamin Assay 203
VII. Luminescence Analyses of Blood Cells 205
VIII. ATP in Blood Plasma 209
IX. Discussion 210
References 211

11. INSTRUMENTATION 219

Philip E. Stanley

I. Introduction 220
II. Reaction Vessel 221
III. Detector 229
IV. Detector Chamber 230
V. Temperature Control 232
VI. Injection 233
VII. Signal 234
VIII. Sensitivity 238
IX. Flow Monitoring 239
X. Microprocessor Control, Automation, and Data Processing 239
Note Added In Proof 241
References 259

Acronyms 261
Author Index 263
Subject Index 275

1

INTRODUCTION

TIMOTHY J. N. CARTER* Wolfson Research Laboratories, Queen Elizabeth Medical Centre, Edgbaston, Birmingham, England

LARRY J. KRICKA† University of Birmingham, Birmingham, England

I. LUMINESCENCE: TERMINOLOGY AND DEFINITIONS

Luminescence is a generic term covering a range of processes which produce light. A classification of the various types of luminescence based on the source of excitation energy is shown in Table 1 (Peng, 1976). This book is concerned exclusively with the applications of

---

*Dr. Carter is currently with Battelle, Centre de Recherche de Genève, Carouge, Genève, Switzerland.

†Dr. Kricka is currently with the University of California at San Diego, La Jolla, California, on a leave of absence from the University of Birmingham.

Table 1 Classification of Luminescence

Type	Origin of excitation energy
Chemiluminescence	Chemical
Bioluminescence	
Electrochemiluminescence	
Photoluminescence	Radiation
Fluorescence	
Phosphorescence	
Sonoluminescence	Thermal
Triboluminescence	
Thermoluminescence	Radiation and thermal

chemiluminescence, and its subdivision bioluminescence, in the clinical laboratory.

Chemiluminescence may be defined as the chemical production of light. It is often confused with fluorescence.The difference between these two processes is the source of the energy producing molecules in an excited state. Chemiluminescence uses the energy of a chemical reaction which promotes product molecules to an excited state. Subsequent decay back to the ground state is accompanied by the emission of light (luminescence). In contrast, incident radiation is the source of the energy which in fluorescence promotes molecules to an excited state.

Bioluminescence is the name given to a special form of chemiluminescence found in biological systems, in which a catalytic protein increases the efficiency of the luminescent reaction. In certain cases the reaction is impossible without the intercession of a protein component. The detailed mechanism of chemiluminescent processes is discussed in detail in Chap. 2.

II. HISTORIC ASPECTS

Luminescence of various types has fascinated humanity for much longer than records show. While it is difficult to identify a definitive first reference in ancient writings because of the use of poetic language, luminescence, especially bioluminescence, was certainly an established fact in the Greek and Roman philosophies (Table 2); Aristotle (384-322 B.C.) described the luminescence of dead fish and fungi in his *De Anima*. References to the phenomenon of luminescence continued to appear intermittently in the literature of the Dark and Middle Ages, but Harvey (1957), in his review, stated categorically that "the end of the Middle Ages left luminescent information in the same condition as during classical times."

With the renaissance of science following the Middle Ages, the study of luminescence became more systematic and in 1555 the first book devoted wholly to the subject was published (Gesner, 1555). During the next 100 years major advances were made in recognizing the categories of physical luminescence, while the field of bioluminescence was ignored. In 1668 Robert Boyle established some of the basic properties of bacterial and fungal systems. In such luminescent systems,

1. Light is generated without perceptible heat.
2. There is susceptibility to inhibition by chemical agents.
3. There is dependence upon air (oxygen).
4. Only a very small pressure of air is sufficient for maximal brightness.
5. The extinction of the light by evacuation of air is reversible.

By the middle of the nineteenth century it had been proved that both bacteria and fungi were the cause of commonly seen luminescent phenomena, and many new species with this property were found in the sea. The physiology, biochemistry, and biophysics of light production continued to make great strides during the nineteenth century, keeping pace with the advancing knowledge of other functional activities of animals. Thus, it was Macaire (1821) who first suggested that some organic compound might be the source of light in the glowworm instead of a phosphorus derivative as had been supposed for many years.

Using this idea, Dubois (1885,1887) at last applied biochemical method to the investigation of bioluminescence by preparing two crude extracts, for both firefly and clam, which when mixed together gave a light-emitting reaction. The extracts, one in hot and the other in cold water, contained, respectively, a heat-stable component, to which he gave the name *luciferine*, and a heat-labile component, to which he gave the name *luciferase*. Luminescence resulted only when luciferine and luciferase of the same species were mixed. He was also able to determine that luciferase was an enzyme system and luciferine its substrate. So great was the impact of this work that these names are still in use today.

After the pioneering work of Dubois, interest in luminescent phenomena increased. Developments came so fast that they became difficult to chronicle. There were, however, important landmarks in

Table 2 Historical Progress in Bioluminescence and Chemiluminescence

350 B.C.	Bioluminescence first described
1555 A.D.	First book wholly devoted to the subject
1668	Robert Boyle establishes basic properties of fungal and bacterial bioluminescence
1821	Realization that organic compounds are involved in bioluminescence
1877	Chemiluminescence of lophine discovered
1885	Description of luciferin and luciferase
1887	Chemiluminescence of pyrogallol described
1889	First use of bioluminescence to detect oxygen
1928	Luminol chemiluminescence described
1935	Lucigenin chemiluminescence described
1947	Substrate and cofactors for firefly bioluminescence described
1948	Firefly bioluminescence used to measure ATP
1953	In vitro bioluminescent bacterial system described

Source: Harvey (1957).

the development of the subject, e.g., the use of luminescent bacteria to detect small amounts of oxygen (Beijerinck, 1889), the extension of the luciferin-luciferase concept to many other bioluminescence systems (Harvey, 1940), the part played by adenosine triphosphate (ATP) and magnesium ions in firefly luminescence (McElroy, 1947), and the isolation of an in vitro bacterial luminescent system (Strehler, 1953).

In contrast to bioluminescence, the history of chemiluminescence in the aqueous phase is recent. Lophine (Radziszewski, 1877) and pyrogallol (Eder, 1887) provided the first examples of chemiluminescence observed in solution (Fig. 1). Soon afterward Dubois (1901), who was interested in all forms of luminescence, discovered the luminescence of aesculin, a glycoside from horse chestnut bark. Further work resulted in the discovery of luminescence of uric acid and asparagine (Guinchant, 1905), and of substances in humus (Weitlaner, 1911) and urine (McDermotte, 1913).

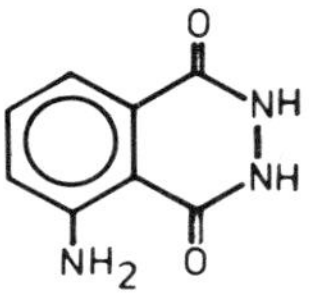

Luminol

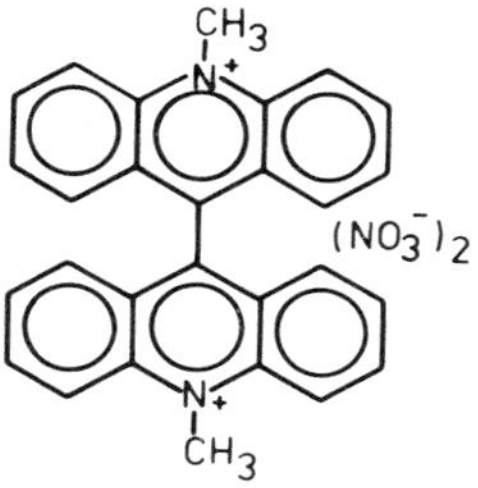

Dimethylbisacridinium Nitrate (Lucigenin)

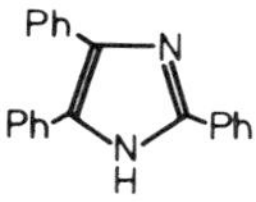

Lophine

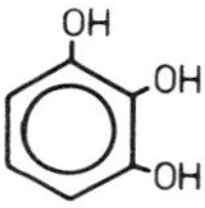

Pyrogallol

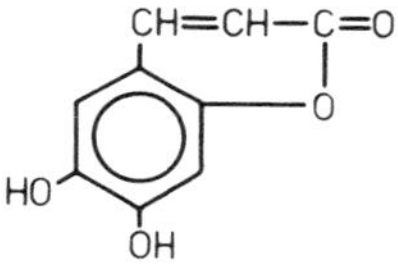

Aesculin

Figure 1 Structure of various chemiluminescent compounds.

Two of the most active luminescent chemicals discovered to date are luminol (Albrecht, 1928) and lucigenin (Gleu and Petsch, 1935). In contrast to these fortuitous discoveries, however, a more systematic approach to the luminescence of organic compounds was recently developed (White and Brundrett, 1973).

For a full review and comprehensive history of luminescence, the unique book by E. N. Harvey (1957) is an established reference work.

III. ROLE OF LUMINESCENCE IN THE CLINICAL LABORATORY

Nowadays the clinical laboratory worker is concerned with the detection and quantitation of a wide range of substances. Many analytical techniques, based on several different physicochemical principles, are utilized. A feature common to all such techniques is an interface between chemistry and physics.

In clinical chemistry, the most common way of achieving such an interface is absorptiometry at both visible and ultraviolet wavelengths. An interface that has so far been relatively unexplored in the clinical laboratory is luminescence.

Analyses based on the measurement of emitted light offer a number of advantages over conventional techniques, namely, high sensitivity, wide linear range, low cost per test, and relatively simple and inexpensive equipment.

A. Sensitivity

The extreme sensitivity of luminescent analyses is the most important advantage of luminescence over other analytical techniques, e.g., the bioluminescent assays for glucose and alcohol have been reported to be 55 and 10 times more sensitive, respectively, than conventional assays (Haggerty et al., 1978). The minimal detectable concentration for an assay ultimately depends on how sensitively light can be detected and on the quantum efficiency (ϕ) of the reaction.

$$\text{Quantum efficiency } (\phi) = \frac{\text{number (or rate) of molecules luminescing}}{\text{number (or rate) of molecules reacting}}$$

In general, bioluminescent reactions are much more efficient than chemiluminescent reactions, typical quantum efficiencies being in the ranges 0.1 to 0.8 and 0.01 to 0.05, respectively. Thus the detection limit for the chemiluminescent assay of hydrogen peroxide is 10^{-9} mol/L (Seitz and Neary, 1974; Bostick and Hercules, 1975; Williams et al., 1976) whereas the detection limits for the more efficient bioluminescent assay of ATP and nicotinamide adenine dinucleotide (NADH) are 10^{-15} (Stanley, 1974) and 10^{-16} (Stanley, 1974) mol/L, respectively.

B. Linear Response

Chemiluminescence is an emission process and hence response is usually linearly proportional to concentration from the minimal detectable concentration up to the point where it is no longer possible to maintain an excess of other reactants relative to the analyte. Thus in the case of ATP assay by the bioluminescent firefly reaction, response is linear over 6 orders of magnitude (Seitz and Neary, 1977).

C. Speed of Analysis

The rapidity with which luminescent assays may be performed depends upon the type of luminescent reaction. In some instances a rapid flash lasting less than a second is obtained, the peak height of which may be related to analyte concentration, while in other cases a more protracted glow occurs with a time course lasting several minutes. In the latter case the integral or partial integral of the light-time curve may be used as a measure of analyte concentration since this is much less sensitive to mixing efficiency. This, however, reduces the speed of analysis.

D. Specificity

Most bioluminescent reactions are specific because they are enzymic processes, but chemiluminescent reactions are nonspecific. For example, luminol will undergo a chemiluminescent reaction with a number of different oxidants, e.g., oxygen, peroxide, superoxide, and iodine. Its reactions are also subject to interference by reducing agents such as uric acid. Specificity may be conferred on luminescent reactions by using them as indicator reactions coupled to intermediates, e.g., peroxide, ATP, NADH, produced enzymatically.

E. Cost

Assays that use sensitive luminescent indicators require much less reagent than conventional assays and are therefore cheaper. Cholesterol oxidase used in the assay of cholesterol is a case in point: the chemiluminescent detection of peroxide is extremely sensitive and the amount of expensive oxidase is a fraction of that normally required.

F. Stability and Toxicity

Chemiluminescent reagents such as luminol, isoluminol, and lucigenin are stable. However, the bioluminescent reagents bacterial luciferase and firefly luciferase suffer from the stability problems encountered with many other enzymes.

To date, no toxic effects have been described for either chemi- or bioluminescent reagents.

It is expected that luminescence will make an impact in two main areas of clinical analysis. First, it may replace conventional spectrophotometric or colorimetric indicator reactions in estimations of the substrates of dehydrogenases and oxidases, where its sensitivity may be used either to quantitate substrates not easily measured by conventional techniques, e.g., prostaglandins and vitamins, or to reduce the quantities of specimen and reagent in the initial enzymic step and therefore the cost of the assay. The second major application of luminescent molecules is to replace radioactive labels in immunoassay. This topic is covered in greater detail in Chap. 9. One of the important features of luminescence as an analytical technique is its applicability to branches of pathology other than clinical chemistry, namely, hematology, immunology, bacteriology, and pharmacology. These applications are discussed in Chaps. 4 to 10.

REFERENCES

Albrecht, H. O. (1928). Uber die Chemiluminescenz des Aminophthalsaurehydrazids. *Z. Phys. Chem. 136*:321-330.

Aristotle (384-322 B.C.). *De Anima, De Sensa, Historia Animalium, Metereologia*. Works, Oxford Edition (1908,1923).

Beijerinck, M. W. (1889). Les bacteries lumineuse dans leur rapports avec l'oxygene. *Arch. Neerl. Sci. 23*:416-427.

Bostick, D. T., and D. M. Hercules (1975). Quantitative determination of blood glucose using enzyme induced chemiluminescence of luminol. *Anal. Chem. 47*:447-452.

Dubois, R. (1885). Note sur la physiologie des pyrophores. *Comp. Rend. Soc. Biol. (Ser. 8) 2*:559-562.

Dubois, R. (1887). Fonction photogenique chez le *Pholas dactylus*. *Comp. Rend. Soc. Biol. (Ser. 8) 3*:564-565.

Dubois, R. (1901). Luminescence obtenue par certains procedes organiques. *Comp. Rend. Ac. Sci. 132*:431.

Eder, J. M. (1887). Phosphorescenzerscheinungen beim Hervorrufen von Gelatineplatten. *Photog. Mitth. 24*:74-82.

Gesner, C. (1555). De raris et admirandis herbis quae sive quod noctu luceant, sive alias ad causas, lunariae nominantur et obiter de alias etiam rebus, quae in tenebris lucent, commentariolus. Tiguri, 1555. Editione hac secunda emendatior, Hafniae.

Gleu, K., and W. Petsch (1935). Die Chemiluminescenz der Dimethyldiaridyliumsalze. *Angew. Chemie 48*:57-59.

Guinchant, J. (1905). Sur la triboluminescence de l'acide arsenieux. *Comp. Rend. Ac. Sci. 140*:1170-1175.

Haggerty, C., E. Jablonski, L. Stav, and M. DeLuca (1978). Continuous monitoring of reactions that produce NADH and NADPH using immobilized luciferase and oxidoreductases from *Beneckea harveyi*. *Anal. Biochem. 88*:162-173.

Harvey, E. N. (1940). *Living Light*. Princeton University Press, Princeton, N.J., pp. 242-248.

Harvey, E. N. (1957). *A History of Luminescence from the Earliest Times Until 1900*. American Philosophical Society, Independence Square, Philadelphia, Pa.

Macaire, J. (1821). Memoire sur la phosphorescence des lampyres. *Ann. Chem. 17*:151-167.

McDermotte, F. A. (1913). Quoted in *A History of Luminescence from the Earliest Times Until 1900* by E. N. Harvey. American Philosophical Society, Philadelphia, Pa., 1957, p. 455.

McElroy, W. D. (1947). The energy source for bioluminescence in an isolated system. *Proc. Natl. Acad. Sci. USA 33*:342-345.

Peng, C. T. (1976). Chemiluminescence. In *Liquid Scintillation: Science and Technology*, A. A. Noujaim, C. Ediss, and L. I. Weibe (Eds.). Academic, New York, p. 315.

Radziszewski, B. (1877). Untersuchungen uber Hydrobenzamid, Amarin und Lophin. *Ber. Chem. Ges. 10*:70-75.

Seitz, W. R., and M. P. Neary (1974). Chemiluminescence and bioluminescence. *Anal. Chem. 46*:188A-202A.

Seitz, W. R., and M. P. Neary (1977). Chemiluminescence and bioluminescence analysis. *Contemp. Topics Anal. Clin. Chem. 1*:49-125.

Stanley, P. E. (1974). Analytical bioluminescent assays using the liquid scintillation spectrometer: A review. In *Liquid Scintillation Counting*, M. A. Crook and P. Johnson (Eds.). Heyden and Son, London, pp. 253-271.

Strehler, B. L. (1953). Luminescence in cell-free extracts of luminous bacteria and its activation by DPN. *J. Amer. Chem. Soc. 75*:1264.

Weitlaner, P. (1911). Weiteres vom Johanniskaferchenlicht, und vom Organismenleuchten uberhaupt. *Verh. Zool. Bot. Ges. Wien 61*:192-202.

White, E. H., and R. B. Brundrett (1973). The chemiluminescence of acylhydrazides. In *Chemiluminescence and Bioluminescence*, M. J. Cormier, D. M. Hercules, and D. Lee (Eds.). Plenum, New York, pp. 231-244.

Williams, D. C., F. G. Huff and W. R. Seitz (1976). Evaluation of peroxylate chemiluminescence for determination of enzyme generated peroxide. *Anal. Chem. 48*:1003-1006.

2

QUANTUM MECHANICAL ASPECTS OF CHEMILUMINESCENCE

PETER H. LLOYD Coulter Electronics Ltd., Luton, Bedfordshire, England

I. INTRODUCTION

When energy is released as a result of a chemical reaction, it normally appears as heat. In a few reactions, however, some part of the energy released appears as visible light. This is the phenomenon of chemiluminescence. Bioluminescence is a special term given to the chemiluminescence of biological systems in which protein catalysts are frequently involved. As far as the discussions here are concerned, bioluminescence is simply a more complex and efficient form of chemiluminescence. While the detailed mechanisms of many chemiluminescent and bioluminescent reactions are still obscure, it is unlikely that one chemical mechanism can be applicable to all types. Therefore, it is not the purpose of this chapter to elucidate chemical mechanisms but rather to discuss in general terms the physicochemical and quantum-mechanical aspects of the phenomenon of chemiluminescence.

II. ENERGETICS OF LUMINESCENT REACTIONS

In order that light be emitted as a result of a chemical reaction, that reaction must generate at least one chemical species in an electronically excited state. Sufficient chemical energy must be available to generate this excited state and there must be some means whereby during the reaction a molecule can pass readily from the ground state to an excited state.

The energy E of a photon is related to the wavelength of the light by the familiar relationship $E = hc/\lambda$, in which h is Planck's constant and c is the velocity of light in vacuo. For visible light, λ lies between approximately 400 and 650 nm and the energies of the corresponding photons lie between approximately 3×10^{-10} and 1.9×10^{-10} eV or 71 and 44 kcal/mol. These energies are high and only the more vigorous chemical reactions can generate such energies. Thus it is not surprising that most, if not all, chemiluminescent reactions are oxidation reactions involving molecular oxygen, hydrogen peroxide, or similar powerful oxidants. The reducing compound must be one which in the process of oxidation can in some way pass to the excited state of the product molecule. There are many examples, of which luminol is probably the most familiar.

It has been stated that the change in free energy of the reaction must be at least as high as the energy of the most energetic photons emitted and these correspond to the short-wavelength end of the emission spectrum. This is not entirely true and the situation is somewhat more complex for reasons which will become more apparent as this discussion proceeds.

Although it is the change in free energy ΔG which determines whether or not a reaction will proceed spontaneously, it is the change in enthalpy ΔH which is available to generate chemically excited states. These two quantities are related by the Gibbs-Duhem equation:

$$\Delta G = \Delta H - T\,\Delta S$$

in which T is the temperature and ΔS is the change in entropy of the system. In most chemiluminescent reactions, however, the change in entropy is likely to be fairly small, so that the two quantities ΔH and ΔG do not differ greatly.

It is also possible that some of the activation energy for the reaction will be available to be absorbed by the excitation. However, as will be seen later, all the energy potentially available could never appear as excitation energy.

Chemiluminescence must take place in two distinct steps. First, the chemical reaction must produce a molecule in an electronically excited state. Second, that molecule must decay to the ground state with the emission of a quantum of radiation. In other words, the product molecule must be fluorescent, and there must not be a readily available path by which the molecule can return to the ground state by means of a radiationless transition. The actual energy of the photon emitted will depend upon complex interactions between the emitting molecule and molecules of the solvent or other species present. Since in condensed phases molecules find themselves in all sorts of rapidly changing environments, the actual energy of the transition, and hence of the emitted photon, depends on the exact environment of the molecule at the moment the transition occurs. In order to understand these points more clearly, it is necessary to discuss in some detail the nature of excited states and the way in which their energy states can be represented.

The energy contained within a diatomic molecule depends solely on the separation of the two atoms involved and can be shown in a simple graph (Fig. 1). The normal separation of the two atoms is represented by the minimum in the curve: as the atoms move closer to one another the energy of the system rises rapidly, i.e., the atoms repel one another; as the atoms move apart the energy also rises, but in such a way that only a finite energy is needed to separate the atoms completely. This is the dissociation energy. Such a graph is a total description of the energy state of a diatomic molecule in the ground (or lowest energy) state. In an excited state, the energy diagram may look similar, but the dissociation energy is generally smaller, and the separation of the atoms at the minimum of the curve somewhat greater, than for the ground state. In some cases the curve may show no minimum at all, so that the molecule is unstable in its excited state and dissociates spontaneously.

In a triatomic molecule, the energy depends on the separations of the atoms in both bonds independently. A three-dimensional graph is therefore required (Fig. 2) in which the energy is plotted vertically and the separations of the atoms in each bond are represented on the other two axes. In fact, such a diagram is not a full description of such a molecule because another variable must be included: the angle between the two bonds. Thus there are three variables on which the energy of the molecule depends: the lengths of the two bonds and the angles between them (or, alternatively, the distance between the nonbonded atoms). In order to represent a triatomic molecule fully, therefore, we require a four-dimensional diagram, three dimensions for the three variables, and one for the energy of the system; and the surface becomes what is called a hypersurface in four dimensions. In general, a molecule of n atoms will have at least n-1 bonds and n-2 angles (more in cyclic molecules although then they will not all be independent). Thus, a molecule of n atoms requires at least 2n-2 dimensions in which to represent its potential energy hypersurface. Of course, there is no way of representing such a hypersurface in any diagram. However, the concept is useful and diagrams of surfaces in three dimensions will be used here to illustrate a number of points. Each of the two-dimensional axes must be taken to represent a

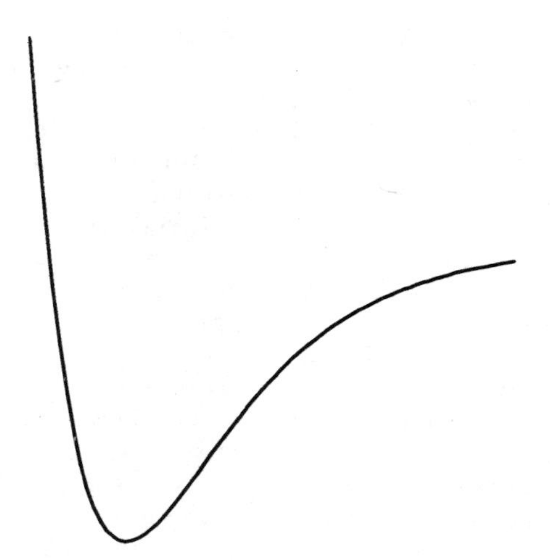

Figure 1 Potential energy curve for a diatomic molecule: (ordinate), energy; (abscissa), separation of the atoms.

composite of many axes in the full hypersurface. It must be understood that the diagrams are purely for purposes of illustration and cannot be taken to represent any particular molecule in reality. In practice, the hypersurfaces are very complex having many local minima separated by hills and saddles. The molecule can be thought of as behaving like a small ball rolling around on its potential energy surface. It will never settle at the bottom of a well; it will continually be knocked about by other molecules and will roll around inside the well. Occasionally it may receive enough energy to roll over a saddle and settle temporarily in another well. This will represent a metastable state from which the molecule will most likely return to its most stable configuration but from which it might suffer a chemical change.

An important point concerning the structure of a molecule in an electronically excited state can be illustrated with potential energy surfaces (Fig. 3). Electronic excitation consists of the promotion of an electron from a bonding molecular orbital to an antibonding orbital (nine combinations are possible involving n, σ, and π orbitals). Such a promotion effectively breaks half a chemical bond, and the molecule has to adjust itself accordingly. Consequently, a molecule in an excited state is structurally different from the same molecule in its ground state. Generally the molecule will be in a looser configuration, although this is not universally true, and this is illustrated in Fig. 3, in which the wells in the potential energy hypersurfaces are not directly above one another.

There are two direct consequences of this difference in the conformations of the molecule in the ground and excited states. In

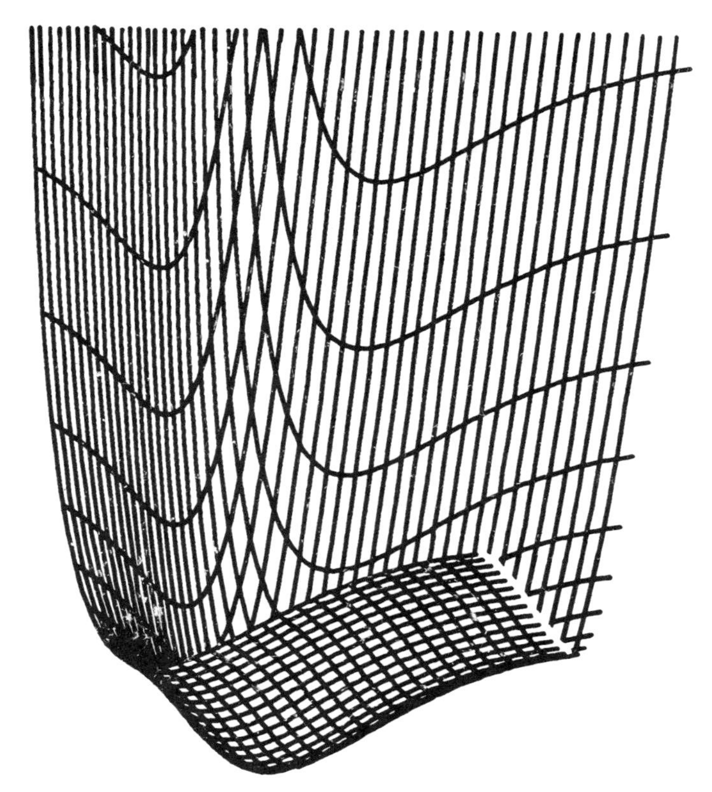

Figure 2 A simplified potential energy surface of a triatomic molecule. The vertical axis is energy, the two horizontal axes are the separations of the atoms along each bond.

general, a molecule spends most of its time close to the bottom of a well in its potential energy hypersurface. If excitation occurs by absorption of a photon, which is a very rapid process compared with molecular vibrations, it will find itself some way away from the bottom of a well in its new hypersurface. It is in fact in a vibrationally excited state. The mean lifetime of a singlet-excited state (about 10^{-9} sec) is very long compared with that of an interatomic vibration (about 10^{-13} sec). Consequently, the molecule will undergo many vibrations while in the electronically excited state, and in condensed phases (e.g., aqueous solution) the molecule will rapidly lose vibrational energy and sink to the bottom of the nearest well in the hypersurface. This is of lower energy and also at a conformation in which the ground state is of higher energy than the molecule had at the moment of excitation. Therefore when the molecule returns to the ground state, the energy change is inevitably smaller than was

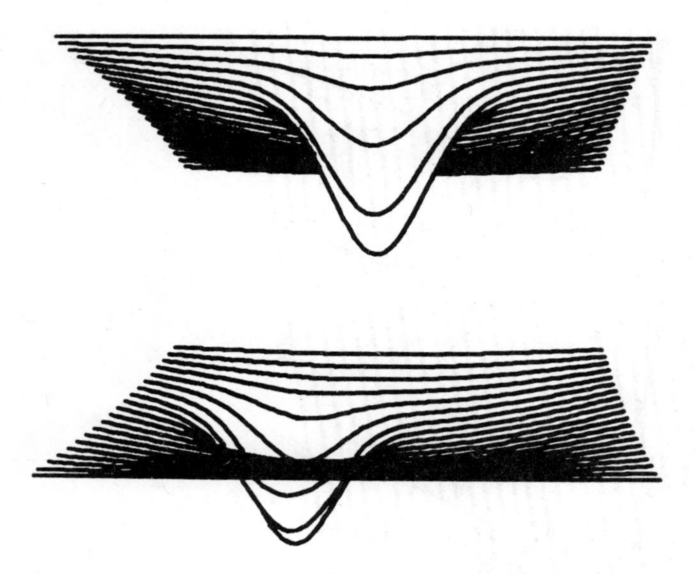

Figure 3 The potential energy surfaces of a molecule in its ground state (lower) and excited state (upper). The vertical axis is energy, the two horizontal axes are composites of various dimensional variables in the molecule. The wells represent the stable forms of the molecule and it can be seen that the excited state is structurally different from the ground state.

required for its excitation and so the emitted light is of longer wavelength. On return to the ground state the molecule again finds itself in a vibrationally excited state and more energy is lost as vibrational energy to the surrounding solvent molecules. In an extreme case, the change in conformation of the molecule while in the excited state may be such that when it returns to the ground state it falls into a well different from the one it left. In this case a photochemical reaction has occurred.

While in the excited state, the molecule may find its way to a position where either the two hypersurfaces cross or at least approach one another closely. Under these circumstances the molecule can return to the ground state in a radiationless transition. If the two hypersurfaces actually cross, the states of the crossing point are said to be degenerate. The molecule is most likely to proceed down the steeper hypersurface which will most probably be the ground state hypersurface.

Since the hypersurfaces are very much more complex than is shown in the diagrams here, it can happen that the course taken by the molecule after excitation will depend upon its exact position on its ground-state surface prior to the excitation. Of the many different paths available, many may take the molecule through configurations from which it may return to the ground state via a radiationless transition, and only a few may lead to the relatively stable configuration required for a fluorescent emission. The quantum yield, defined as the proportion of excited molecules which emit a photon on return to the ground state, is consequently very frequently less than 100%.

The yields in chemiluminescent reactions, usually defined as the number of photons emitted divided by the number of molecules reacting, are invariably small (although *bio*luminescent reactions have very high yields). For example, the best known compound, luminol (5-amino-2,3-dihydro-1,4-phthalazinedione), has a yield of only about 1%. Luminol is quantitatively oxidized to 3-aminophthalate (White et al., 1964), a molecule which is fluorescent. The fluorescence yield of aminophthalate is about 30% (Lee and Seliger, 1972) and it is established that the emission of light from the oxidation of luminol comes from the decay of the first singlet-excited state of 3-aminophthalate (White and Bursey, 1964). From the figures given above for the quantum yields, it can be deduced that only about 4% of the aminophthalate is generated in the excited state.

Although very general mechanisms have been proposed to account for the chemical generation of molecules in excited states, as far as the present author is aware, only one detailed chemical and quantum-mechanical mechanism has been described (Michl, 1977) which shows in a direct way how such excitation can occur. It is generally (though not universally) accepted that the oxidation of luminol to aminophthalate proceeds via the dianion of cyclic aminophthaloyl peroxide (Gundermann, 1968; White and Rosewell, 1970; Michl, 1975). Michl (1975,1977) pointed out that conversion of cyclic aminophthaloyl peroxide to aminophthalate involves only the stretching of the (already weak) O-O bond, but that although the two molecules are geometrically similar they are not electronically identical. The peroxide corresponds to the doubly-excited $n,n \rightarrow \pi^*,\pi^*$ state of the aminophthalate, and the aminophthalate corresponds to the doubly-excited $\pi,\pi \rightarrow \sigma^*,\sigma^*$ state of the peroxide. What this means is that as the O-O bond breaks, the electrons do not have to be promoted to the higher energy state, *they are already there*. The orbitals on the oxygen atoms which are contributing to the σ bond linking the two oxygen atoms in the peroxide have the same spatial disposition as the antibonding π^* orbitals in the aminophthalate. The σ orbital is occupied by two electrons in the peroxide and simply becomes the π^* orbital of the phthalate. The fact that this mechanism would generate a doubly-excited molecule is not in conflict with the fact that the emission is from the singlet-excited state. There are simply two electrons in the higher molecular orbital, and it is intriguing to speculate that this implies a potential quantum yield of 200%.

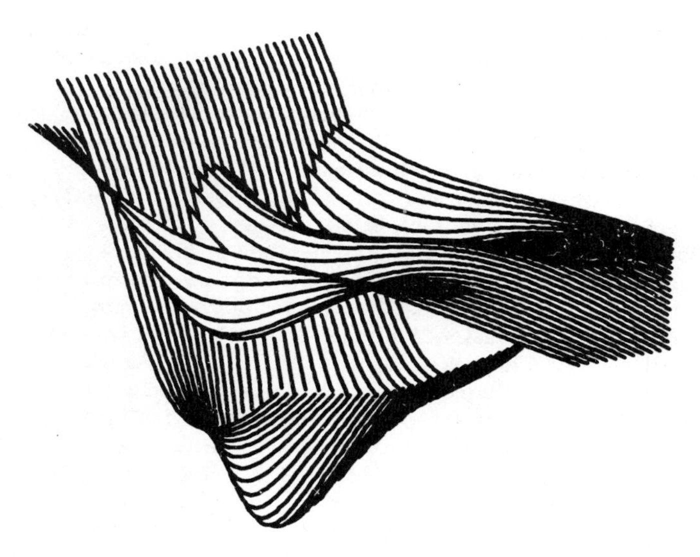

Figure 4 A possible form for the potential energy surfaces involved in the transition between cyclic 3-aminophthaloyl peroxide and 3-aminophthalic acid. The diagram was drawn to illustrate the discussion in the text and is not the result of quantum mechanical calculations.

Looked at in terms of potential energy surfaces, the mechanism is illustrated in Fig. 4. The potential energy well of the peroxide is separated by a low saddle from the well of the excited aminophthalate. However, the groundstate surface of the aminophthalate must cross this former surface because it must be of lower energy in the geometry of the aminophthalate and of higher energy in the geometry of the peroxide. (The aminophthalate, remember, has the structure of an excited state of the peroxide.) If during its passage over the barrier between peroxide and excited aminophthalate the molecule gets close to the surface of the ground state aminophthalate, it may cross over and end up in the ground state. However, if it takes a route which keeps it well away from the ground state surface, then it will pass successfully to the excited state of the product.

The likelihood of a successful transition will depend upon where, within the saddle region, the ground state surface crosses the other surface. It must be remembered that the molecule will not always take the lowest path over the barrier. The analogy of a small ball rolling about on the surface is useful here. As it rolls around within the potential energy well of the peroxide, the final push which sends it over the saddle into the well of the aminophthalate may very well occur in such a direction that it rolls well up the side of the saddle as it passes over. This mechanism provides a ready explanation of how the reaction can produce a molecule in its excited state and also why it may do so with less than 100% yield. The function of the enzymes in bioluminescence may be to move the surfaces in such a way that the molecule never finds the region of crossover and consequently always reaches the excited state of the product.

It must also be pointed out that the molecule has reached the excited state surface by a completely different route from that taken during fluorescence. Consequently it will arrive on that surface at possibly quite a different point, and consequently what happens to it while it is on that surface may also be different. There is no a priori reason to suppose that the chances of its reaching a region of the surface from which it can make a radiationless transition to the ground state is the same in both cases. In other words, the "fluorescence yield" may be quite different in the two cases. It could even be that the molecules pass to the excited state with great efficiency but in a molecular configuration in which they are very likely to find a ready transition to the ground state.

In the case of luminol itself, there is a perfectly possible mechanism whereby the preceding intermediate to the peroxide (an azo peroxide) could decompose directly to the aminophthalate (Michl, 1977) without passing through the peroxide at all. Probably both mechanisms are possible which would further reduce the yield of chemiluminescence.

III. SUMMARY

There are a number of conditions which must be satisfied before chemiluminescence can be observed. Since the overall yield of chemiluminescence will be the product of the yields in each step, it is not surprising that quantum yields are universally so low. Undoubtedly, the role of protein catalysts in bioluminescence is to control and direct the reactions in such a way as to achieve near 100% yield at each step. In order to achieve high overall yields, the following conditions at least have to be satisfied:

1. The chemical reaction must be able to generate *at least* 40 kcal/mol.
2. A chemical mechanism must exist capable of generating product molecules in an electronically excited state.
3. There must be no competing mechanisms generating the product in its ground state.
4. There must be no side reaction producing other products.
5. There must be no means whereby the excited product molecule can return to the ground state via a radiationless transition.
6. A product molecule must be sufficiently stable in its excited state to survive long enough for emission to occur before any possible chemical change.

7. There must be no means whereby the excited molecule can transfer its energy to other molecules present which can in turn return to the ground state via radiationless transitions.
8. The medium must be transparent at the wavelength of emission so that the light can be detected from outside.

Nature has shown that these conditions can be satisfied by using complex catalysts along with relatively simple organic compounds. It remains an exciting challenge to organic chemists to achieve similar results without the complex catalysts.

ACKNOWLEDGMENTS

I am grateful to Professor T. P. Whitehead, who allowed me to use the computer in his department, and to Mrs. A. M. Peters and Mr. I. Clarke for their constant help and encouragement during the writing of the programs with which my diagrams were drawn.

REFERENCES

Gundermann, K. D. (1968). *Chemiluminiszenz Organischer Verbindungen*. Springer-Verlag, Berlin, p. 84.

Lee, J., and H. H. Seliger (1972). Quantum yields of the luminol chemiluminescence reaction in aqueous and aprotic solvents. *Photochem. Photobiol. 15*:227-237.

Michl, J. (1975). In *Physical Chemistry*, Vol. 3, H. Eyring, D. Henderson, and W. Jost (Eds.). Academic, New York, pp. 125-169.

Michl, J. (1977). The role of biradicaloid geometries in organic photochemistry. *Photochem. Photobiol. 25*:141-148.

White, E. H., and M. M. Bursey (1964). Chemiluminescence of luminol and related hydrazides: The light emission step. *J. Amer. Chem. Soc. 86*:941-942.

White, E. H., and D. F. Rosewell (1970). The chemiluminescence of organic hydrazides. *Acc. Chem. Res. 3*:54-62.

White, E. H., O. Zafiriou, H. M. Kagi and J. M. M. Hall (1964). Chemiluminescence of luminol: The chemical reaction. *J. Amer. Chem. Soc. 86*:940-941.

3

CHEMILUMINESCENCE

GARY H. G. THORPE Wolfson Research Laboratories, Queen Elizabeth Medical Centre, Edgbaston, Birmingham, England

LARRY J. KRICKA* University of Birmingham, Birmingham, England

TIMOTHY J. N. CARTER† Wolfson Research Laboratories, Queen Elizabeth Medical Centre, Edgbaston, Birmingham, England

I. Introduction 21

II. Liquid Phase Chemiluminescence 23
   A. Peroxide Decomposition 23
   B. Electron Transfer Chemiluminescence 33
   C. Singlet Oxygen Chemiluminescence 34

III. Gas Phase Chemiluminescence 34

IV. Solid Phase Chemiluminescence 36

V. Conclusion 37

References 37

I. INTRODUCTION

Although the study of luminescence in living systems originated in antiquity, it was not realized until the late nineteenth century that the phenomenon could be associated with relatively simple organic

---

*Dr. Kricka is currently with the University of California at San Diego, La Jolla, California, on a leave of absence from the University of Birmingham.
†Dr. Carter is currently with Battelle, Centre de Recherche de Genève, Carouge, Genève, Switzerland.

reactions. Today many chemiluminescent systems are known, including reactions in the liquid, gas, and solid phases (Isacsson and Wettermark, 1974).

Intensive studies of the detailed mechanisms involved are in progress (Schuster et al., 1979; McCapra and Leeson, 1979; Hart and Cormier, 1979), but as yet many aspects of the mechanisms are not well understood. Chemiluminescence can be defined as the emission of light from chemical reactions at ordinary temperatures. Chemiluminescent reactions produce a reaction intermediate or product in an electronically excited state and radiative decay of the excited state results in light production. The radiation produced may be in the ultraviolet, infrared, or visible region of the spectrum. When the excited state produced is a singlet, then the radiative process is identical to fluorescence, and when the excited state is a triplet, phosphorescent emission results.

Chemiluminescent reactions are relatively uncommon as most chemical reactions release energy as heat via vibrational excitation of ground state products, and even when electronically excited states are produced, most of them decay via nonradiative processes (Rauhut, 1979).

Several mechanisms have been described to explain the manner by which chemical energy is provided for chemiluminescent reactions; these include reactions involving peroxide decomposition, singlet oxygen, ion radicals, and chemically initiated electron exchange.

In theory, a single molecule of chemiluminescent reactant could produce one electronically excited molecule which could then decay to emit one photon of light. Chemiluminescent reactions are not, however, this efficient, and although a quantum yield (i.e., the ratio of the number of emitted photons to the number of reacting molecules) of 1 Einstein/mol (or 100%) is theoretically possible, it is seldom greater than 1% of this (Gorus and Schram, 1979). As chemiluminescence is a multistep process (Scheme 1), the final quantum yield is the product of

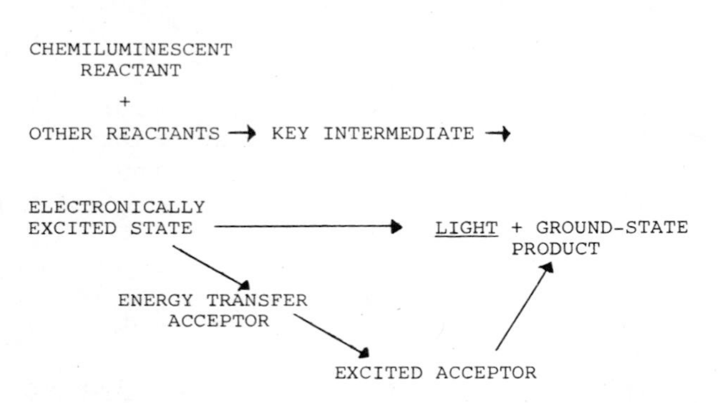

Scheme 1 Chemiluminescent reaction sequences.

the chemical and physical efficiencies of the processes involved (Brundrett et al., 1972). Efficient chemiluminescence requires selective production of a key intermediate and efficient conversion of the key intermediate to the singlet-excited state of a highly fluorescent product. Competitive side reactions can decrease formation of the key intermediate, and products may even obscure the chemiluminescent reaction. Efficient excitation requires a large instantaneous energy release in a single excitation step (Rauhut, 1979; Rauhut et al., 1965; Sheehan, 1975). Two-step reactions in which energy is released at each step cannot be chemiluminescent in solution, as the energy released in the first step is lost as vibrational energy to the solvent before the second step can raise the energy level of the intermediate to the required level. Chemiluminescent quantum yields are also reduced because the fluorescent quantum yields of most molecules are low and phosphorescence is inefficient in liquids (see Chap. 2).

Classification of chemiluminescent reactions may be based either on the class of compound, or on the mechanism of the luminescent reaction, or on the phase in which luminescence occurs, i.e., gas, liquid, or solid. In this chapter the last type of classification has been adopted and emphasis has been placed on liquid phase reactions because these are of greatest importance in analysis.

II. LIQUID PHASE CHEMILUMINESCENCE

Many liquid phase chemiluminescent reactions involving peroxides proceed through the decomposition of reaction intermediates with the simultaneous formation of two carbonyl groups. The concerted formation of these groups meets the energy requirements for the chemiluminescent reaction. The new carbonyl groups need not be part of the excited product but must be formed synchronously with the excited state (Rauhut, 1979). Alternative mechanisms involving electron transfer (Sec. II.B) and singlet oxygen (Sec. II.C) have also been proposed for liquid phase chemiluminescence.

A. Peroxide Decomposition

1, 2-Dioxetanes

1,2-Dioxetanes have recently been isolated, although they were expected to be thermally unstable. Some are stable at room temperature but most decompose below 80°C and require careful synthesis and handling. Simple dioxetanes [(1), Scheme 2] decompose to produce carbonyl compounds in excited states, including singlet states, and thus decomposition is accompanied by light emission (Wilson, 1976). Both photochemical and chemiluminescent methods have been used to determine the relative yields of the excited states (Lechtken et al.,

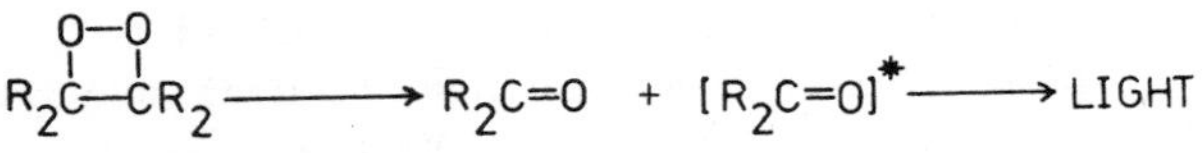

$$R_2C(-O-O-)CR_2 \longrightarrow R_2C{=}O + [R_2C{=}O]^* \longrightarrow \text{LIGHT}$$

(1)

Scheme 2 Chemiluminescent decomposition of dioxetanes. (From Rauhut, 1979.)

1973; Wilson and Schaap, 1971): in general, the yield of triplet-excited states is higher than that of the singlet state although these vary with the substituent groups of the dioxetanes.

Tetraalkyldioxetanes. The highest yield of excited product in dioxetane decomposition reactions is produced from the decay of tetramethyl-1,2-dioxetane, where 50% of the acetone produced is in the triplet form; the singlet state is formed only in low yields (0.1%). As both fluorescence and phosphorescence from acetone are poor, the addition of fluorescent or triplet acceptors to the system results in increased light production (Rauhut, 1979).

Higher yields of singlet-excited states are produced by other dioxetanes. That prepared by McCapra and co-workers based on the acridan system (2) produced singlet-excited states with a yield of up to 30% (McCapra et al., 1975; Hart and Cormier, 1979).

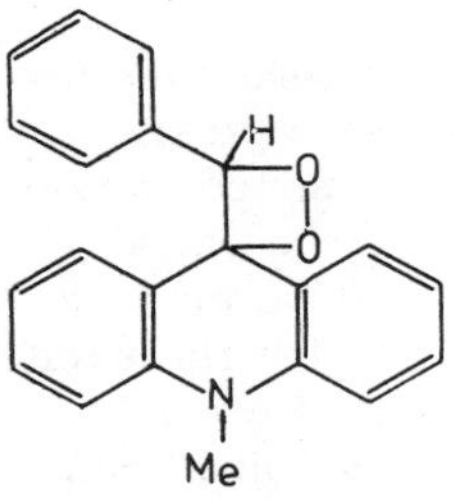

(2)

Many chemiluminescent reactions, such as those of imidazoles and acridine derivatives, may proceed through unstable dioxetane intermediates.

Imidazoles. The chemiluminescent oxidation of 2,4,5-triphenylimidazole [lophine (3), Scheme 3] was first reported by Radziszewski in 1877. Suitable oxidizing agents include oxygen in aqueous alkaline dimethyl sulfoxide (Scheme 3), hydrogen peroxide/sodium hypochlorite, and hydrogen peroxide/potassium ferricyanide. An unstable di-

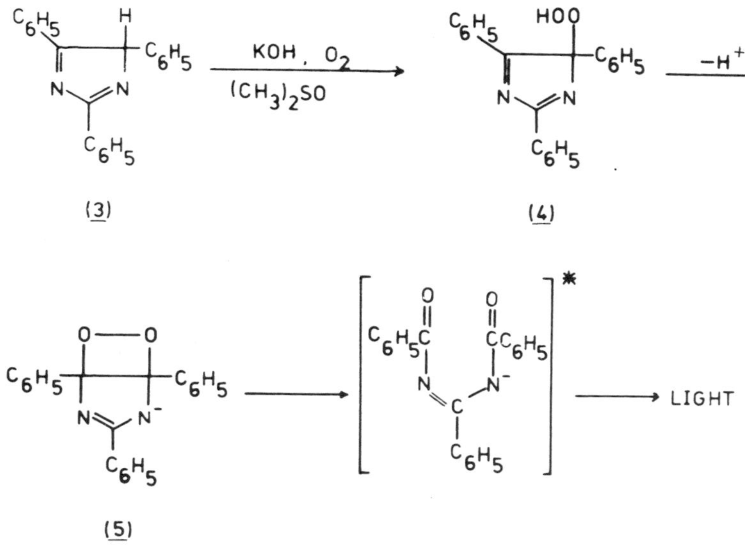

Scheme 3 Chemiluminescent oxidation of lophine. (From Rauhut, 1979.)

oxetane intermediate (5) has been implicated in the mechanism of the reaction, and the hydroperoxide intermediate (4) has been isolated and shown to emit light when treated with basic ethanol. Various metal ions (e.g., $AuCl_4^-$, Cr(III), MnO_4^-) enhance the chemiluminescent oxidation of lophine, which has been utilized in the determination of such ions (MacDonald et al., 1979).

Acridine Derivatives. N,N'-dimethyl-9,9'-bisacridine [lucigenin (6)] is one of the classic organic chemiluminescent agents. Chemiluminescence is obtained upon oxidation by hydrogen peroxide in basic aqueous solution (Scheme 4). The quantum yield of the reaction is about 0.5% based on lucigenin but 1.6% based on the low-yield product of the reaction, N-methylacridone (8) (Rauhut, 1979). The dioxetane intermediate (7) proposed in the chemiluminescent reaction has been prepared and also has a quantum yield of 1.6% (Lee et al., 1976). The reaction between lucigenin and hydrogen peroxide to produce light proceeds without catalysis, although it is accelerated by metal ions leading to an intensification of chemiluminescence. Light emission occurs on the addition of numerous nucleophiles and reducing agents to solutions of lucigenin (Maskiewicz et al., 1979a,b; Veazey and Nieman, 1979). NADH is also capable of light production under certain circumstances (Totter et al., 1960).

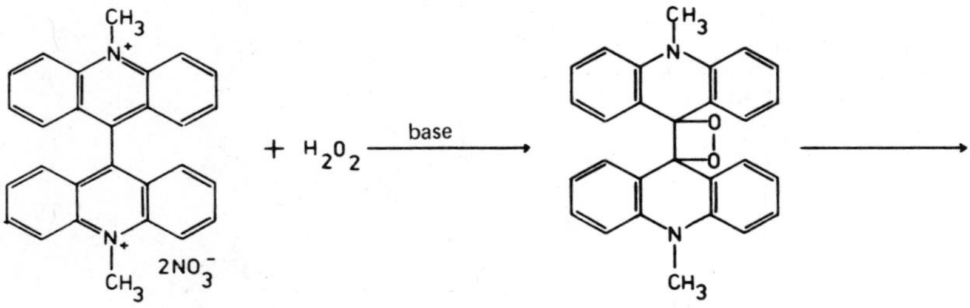

(6) (7) 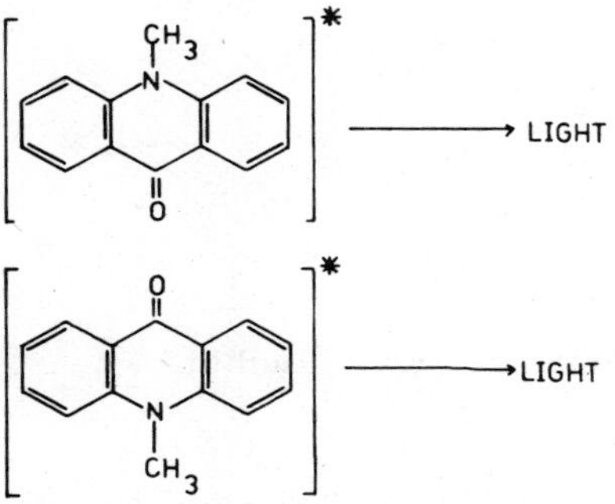 (8)

Scheme 4 Chemiluminescent oxidation of lucigenin by hydrogen peroxide in basic aqueous solution.

Novel acridinium phenyl carboxylates, e.g., p-acetylphenyl-10-methylacridinium-9-carboxylate fluorosulfonate (9), which undergo chemiluminescent oxidation with hydrogen peroxide have been described and applied to clinical analyses (McCapra et al., 1977). The

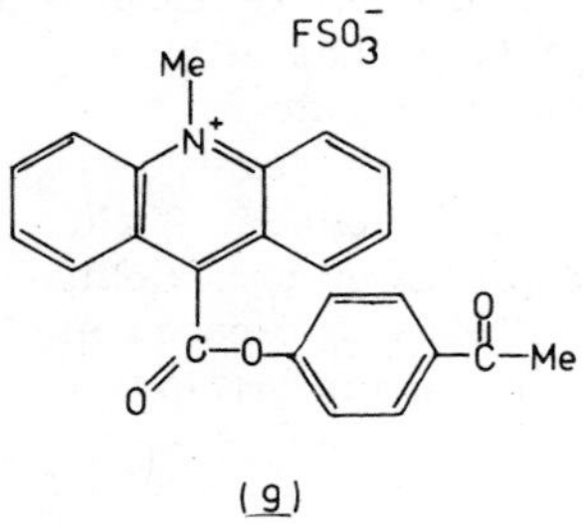

(9)

optimal pH of oxidation of these substances depends on the nature of the substituent in the phenyl group, thus allowing an acridinium salt to be tailored to fit particular pH requirements. Their shelf-life and susceptibility to interferences are claimed to be superior to other chemiluminescent reagents (McCapra et al., 1978).

Tetraminoethylenes. Tetrakis(dimethylamino)ethylene (TMAE) reacts spontaneously with oxygen to produce light and has been proposed as a chemical light source. The products of the reaction are tetramethylurea and tetramethyloxamide, both of which quench chemiluminescence, with TMAE itself being the emitting excited state. An energy transfer process from an excited state produced by dioxetane decomposition and a mechanism involving chemical dimerization and regeneration of excited TMAE through decomposition of the dimer have been proposed for the reaction (Rauhut, 1979).

1,2-Dioxetanones

1,2-Dioxetanones (α-peroxylactones) were proposed as key intermediates in several bioluminescent reactions (Hopkins et al., 1967; White et al., 1969) and have subsequently been prepared (using low-temperature techniques) and shown to decompose (Scheme 5) under ambient conditions to produce aldehydes and ketones (Adam, 1975). This decomposition reaction is only weakly chemiluminescent since the products are poorly fluorescent. Addition of a fluorescer increases light emission (Adam et al., 1974,1978; Schmidt and Schuster, 1978). At low concentrations of fluorescer the rate of decomposition remains unaltered but at high concentrations the fluorescer acts catalytically and the rate of decomposition of 1,2-dioxetanones is increased. The rate of decomposition is also affected by the ionization potential of the fluorescer (Hart and Cormier, 1979).

The chemiluminescent decomposition of dioxetanones in the presence of a fluorescer is thought to proceed via a chemically initiated electron exchange reaction (Hart and Cormier, 1979).

1,2-Dioxetanediones

Esters of oxalic acid such as bis(2,4,6-trichlorophenyl)oxalate [TCPO (10), Scheme 6] react with peroxide in the presence of suitable fluorescers, e.g., rubrene, in some of the most efficient non-

$$(C_6H_5)_2\overset{O-O}{\overset{|\quad\;\, |}{C-C}}{=}O \;+\; FLR \longrightarrow (C_6H_5)_2C{=}O \;+\; CO_2 \;+\; [FLR]^* \longrightarrow LIGHT$$

Scheme 5 An example of the chemiluminescent decomposition of 1,2-dioxetanes in the presence of a fluorescer (FLR). (From Rauhut, 1979.)

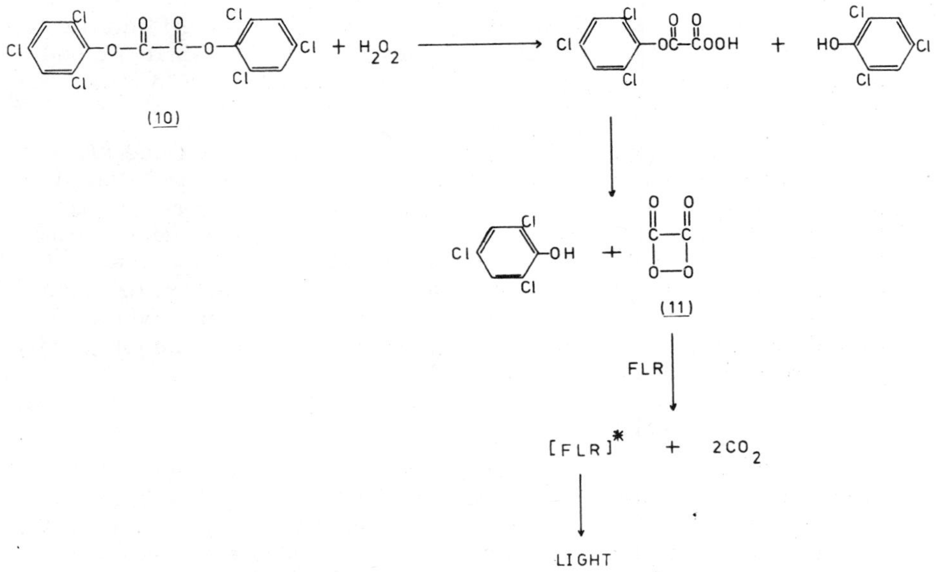

(10) + H_2O_2 → ... (11) → FLR → $[FLR]^*$ + $2CO_2$ → LIGHT

Scheme 6 The chemiluminescent reaction of bis(2,4,6-trichlorophenyl) oxalate with hydrogen peroxide (FLR = fluorescer).

enzymic chemiluminescent reactions known. Quantum yields may be as high as 27% and a tentative reaction mechanism involving catalytic decomposition of 1,2-dioxetanedione (<u>11</u>) by the fuorescer in the excitation step is outlined in Scheme 6. The reactions are very efficient because (1) the dioxetanedione-fluorescer complex generates the singlet-excited state of the fluorescer in high yield, (2) the fluorescers used are highly efficient, (3) the key intermediate is metastable toward unimolecular decomposition, and (4) unwanted nonluminescent side reactions can be reduced by suitable choice of reaction conditions. The chemiluminescent intensity and lifetime can also be altered by the addition of weak acids, bases, and certain salts (Rauhut, 1979). The efficiency of an oxalate chemiluminescent reaction also depends on the nature of the substituents. Good "leaving" groups facilitate ring closure to the dioxetanedione intermediate and thus increase chemiluminescent efficiency (Rauhut et al., 1967). Efficient chemiluminescent oxalate derivatives include electronegatively substituted aliphatic and aromatic esters, amides, sulfonamides, oxalyl chloride, O-oxalyl hydroxylamine derivatives, and mixed oxalic-carboxylic anhydrides.

Other chemiluminescent reactions, e.g., the reaction of chlorinated esters and ethers with hydrogen peroxide and a fluorescer,

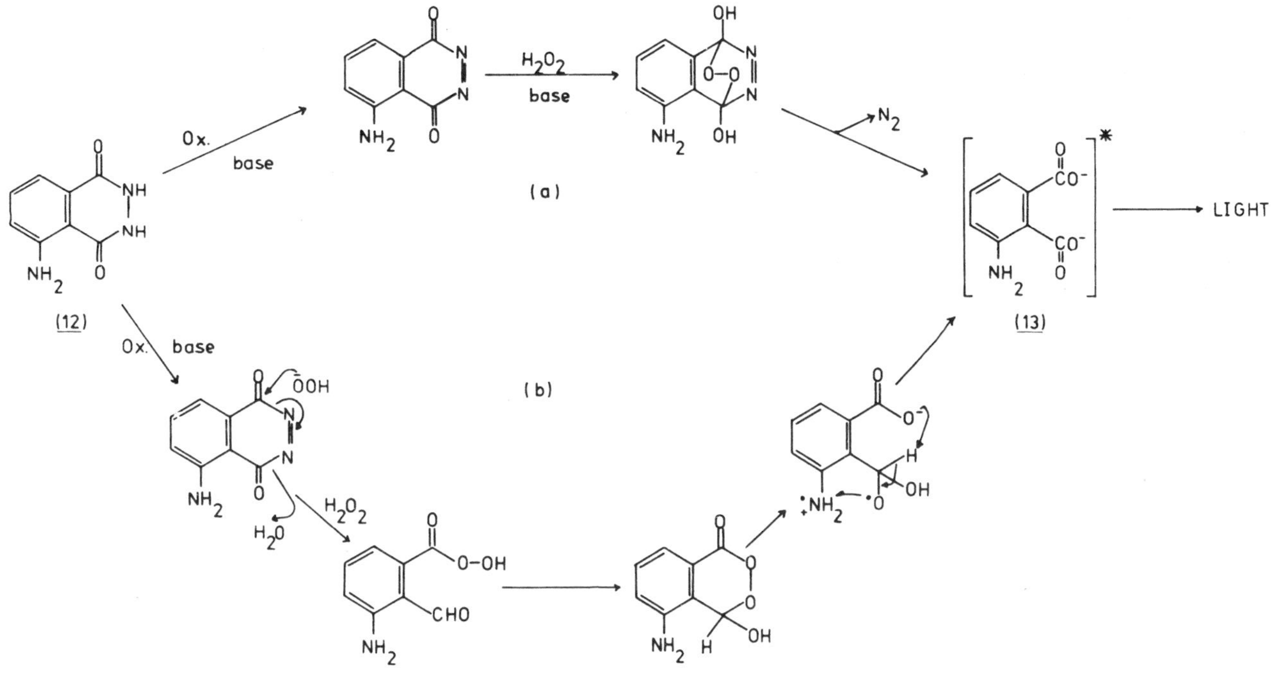

Scheme 7 Proposed mechanisms for the chemiluminescent oxidation of luminol.

appear to be analogous to oxalate chemiluminescence, although the mechanistic details may vary (Maulding and Roberts, 1972).

The oxalate chemiluminescent reaction has been exploited as a lighting system (Bollyky and Rauhut, 1971) and recently bis(2,4,5-trichloro-6-carbopentoxylphenyl)oxalate has been advocated for use in the chemiluminescent measurement of peroxide generated in enzyme-coupled assays (Scott et al., 1980).

Acylhydrazides

The chemiluminescent reactions of cyclic diacylhydrazides such as luminol [5-amino-2,3-dihydro-1,4-phthalazinedione (12) Scheme 7] have been widely used in chemical analysis (Whitehead et al., 1979; Isacsson and Wettermark, 1974) and extensively studied (Roswell and White, 1978).

In aprotic solvents, such as dimethyl sulfoxide, luminol reacts with oxygen in the presence of a strong base to produce light. Similarly, in aqueous solution many oxidants in the presence of suitable catalysts or cooxidants produce light (Scheme 7). In general, hydrogen peroxide is the most commonly used oxidant, in the presence of catalysts such as $Fe(CN_6)^{3-}$, Cu^{2+}, or hemin; other oxidants include hypochlorite, iodine, permanganate, and oxygen. The quantum yield is approximately 1%, with the efficiency, wavelength, and pH optimum of light emission depending largely on the reaction conditions. The exact manner in which chemiluminescence is produced remains in doubt. Mechanisms involving the intermediary role of diazoquinones and dioxetanes (Scheme 7a) and electron transfer (Scheme 7b) (Roswell and White, 1978; Rauhut, 1979; McCapra and Leeson, 1979; Gundermann, 1978) have been proposed. The aminophthalate dianion (13) has been shown to be the emitting fluorescer (White and Bursey, 1964).

Chemical coupling of luminol to other materials reduces the chemiluminescent efficiency (Simpson et al., 1979; Hersh et al., 1979). To achieve higher quantum yields and obtain more efficient chemiluminescent-labeled conjugates for monitoring competitive binding reactions, many derivatives of luminol have been investigated (Gundermann, 1978). Several derivatives [e.g., (14)] have been pro-

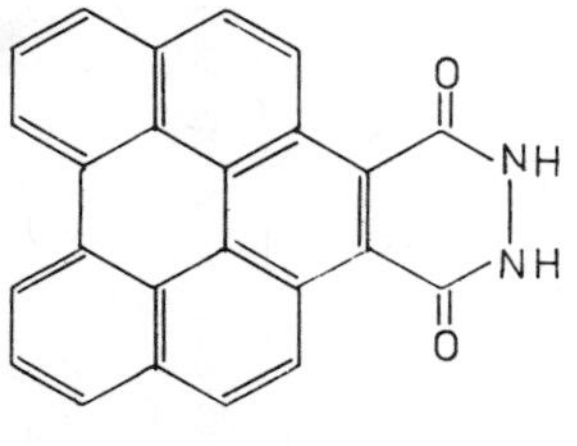

(14)

duced which have chemiluminescent quantum yields substantially higher than luminol (McCapra, 1966; Gundermann, 1978). Electron-withdrawing substituents in the luminol benzene ring decrease luminescence whereas electron-donating substituents increase light yields, with substitution at positions 5 and 8 being more effective than at 6 and 7. A complete loss of light emission occurs if the heterocyclic ring is substituted (Schroeder and Yeager, 1978). A shift in the position of the amino group reduces efficiency, isoluminol (6-amino-2,3-dihydrophthalazine-1,4-dione) (15) being only 10% as efficient as luminol. Alkylation of the amino group of isoluminol derivatives leads to a several-fold increase in chemiluminescence whereas similar substitution of luminol produces steric hindrance resulting in a significant decrease in chemiluminescence (Schroeder and Yeager, 1978). These observations are of considerable importance in the preparation of conjugates (see Chap. 9).

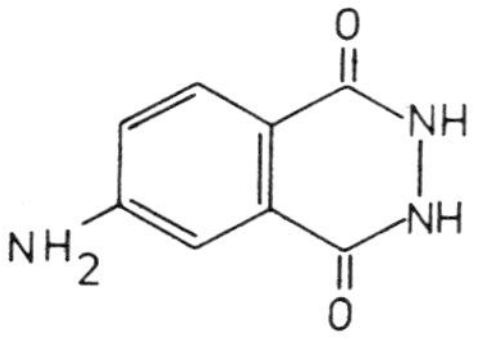

(15) 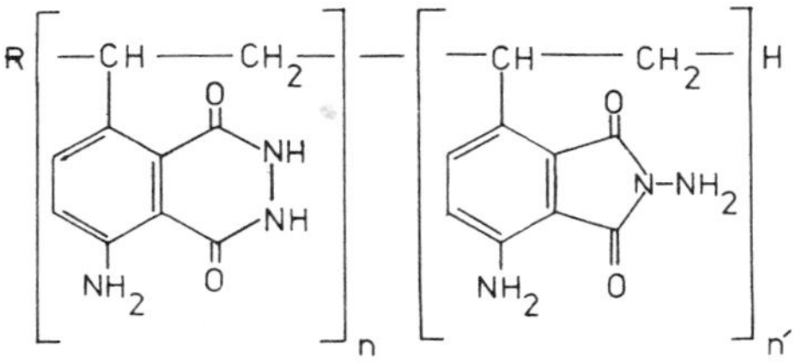 (16)

Polymers of luminol (16) exhibit very low quantum yields (Gundermann, 1978). Oligomers containing luminol subunits have been prepared by treating an oligomer of dimethyl-3-nitro-6-vinylphthalate with hydrazine. The product, which contained luminol and N,3-diaminophthalimide units, gave 0.5% of the chemiluminescence yield given by luminol. The poor efficiency of the luminol oligomer was attributed to quenching by neighboring phthalimide units, as quantum yields approaching that of luminol were obtained if the distance between the luminol units was increased (Gundermann and Giesecke, 1979).

Noncyclic monoacylhydrazides are chemiluminescent under similar conditions to cyclic hydrazides and quantum yields of 0.3% have been reported (Roswell and White, 1978).

Organometallics

Several organometallics react with oxygen to produce chemiluminescence. Aryl magnesium halides, especially bromides such as

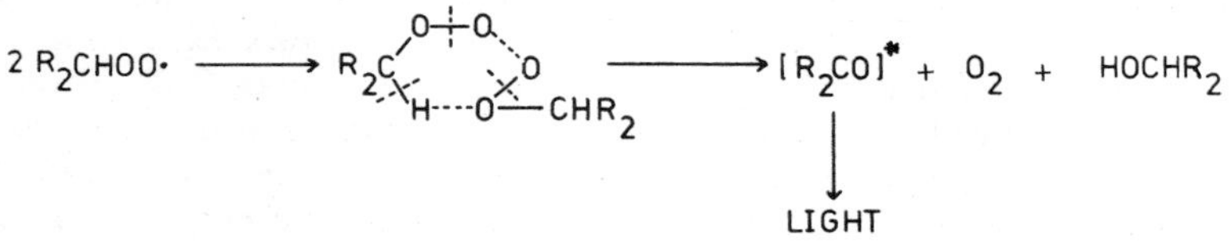

Scheme 8 The chemiluminescent termination step for secondary peroxy radicals. (From Rauhut, 1979.)

p-chlorophenylmagnesium bromide, react with oxygen in ether to generate light. Although the exact mechanism is obscure, it appears probable that free radicals are involved and ArOOMgBr is an intermediate: brominated biphenyls are the emitting species (Walling and Buckler, 1955; Bolton and Kearns, 1974). Aryl Grignard reagents also produce light in reactions with benzoyl peroxide and with nitro compounds. Weak chemiluminescence is produced when alkyl Grignard reagents, and lithium diphenylphosphide or related organophosphides, react with oxygen (Rauhut, 1979).

Oxidation Reactions

Many organic compounds react very slowly with atmospheric oxygen at room temperature to produce weak chemiluminescence (Mendenhall, 1977; Gundermann, 1978). Chemiluminescence emission occurs from papers, plastics, rubber, edible materials, resins, cloth, and many other nonmetallic articles (Mendenhall, 1977). In many autoxidation reactions, light is produced by the combination of peroxyl radicals (Scheme 8), which are primarily responsible for the propagation and termination of the autoxidation chain reaction (Rauhut, 1979; Gundermann, 1978). Tertiary peroxyl radicals also produce chemiluminescence, although with lower efficiencies.

A large number of liquid phase oxidation reactions are chemiluminescent. These include the reactions of carbenes or phenanthrene quinone with oxygen, and coumarin derivatives or nitriles with hydrogen peroxide. Several chemiluminescent reactions, e.g., the oxidation of iron(II) with H_2O_2 and $KHCO_3$, produce electron-accepting radicals such as $HO^{\cdot}$ in the presence of carbonate ions (Stauff et al., 1973). The reactions appear to involve the carbonate radical anion as a key intermediate and this may also account for many observations of weak chemiluminescence in oxidation reactions (Rauhut, 1979). Sulfite is oxidized to sulfate by hypochlorite or H_2O_2 in reactions which also emit light. The intermediate reaction steps and the nature of the emitting species are not well established, although an excited S_2 molecule has been proposed (Burguera and Townshend, 1980).

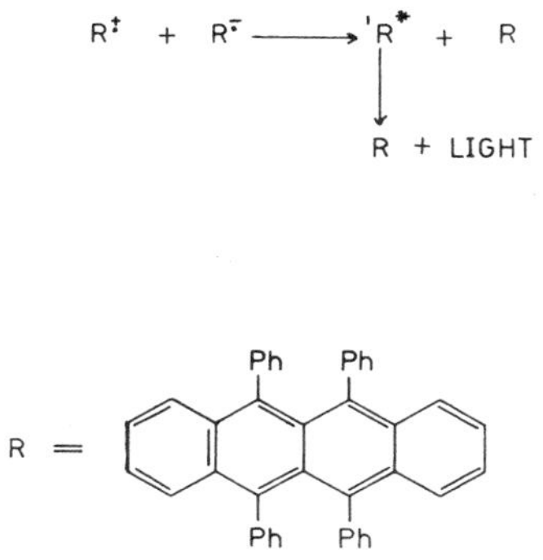

$$R^{\dot{+}} + R^{\dot{-}} \longrightarrow {}^{1}R^{*} + R$$

$$ {}^{1}R^{*} \longrightarrow R + \text{LIGHT}$$

R = (rubrene: tetracene core with four Ph substituents)

Scheme 9 Electron transfer chemiluminescence from rubrene radicals (R).

B. Electron Transfer Chemiluminescence

Electron transfer reactions are capable of producing excited products which may decay to the ground state with the emission of light (Faulkner, 1976; Schuster et al., 1979). A typical example of such reactions is the mutual annihilation of rubrene anion and cation radicals to produce a rubrene molecule in its first excited singlet state (Scheme 9). This type of process is quite general among aromatic molecules and their radical ions (Faulkner, 1976).

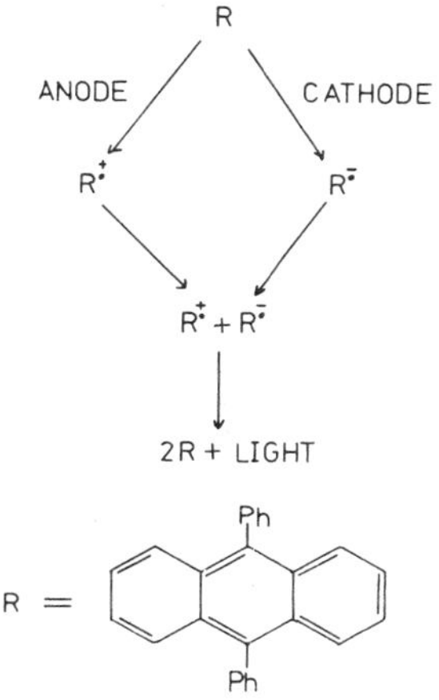

$$R \xrightarrow{\text{ANODE}} R^{\dot{+}} \qquad R \xrightarrow{\text{CATHODE}} R^{\dot{-}}$$

$$R^{\dot{+}} + R^{\dot{-}} \longrightarrow 2R + \text{LIGHT}$$

R = (anthracene core with two Ph substituents)

Scheme 10 Electrogenerated chemiluminescence from diphenyl anthracene (R).

Chemiluminescence in both solid and liquid phases has been described. In the liquid phase chemiluminescence may occur during the recombination of anion and cation radicals generated by electrochemical methods (Schuster et al., 1979; Rauhut, 1979), e.g., reaction of the anion and cation radical forms of 9,10-diphenylanthracene (Scheme 10). Quantum yields of 8-10% have been reported for the electrochemiluminescence of 9,10-diphenylanthracene (Rauhut, 1979) and up to 20% for that from the recombination of the 9,10-diphenylanthracene anion and the thianthrene cation (Bard et al., 1973; Keszthelyi et al., 1975).

C. Singlet Oxygen Chemiluminescence

Reactions involving the electronically excited singlet states of oxygen can produce chemiluminescence and singlet oxygen has been implicated in several chemiluminescent reactions (Rauhut, 1979; Gundermann, 1978). In solution the excited state may be produced by reaction of hydrogen peroxide with hypochlorite ions (Foote and Wexler, 1964) and by thermal decomposition of the complex formed between triphenylphosphite and ozone (Murray and Kaplan, 1968,1969). Singlet oxygen is thought to be involved in several chemiluminescent reactions such as the reaction between alkaline pyrogallol, formaldehyde, and hydrogen peroxide, and the decomposition reactions of hydroperoxides and secondary dialkyl peroxides (Rauhut, 1979).

III. GAS PHASE CHEMILUMINESCENCE

Perhaps the earliest recorded example of gas phase chemiluminescence is the emission of light from the oxidation of white phosphorus in moist air. The chemiluminescence actually occurs from phosphorus vapor present just above the surface of the solid. The reaction has been studied using phosphorus vapor (P_4) and moist air in a nitrogen carrier (Van Zee and Khan, 1976), although the complex mechanism of this reactions remains as yet unresolved (Campbell and Baulch, 1978).

Gas phase chemiluminescence has found wide applicability in the monitoring of air pollution and has been studied in simple gases, flames, metal vapors, chemical lasers, molecular beams, and recombi-

$$NO + O_3 \longrightarrow O_2 + [NO_2]^* \longrightarrow \text{LIGHT}$$

$$O + O + SO_2 \longrightarrow O_2 + [SO_2]^* \longrightarrow \text{LIGHT}$$

Scheme 11 Examples of gas phase chemiluminescence from ozone and atomic oxygen.

$$K + SO_2 \longrightarrow [K^{+} \cdot SO_2^{-}] \longrightarrow SO_2 + [K]^{*} \longrightarrow \text{LIGHT}$$

Scheme 12 An example of gas phase chemiluminescence from alkali metals and gases.

nation reactions (Campbell and Baulch, 1978; Seitz and Neary, 1977; Cormier et al., 1973).

Ozone and atomic oxygen can react with other gases such as oxides of nitrogen, olefins, sulfur dioxide, and hydrogen sulfide to emit light (Seitz and Neary, 1977). Proposed mechanisms are given in Scheme 11. Numerous other examples are known (Campbell and Baulch, 1978) and gas phase chemiluminescence involving the formation and decomposition of unsubstituted 1,2-dioxetane has been observed when oxygen reacts with C_2H_4 in a heated flow reactor (Bogan et al., 1976).

Chemiluminescent reactions between alkali metal vapors and gases such as N_2, CO_2, SO_2, and several hydrocarbons have been observed (Herschbach, 1973). The reactions involve no net chemical change and produce light via excited metal atoms (Scheme 12).

Chemical reactions in flames can produce electronically excited species and hence chemiluminescence. The technique has been used in air pollution monitoring and detection systems for gas chromatography (Seitz and Neary, 1977). The excited species need not be directly involved in a chemical reaction as it can receive some of the energy liberated from highly exothermic reactions occurring in the flame. In a sodium chloride cool flame, light emission results from the production of excited sodium atoms (Strive et al., 1971) (Scheme 13). In cool hydrogen-air flames chemiluminescent emission is obtained from carbon (as CH_2 and CH), nitrogen (as HNO), phosphorus (as POH), sulfur (as S_2), selenium (as Se_2), and several halogens (Seitz and Neary, 1977). For sulfur, the reaction sequence proposed is shown in Scheme 14. Gas phase chemiluminescence reactions, such as the emission of SmF in the reaction of Sm atoms with NF_3 in an argon carrier at high pressure, have been investigated in the search for chemical laser systems (Eckstrom et al., 1975).

Combination reactions in the gas phase (Scheme 15), such as those that follow the dissociation of molecules by an electric or micro-

$$Na + Cl_2 \longrightarrow NaCl + Cl\cdot$$

$$Na_2 + Cl\cdot \longrightarrow NaCl + [Na]^{*} \longrightarrow \text{LIGHT}$$

Scheme 13 Flame chemiluminescence from excited sodium atoms in a sodium chloride cool flame. (From Rauhut, 1979.)

$$H + H + S_2 \longrightarrow H_2 + [S_2]^* \longrightarrow \text{LIGHT}$$

Scheme 14 Sulfur chemiluminescence in a cool hydrogen-air flame.

wave discharge, may produce chemiluminescence (Rauhut, 1979; Thrush and Golde, 1973; Ogryzlo and Pearson, 1968). These types of reaction are often called "afterglows."

IV. SOLID PHASE CHEMILUMINESCENCE

Light emission from solids, produced by processes such as electroluminescence, has been widely investigated (Blasse, 1978) and has found numerous applications for example in cathode ray tube phosphors and x-ray screens (Williams, 1978). The number of solids which undergo chemical oxidation reactions leading to chemiluminescence is, however, small. Perhaps the best illustration of this type of reaction is the chemiluminescence produced from the oxidation of siloxene. This compound and its derivatives are highly polymerized, with a permutoid structure, and can be readily produced from calcium silicide and hydrochloric acid. The basic formula is $(Si_6H_6O_3)n$ (17) but hydroxyl

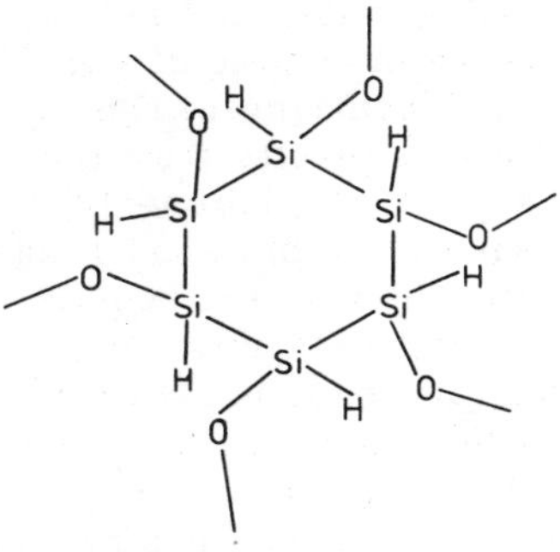

(17)

and chloride groups may be substituted in the basic structure. Oxidation of siloxene with hydrogen peroxide, permanganate, ceric sulfate, chromic acid, nitric acid, or other strong oxidants results in a

$$N\cdot + N\cdot \longrightarrow [N_2]^* \longrightarrow \text{LIGHT}$$

Scheme 15 The nitrogen afterglow; an example of combination chemiluminescence. (From Rauhut, 1979.)

red chemiluminescence (Isacsson and Wettermark, 1974; Rauhut, 1979). An energy transfer process, after oxidation, to hydroxysiloxene present in the structure is believed to be responsible for the emission (Isacsson and Wettermark, 1974). Solid lithium organophosphides also react with oxygen to produce light (Strecker et al., 1973).

Electron transfer reactions have been reported to produce chemiluminescence in solid phase reactions. Examples include the luminescence from anthracene crystals subjected to an alternating current (Helfrich and Schneider, 1966) and the luminescence produced from the recombination of an electron and the carbazole free radical in a frozen glass matrix (Linschitz et al., 1954).

A solid-state chemiluminescent indicator in the form of paper impregnated with 0.1% luminol, $CuSO_4$, and Na_2CO_3 has been described in the Russian literature and used for hydrogen peroxide determination (Arganov and Reiman, 1979).

V. CONCLUSION

Chemiluminescent reactions occur in solid, liquid, and gas phases and involve a diversity of organic and inorganic molecules. In many instances the mechanism of the light-producing reaction is incompletely understood and the subject of some controversy. The analytical potential of most chemiluminescent reactions is unexplored and only liquid phase reactions have found widespread application in clinical analysis, e.g., the reaction of luminol or lucigenin with peroxide.

REFERENCES

Adam, W. (1975). Biological light, α-peroxylates as bioluminescent intermediates. *J. Chem. Educ. 52*:138-145.

Adam, W., O. Cueto and F. Yany (1978). On the mechanism of the rubrene-enhanced chemiluminescence of α-peroxylactones. *J. Amer. Chem. Soc. 100*:2587-2589.

Adam, W., G. A. Simpson and F. Yany (1974). Mechanism of direct and rubrene enhanced chemiluminescence during α-peroxylactone decarboxylation. *J. Phys. Chem. 78*:2559-2569.

Arganov, K., and L. V. Reiman (1979). Chemiluminescence analysis of hydrogen peroxide using a solid phase indicator. *Zh. Anal. Khim. 36* (8): 1533-1538.

Bard, A. J., P. Csaba, C. P. Keszthelyi, H. Tachikawa and N. E. Tokel (1973). On the efficiency of electrogenerated chemiluminescence. In *Chemiluminescence and Bioluminescence*, M. J. Cormier, D. M. Hercules, and J. Lee (Eds.). Plenum, New York, pp. 193-209.

Blasse, G. (1978). Materials science of the luminescence of inorganic solids. In *Luminescence of Inorganic Solids*, B. DiBartolo (Ed.). Plenum, New York, pp. 457-494.

Bogan, D. J., R. S. Sheinson and F. W. Williams (1976). Gas phase dioxetane chemistry. Formaldehyde (A → X) chemiluminescence from the reaction of O_2 ('Δg) with ethylene. *J. Amer. Chem. Soc. 98*: 1034-1036.

Bollyky, L. J., and M. M. Rauhut (1971). U.S. Patent 3,597,367.

Bolton, P. M. and D. R. Kearns (1974). The molecular oxygen induced chemiluminescence of aryl Grignard reagents. *J. Amer. Chem. Soc. 96*:4651-4654.

Brundrett, R. B., D. F. Roswell and E. H. White (1972). Yields of chemically produced excited states. *J. Amer. Chem. Soc. 94*:7536-7541.

Burguera, J. L., and A. Townshend (1980). Determination of ng/ml levels of sulphide by a chemiluminescent reaction. *Talanta 27*:309-314.

Campbell, I. M., and D. L. Baulch (1978). Chemiluminescence in the gas phase. *Gas Kin. Energ. Trans. 3*:42-81.

Cormier, M. J., D. M. Hercules and J. Lee (Eds.) (1973). *Chemiluminescence and Bioluminescence*. Plenum, New York.

Eckstrom, D. J., S. A. Edelstein, D. L. Huestis, and S. W. Benson (1975). Chemiluminescence studies. IV. Pressure-dependent photon yields for Ba, Sm and Eu reactions with N_2O, O_3, O_2, F_2 and NF_3. *J. Chem. Phys. 63*:3828-3834.

Faulkner, L. R. (1976). Chemiluminescence in the liquid phase: Electron transfer. In International Review of Science, Physical Chemistry Series Two, Vol. 9, *Chemical Kinetics*, D. R. Herschbach (Ed.). Butterworths, London, pp. 213-263.

Foote, C. S., and S. Wexler (1964). Olefin oxidations with excited singlet molecular oxygen. *J. Amer. Chem. Soc. 86*:3879-3880.

Gorus, F., and E. Schram (1979). Applications of bio- and chemiluminescence in the clinical laboratory. *Clin. Chem. 25*:512-519.

Gundermann, K. D. (1978). Chemiluminescence: Current status. In *Proceedings of the International Symposium on Analytical Applications of Bioluminescence and Chemiluminescence*, E. Schram and P. Stanley (Eds.). State Printing and Publishing, Westlake Village, Calif., pp. 37-86.

Gundermann, K. D., and H. Giesecke (1979). Constitution and chemiluminescence. XI. Oligomers of 5-amino-8-vinylphthalazine-1,4(^{2}H,^{3}H)-dione. *Liebigs Ann. Chem. 8*:1085-1093.

Hart, R. C., and M. J. Cormier (1979). Recent advances in the mechanisms of bio- and chemiluminescent reactions. *Photochem. Photobiol. 29*:209-215.

Helfrich, W., and W. G. Schneider (1966). Transients of volume-controlled current and recombination radiation in anthracene. *J. Chem. Phys. 44*:2902-2909.

Herschbach, D. R. (1973). Transvibronic reactions in molecular beams. In *Chemiluminescence and Bioluminescence*, M. J. Cormier, D. M. Hercules, and J. Lee (Eds.). Plenum, New York, pp. 29-43.

Hersch, L. S., W. P. Vann, and S. A. Wilhelm (1979). A luminol-assisted, competitive-binding immunoassay of human immunoglobulin G. *Anal. Biochem. 93*:267-271.

Hopkins, T. A., H. H. Seliger, E. H. White, and M. W. Cass (1967). The chemiluminescence of firefly luciferin. A model for the bio-luminescent reaction and identification of the product excited state. *J. Amer. Chem. Soc. 89*:7184-7150.

Isacsson, U. and G. Wettermark (1974). Chemiluminescence in analytical chemistry. *Anal. Chim. Acta 68*:339-362.

Keszthelyi, C. P., N. E. Tokel-Takyoryan, and A. J. Bard (1975). Electrogenerated chemiluminescence: Determination of the absolute luminescence efficiency in electro-generated chemiluminescence; 9,10-diphenylanthracene-thianthrene and other systems. *Anal. Chem. 47*(2):249-256.

Lechtken, P., A. Yekta, and N. J. Turro (1973). Tetramethyl-1,2-dioxetane. A mechanism for an autocatalytic decomposition. Evidence for a quantum chain reaction. *J. Amer. Chem. Soc. 95*: 3027-3028.

Lee, K., L. A. Singer and K. D. Legg (1976). Chemiluminescence from the reaction of singlet oxygen with 10,10'-dimethyl-9-9'-biacridylidene. A reactive 1,2-dioxetane. *J. Org. Chem. 41*:2685-2688.

Linschitz, H., M. G. Berry and D. Schweitzer (1954). The identification of solvated electrons and radicals in rigid solutions of photoxidized organic molecules. Recombination luminescence in organic phosphors. *J. Amer. Chem. Soc. 76*:5833-5839.

MacDonald, A., K. W. Chan and T. A. Nieman (1979). Lophine chemiluminescence for metal ion determinations. *Anal. Chem. 51*: 2077-2082.

Maskiewicz, R., D. Sogah and T. C. Bruice (1979a). Chemiluminescent reactions of lucigenin. 1. Reactions of lucigenin with hydrogen peroxide. *J. Amer. Chem. Soc. 101* (18):5347-5354.

Maskiewicz, R., D. Sogah and T. C. Bruice (1979b). Chemiluminescent reactions of lucigenin. 2. Reactions of lucigenin with hydroxide ion and other nucleophiles. *J. Amer. Chem. Soc. 101*(18):5355-5364.

Maulding, D. R., and B. G. Roberts (1972). The chemiluminescence of tetrachloroethylene carbonate and related compounds. *J. Org. Chem. 37*:1458-1459.

McCapra, F. (1966). The chemiluminescence of organic compounds. *Q. Rev. (London) 20*:485-510.

McCapra, F., and P. D. Leeson (1979). Possible mechanism for bacterial bioluminescence and luminol chemiluminescence. *J. Chem. Soc. Chem. Commun.* 3:114-117.

McCapra, F., K. Zaklika and J. Gilmore (1975). NTIS Report AD A011031.

McCapra, F., D. E. Tutt and R. M. Topping (1977). Assay method utilizing chemiluminescence. British Patent 1,461,877.

McCapra, F., D. Tutt and R. M. Topping (1978). The chemiluminescence of acridinium phenyl carboxylates in the assay of glucose and hydrogen peroxide. In *Proceedings of the Internal Symposium on Analytical Applications of Bioluminescence and Chemiluminescence*, E. Schram and P. Stanley (Eds.). State Printing and Publishing, Westlake Village, Cal., p. 221.

Mendenhall, G. D. (1977). Analytical applications of chemiluminescence. *Angew. Chem. 16*:225-232.

Murray, R. W., and M. L. Kaplan (1968). Gas-phase reactions of singlet oxygen from a chemical source. *J. Amer. Chem. Soc. 90*: 4161-4162.

Murray, R. W., and M. L. Kaplan (1969). Singlet oxygen sources in ozone chemistry. Chemical oxygenations using the adducts between phosphite esters and ozone. *J. Amer. Chem. Soc. 91*:5358-5364.

Ogryzlo, E. A., and A. E. Pearson (1968). Excitation of violanthrone by singlet oxygen. A chemiluminescence mechanism. *J. Phys. Chem. 72*:2913-2916.

Rauhut, M. M. (1979). Chemiluminescence. In *Kirk-Othmer Encycl. Chem. Technol. 3rd Ed., Vol. 5*, M. Graysen and D. Echroth (Eds.). Wiley, New York, pp. 416-450.

Rauhut, M. M., D. Sheehan, R. A. Clarke and A. M. Semsel (1965). Structural criteria for chemiluminescence in acyl peroxide decomposition reactions. *Photochem. Photobiol. 4*(6):1097-1110.

Rauhut, M. M., L. J. Bollyky, B. G. Roberts, M. Loy, R. H. Whitman, A. V. Iannotta, A. M. Semsel, and R. A. Clarke (1967). Chemiluminescence from reactions of electronegatively substituted aryl oxalates with hydrogen peroxide and fluorescent compounds. *J. Amer. Chem. Soc. 89*:6515-6522.

Roswell, D. F., and E. H. White (1978). The chemiluminescence of luminol and related hydrazides. In *Methods in Enzymology, Vol. 57, Bioluminescence and Chemiluminescence*, M. A. DeLuca (Ed.). Academic, New York, pp. 409-423.

Schmidt, S. P., and G. B. Schuster (1978). Kinetics of unimolecular dioxetanone chemiluminescence. Competitive parallel reaction paths. *J. Amer. Chem. Soc. 100*:5559-5561.

Schroeder, H. R., and F. M. Yeager (1978). Chemiluminescence yields and detection limits of some isoluminol derivatives in various oxidation systems. *Anal. Chem. 50*(8):1114-1120.

Schuster, G. B., B. Dixon, J. Koo, S. P. Schmidt and J. P. Smith (1979). Chemical mechanisms of chemi- and bioluminescence. Reactions of high energy content organic compounds. *Photochem. Photobiol. 30*:17-26.

Scott, G., W. R. Seitz and J. Ambrose (1980). Improved determination of hydrogen peroxide by measurement of peroxyoxalate chemiluminescence. *Anal. Chim. Acta 115*:221-228.

Seitz, W. R., and M. P. Neary (1977). Chemiluminescence and bioluminescence analysis. *Contemp. Topics Anal. Clin. Chem. 1*:49-125.

Sheehan, D. (1975). U.S. Patent 3,914,255.

Simpson, J. S. A., A. K. Campbell, M. E. T. Ryall and J. S. Woodhead (1979). A stable chemiluminescent-labelled antibody for immunological assays. *Nature 279*:646-647.

Stauff, J., U. Sander and W. Jaeschke (1973). Chemiluminescence of perhydroxyl- and carbonate radicals. In *Chemiluminescence and Bioluminescence*, M. J. Cormier, D. M. Hercules, and J. Lee (Eds.). Plenum, New York, pp. 131-143.

Strecker, R. A., J. L. Snead and G. P. Sollot (1973). Chemiluminescence of lithium phosphides. *J. Amer. Chem. Soc. 95*:210-214.

Strive, W. S., T. Kitagawa and D. R. Herschbach (1971). Chemiluminescence in molecular beams: Electronic excitation in reactions of C^1 atoms with Na_2 and K_2 molecules. *J. Chem. Phys. 54*:2759-2761.

Thrush, B. A., and M. F. Golde (1973). The nitrogen afterglow. In *Chemiluminescence and Bioluminescence*, M. J. Cormier, D. M. Hercules, and J. Lee (Eds.). Plenum, New York, pp. 73-83.

Totter, J. R., V. J. Medina and J. L. Scoseria (1960). Luminescence during the oxidation of hypoxanthine by xanthine oxidase in the presence of dimethylbiacridylium nitrate. *J. Biol. Chem. 235*:238-241.

Van Zee, R. J., and A. V. Khan (1976). Transient emitting species in phosphorus chemiluminescence. *J. Chem. Phys. 65*:1764-1772.

Veazey, R. L., and T. A. Nieman (1979). Chemiluminescent determination of clinically important organic reductants. *Anal. Chem. 51* (13):2092-2096.

Walling, C., and S. A. Buckler (1955). The reaction of oxygen with organometallic compounds. A new synthesis of hydroperoxides. *J. Amer. Chem. Soc. 77*:6032-6038.

White, E. H., and M. M. Bursey (1964). Chemiluminescence of luminol and related hydrazides: The light emission step. *J. Amer. Chem. Soc. 86*:941-942.

White, E. H., E. Rapaport, T. A. Hopkins, and H. H. Seliger (1969). Chemi- and bioluminescence of firefly luciferin. *J. Amer. Chem. Soc. 91*:2178-2180.

Whitehead, T. P., L. J. Kricka, T. J. N. Carter and G. H. G. Thorpe (1979). Analytical luminescence: Its potential in the clinical laboratory. *Clin. Chem. 25*(9):1531-1546.

Williams, F. (1978). Applications of Luminescence. In *Luminescence of Inorganic Solids*, B. DiBartolo (Ed.). Plenum, New York and London, pp. 539-546.

Wilson, T. (1976). Chemiluminescence in the liquid phase: Thermal cleavage of dioxetanes. In *International Review of Science, Physical Chemistry Series Two*, Vol. 9, *Chemical Kinetics*, D. R. Herschbach (Ed.). Butterworths, London-Boston, pp. 265-322.
Wilson, T., and A. P. Schaap (1971). The chemiluminescence from cis-diethoxy-1,2-dioxetane. An unexpected effect of oxygen. *J. Amer. Chem. Soc.* *93*:4126-4136.

4

ANALYTICAL APPLICATIONS OF BIOLUMINESCENCE: THE FIREFLY SYSTEM

ARNE LUNDIN* Arrhenius Laboratory, University of Stockholm, Stockholm, Sweden

I.	Introduction	44
II.	Recent Improvements of the Kinetics of the Firefly Reaction	44
III.	Technical Considerations	49
	A. Instrumentation	49
	B. Reagents	50
	C. Buffers	51
	D. Linear Range	52
	E. Analytical Interference from Sample	53
IV.	Applications	54
	A. Biomass Determinations	54
	B. Quantitation of Antibiotic and Vitamin Effects in Microorganisms	56
	C. Applications Involving Changes in Cellular ATP Concentrations	57
	D. Determination of Enzymes and Metabolites	59
V.	Conclusion	66
	References	66

*Dr. Lundin is presently with the LKB-Produkter AB, Bromma, Sweden.

This study was in part supported by the Swedish Board for Technical Development and by the Swedish Natural Science Research Council.

I. INTRODUCTION

The requirement for ATP in the firefly luciferase reaction was shown by McElroy in 1947 (McElroy, 1947). The potential analytical usefulness of the firefly reaction not only for sensitive assays of ATP but for assays of all metabolites and enzymes participating in ATP-converting reactions was shown by Strehler and Totter (1952). Today more than 1000 papers on analytical applications of the firefly reaction have been published. These applications represent a wide variety of disciplines such as biochemistry (e.g., bioenergetics), biology (e.g., microbiology, immunology, oceanography, limnology), medicine (e.g., clinical chemistry and bacteriology), and space research (life detection on other planets).

Although many research laboratories have used the firefly reaction extensively for analytical purposes, not a single application has gained a general acceptance in routine laboratories, i.e., in clinical chemistry or clinical bacteriology. One of the purposes of this chapter is to illustrate how recent improvements of the kinetics of the firefly reaction will, in all probability, change this picture in the near future. The use of firefly luciferase in assays based on immobilized enzymes (Lee et al., 1977) and luminescence immunoassays (Carrico et al., 1976) represent highly interesting aspects of the development of the firefly assay. Since these new techniques are discussed elsewhere in this book, they will not be considered here.

II. RECENT IMPROVEMENTS OF THE KINETICS OF THE FIREFLY REACTION

The mechanism of the firefly luciferase reaction has been described elsewhere (DeLuca, 1976). However, some kinetic aspects of particular analytical importance will be discussed in this chapter.

The present knowledge of the firefly reaction as described in a recent review by DeLuca (1976) is summarized in Fig. 1. The initial activation of D-luciferin (L_{red}) with ATP (or rather, MgATP) results in the formation of enzyme-bound D-luciferyl adenylate (L_{red}-AMP) and free pyrophosphate (PP_i). The reaction is reversible and pyrophosphate is an inhibitor of the forward reaction (McElroy et al., 1953). Under normal reaction conditions, i.e., low levels of luciferase and ATP), inhibitory concentrations of pyrophosphate will not be formed because only a minor portion of the ATP is degraded in the luciferase reaction (Lundin et al., 1976).

Before the oxidative decarboxylation of D-luciferyl adenylate to AMP, CO_2, and electronically excited oxyluciferin, there are at least two rate-limiting steps (DeLuca and McElroy, 1974). These steps result in a lag of 0.025 sec before any light is emitted and a time to reach peak light intensity of 0.25 to 0.3 sec. When the electronically

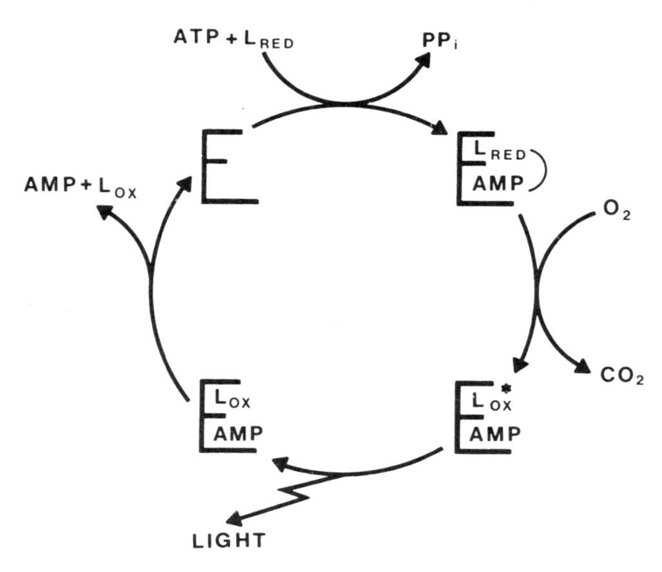

Figure 1 Schematic representation of the firefly luciferase reaction (symbols are explained in the text).

excited oxyluciferin goes to the ground state, a photon is emitted with an overall quantum yield of 0.88 (Seliger and McElroy, 1960).

Before luciferase can participate in a second reaction cycle, oxyluciferin and AMP must be released from the active site of the enzyme. However, the enzyme-product complex is very stable. Gates and DeLuca (1975) were able to isolate enzyme-product complexes containing oxyluciferin and, in the presence of pyrophosphatase, AMP by gel filtration on Sephadex G-25. A completely stable enzyme-product complex should result in a rapid inactivation of luciferase and a rapid decay of the light emission. In addition, AMP and oxyluciferin are competitive inhibitors of luciferase with K_i values of 2.4×10^{-4} mol/L (Lee et al., 1970) and 2.3×10^{-7} mol/L (Goto et al., 1973), respectively.

Lundin et al. (1976) showed that it is possible to define analytical conditions to avoid product inhibition. Under these conditions ATP concentrations below 10^{-6} mol/L produce a stable light emission proportional to the ATP concentration. Thus the ATP concentration in the sample can be monitored simply by continuously measuring the light emission as shown in Fig. 2. This new technique, which has been called bioluminescent ATP monitoring, has several advantages over previously used assays based on the firefly reaction. As shown in Table 1, these advantages cover all types of assays based on the firefly reaction, i.e., determination of ATP in biological extracts, endpoint determination of metabolites participating in ATP-converting

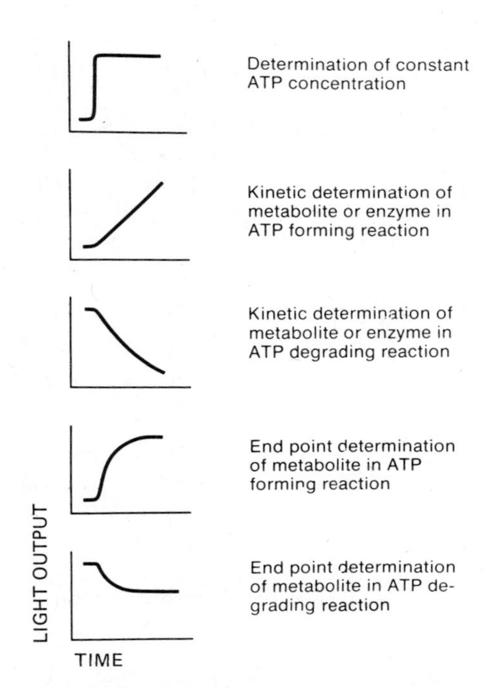

Figure 2 Light emission kinetics in different types of assays obtainable by ATP monitoring.

reactions, and kinetic determination of metabolites and enzymes participating in ATP-converting reactions.

In order to obtain conditions suitable for ATP monitoring, there are five requirements that should be fulfilled:

1. There should be no accumulation of inhibitors or enzyme-product complex in the firefly reaction. This means that the reagent composition should be carefully optimized. Furthermore, the ATP concentration should be low, i.e., below 10^{-6} mol/L, in order to have a low rate of product formation. An ATP concentration below 10^{-6} mol/L is also required to produce a linear relation between light emission and ATP concentration. This may be inferred from the Michaelis-Menten equation $v/V_{max} = S/(K_m + S)$ using a K_m value of 2.5×10^{-4} mol/L ATP (Lee et al., 1970).
2. There should be essentially no degradation of ATP. The degradation of ATP in the luciferase reaction is less than 1%/min, even in the reagent with the highest activity commercially available (LKB-Wallac). However, in crude reagents there may be enzyme systems degrading ATP (Lundin and Thore, 1975a).

Table 1 Advantages of ATP Monitoring as Compared to Assays Using "Flash-Type" Firefly Reagents

Type of assay	Advantage	Explanation
A. Determination of constant ATP concentration	Accuracy	Internal calibration can be done in each single assay by measuring the light signal before and after addition of a known amount of ATP. Interference with luciferase reaction disturbing kinetics of light emission is easily observed.
	Sensitivity	Light signal can be integrated for any desired period of time without loss of accuracy.
	Convenience	Stable light emission allows convenient mixing of reagent and sample, e.g., outside instrument. No special equipment needed to capture peak light emission.
B. Kinetic determination of metabolite or enzyme	Accuracy	Continuous monitoring gives more analytical information than two-point measurement. Interference with ATP-converting reaction easily observed. See also section A in this table.
	Sensitivity	Measuring time is not fixed in each series of assays (as in two-point measurements), but can be adjusted individually to the rate of the reaction studied.
	Convenience	Rate of formation or degradation of ATP determined in a single assay, i.e., less labor and less reagent.

Table 1 (Continued)

Type of assay	Advantage	Explanation
C. Endpoint determination of metabolite	Accuracy	Continuous monitoring is attained. See also sections A and B in this table.
	Sensitivity	Even a small increase of the ATP concentration can be detected, since the assay is performed in a single curette. See also Sec. A in this table.
	Convenience	By eliminating the need for multiple sampling from reaction mixture (to assure endpoint attainment), less labor and reagent are used.

3. The luciferase reagent should be highly purified and contain no enzymes or metabolites participating in ATP-converting reactions. Since samples may contain enzymes or metabolites in addition to those being studied, reactions with impurities in the luciferase reagent may effect the ATP concentration. Thus the reagent should be as pure as possible and levels of contaminants should be specified.
4. Reaction conditions should be chosen to avoid a continuous inactivation of luciferase (or luciferin) during the measurements. This means that extremes in pH (pH 6 to 8 is appropriate) or high temperatures should be avoided. Compounds denaturating or inactivating luciferase should also be avoided.
5. The temperature should be constant while measuring. The luciferase reaction, as any enzymatic reaction, is temperature-dependent, and a change in the temperature affects the rate of the reaction, i.e., the intensity of the light emitted.

There have been some previous attempts to use the firefly reaction in coupled enzyme systems to study ATP-converting reactions (Strehler and Totter, 1952; Lemasters and Hackenbrock, 1973,1976; Witteveen et al., 1974). However, the term *ATP monitoring* will be used only in connection with applications in which the requirements given above were fulfilled. A number of such applications have been published and will be given later in the chapter. From Table 1 it is clear that the advantages introduced by the ATP-monitoring technique considerably improve the applicability of the firefly reaction, particularly in the study of ATP converting reactions. A firefly reagent suitable for ATP monitoring is now commercially available (LKB-Wallac), and the number of applications of ATP monitoring will undoubtedly increase rapidly in the near future.

III. TECHNICAL CONSIDERATIONS

A. Instrumentation

The ideal light-measuring instrument or luminometer should meet the following requirements (see also Chap. 11):

1. The signal-to-noise ratio should be as high as possible to allow measurements of low ATP concentrations.
2. The spectral response of the photomultiplier used should be as flat as possible in the wavelength region for light emission in the luciferase reaction (500 to 700 nm). This would make changes in reaction conditions less critical. The spectrum of the light emitted is changed from yellow-green to red at low pH and in the presence of Zn^{2+}, Cd^{2+}, and Hg^{2+} (McElroy et al., 1969).
3. A luminometer should be capable of maintaining a constant temperature in the reaction mixture during measurement. The rate of the luciferase reaction is temperature-dependent with a maximum at approximately 25°C (Fig. 3) and a changing temperature will affect the intensity of the emitted light.
4. It should be possible to mix reagents and sample in a light-sealed reaction chamber. With firefly reagents, which produce a flash of light, it is necessary to inject the reagent directly to the sample in this chamber in order to assure detection of the peak light emission giving the most accurate results (Lundin and Thore, 1975a; Riemann, 1979). In ATP monitoring it is often advantageous to make low-volume additions of reagents in the light-sealed chamber. This necessitates facilities for reagent injection and for stirring of the reaction mixture.
5. Alternative outputs are advantageous. In studying the kinetics of the firefly reaction or ATP monitoring of ATP-converting reactions, a potentiometric recorder or other time-resolving recording unit is required. In measurements of constant ATP levels, i.e., biological extracts, a digital display or printer may be more convenient.
6. In the study of photophosphorylation by ATP monitoring it is necessary to be able to illuminate the reaction mixture while measuring (Lundin and Baltscheffsky, 1978; Lundin et al., 1979a).

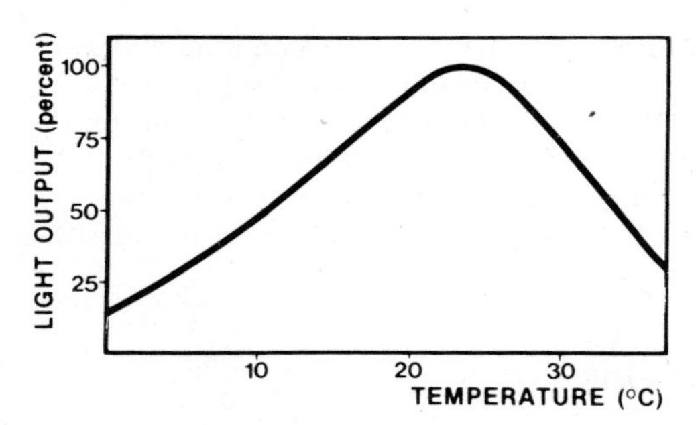

Figure 3 Temperature dependence of firefly luciferase reaction. (From Myhrman et al., 1978.)

No commercially available luminometer conforms to all the requirements given above. Thus it is necessary to make a suitable compromise for each particular application.

B. Reagents

A highly purified reagent (preferably fulfilling the requirements enumerated earlier in this chapter) will give the most accurate results (cf. Table 1). Contaminating ATP-converting enzymes in reagents can cause analytical interference and complex kinetics in the light emission of the firefly reaction (Rasmussen and Nielsen, 1968; Lundin and Thore, 1975a). Furthermore, calibration by addition of a known amount of ATP to the reaction mixture while measuring will not be possible. This means that analytical interference in individual samples may not be easily compensated.

Several methods for the purification of firefly luciferase have been published (Green and McElroy, 1956; Nielsen and Rasmussen, 1968; Bény and Dolivo, 1976; Lundin, 1978). The method by Lundin (1978) is based on ammonium sulfate precipitation followed by isoelectric focusing and gives a luciferase appearing as a single band both in SDS gel electrophoresis and analytical isoelectric focusing (Lundin et al., 1981). This reagent is now commercially available (LKB-Wallac).

In order to obtain the inherent sensitivity and specificity of the firefly assay, it is important that all reagents be highly purified. Inhibitors may decrease the sensitivity or affect the kinetics of the firefly reaction. Contaminating enzymes or metabolites may cause interfering side reactions which decrease the specificity of the technique. Two examples of such interference, taken from the development of the creatine kinase assay by ATP monitoring (Lundin and Styrelius, 1978; Lundin et al., 1979c; Lundin et al., 1982), are as follows:

1. Commercially available ADP is contaminated by approximately 2% ATP. Using the optimum ADP concentration in the assay (0.5 mol/L ADP) results in 10^{-5} mol/L ATP, i.e., far greater than the upper limit of the linear range of the firefly assay (10^{-6} mol/L ATP). An ADP preparation of adequate purity can be obtained by ion exchange chromatography (Lundin, 1978).
2. Some batches of commercially available creatine phosphate are contaminated by a potent inhibitor of the firefly reaction (Lundin et al., 1979c). A suitable purification method has not yet been found. Creatine phosphate of adequate quality can, however, be obtained by a selection of appropriate batches of commercially available preparations.

C. Buffers

The pH optimum of the firefly reaction is 7.75 (Lundin et al., 1976). In ATP monitoring, an interval of pH 6 to 8 may be used, provided that the buffering capacity is such that pH can be carefully controlled. However, luciferase is extremely sensitive to changes in pH and the activity at pH 6 is only 8% of the activity obtainable at pH 7.5 (uncorrected for the lower spectral response of the photomultiplier for red light obtained at low pH).

The inhibition of the luciferase reaction by salts is due to an increase in the K_m for MgATP occurring at high ionic strength. Furthermore, there is a specific anion inhibition resulting, presumably, from a small conformational change in the active site upon the binding of the anion to luciferase (Denburg and McElroy, 1970). The order of effectiveness of inhibition by anions has been described by Denburg and McElroy (1970) as $SCN^- > I^- \sim NO_3^- > Br^- > Cl^-$ and by Gilles et al. (1976) as $ClO_4^- > I^- > Cl^- >$ acetate.

In a particular application the choice of buffer should be made according to the following criteria:

1. The buffer should have a high buffering capacity at the pH used in the assay, thereby enabling use of the buffer at a low ionic strength.
2. The anion of the buffer should be chosen to give a low inhibition of the firefly reaction, e.g., acetate.
3. There should be no time-dependent inactivation or activation of luciferase. This is of particular importance in ATP monitoring.

The choice of buffers in ATP monitoring causes particular problems, since firefly luciferase (in contrast to conventionally used auxiliary enzymes in coupled enzymatic assays) is not present in excess concentration. An example may be given from the development of the assay of creatine kinase by ATP monitoring. The pH optima for the creatine kinase and firefly reactions are 6.7 and 7.75, respectively. The pH optimum for the coupled reaction, i.e., the rate of increase

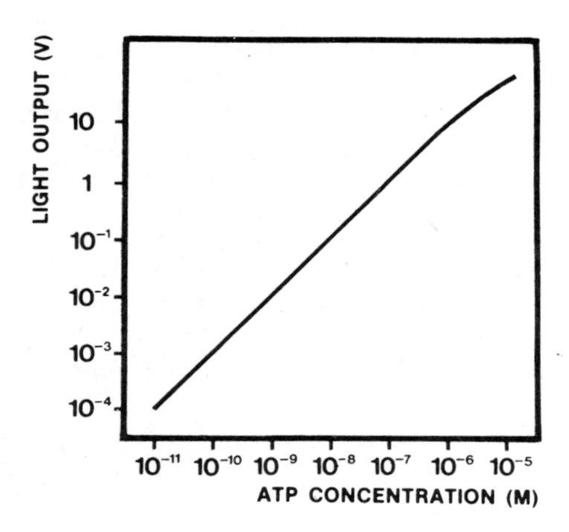

Figure 4 Linearity of firefly assay in 0.1 mol/L tris acetate buffer, pH 7.75. (From Myhrman et al., 1978.)

of the light emission, is pH 7.5. In this particular assay 0.1 mol/L imidazole acetate, pH 6.7, was used because it is the buffer generally accepted for the conventional spectrophotometric assay of creatine kinase. It was assumed that the use of other buffers might cause difficulties in the interpretation of assay results by the clinician. In assays of other enzymes it may be preferable to choose buffers on entirely different grounds.

D. Linear Range

The range of a linear relation between light emission and ATP concentration goes from the detection limit of the assay (approximately 10^{-11} mol/L) to an upper limit, which can be calculated from the K_m of luciferase for ATP (or rather MgATP) using the Michaelis-Menten equation $v/V_{max} = S/(K_m + S)$. The K_m value for ATP increases with pH, ionic strength, and decreases with luciferin concentration (Lee et al., 1970). Under optimal analytical conditions the K_m is 2.5×10^{-4} mol/L corresponding to an upper limit of the linear range of approximately 10^{-6} mol/L ATP. However, under other analytical conditions the K_m can be as low as 10^{-4} mol/L ATP or as high as 10^{-3} mol/L ATP (Lee et al., 1970). Furthermore, the presence of competitive inhibitors such as AMP, ADP, and other nucleotides may increase the apparent K_m value, $K'_m = K_m(1 + I/K_i)$, since K_i values are often in the interval 2 to 80×10^{-4} mol/L (Lee et al., 1970). Thus the upper limit of the linear range of the firefly assay is normally approximately 10^{-6} mol/L ATP (Fig. 4) but may under certain analytical conditions be even higher. In each particular application, therefore, it is worthwhile to determine the linear range of the assay.

E. Analytical Interference from Sample

Interference caused by nucleotide-converting reactions, inhibitors, inactivators, and variations in pH and ionic strength have already been mentioned.

Sample pretreatment is an important step in the minimizing of analytical interference caused by the biological sample as such. Furthermore, agents used in, for example, extraction of adenine nucleotides from biological material are often strongly denaturing or inhibiting, causing serious analytical interference. A general discussion on pretreatment of all types of biological samples is beyond the scope of this chapter. However, some important aspects are illustrated in a study of methods for extraction of bacterial adenine nucleotides, determined by the firefly assay (Lundin and Thore, 1975b).

In the above mentioned study five bacterial species were extracted using 10 different methods and the levels of ATP, ADP, and AMP were determined by the firefly assay. Analytical interference by bacterial enzymes not inactivated during the extraction was found to be a major problem. However, this problem could be solved by inclusion of ethylenediaminetetraacetate in the extraction media.

The extraction methods were compared with respect to yield of adenine nucleotides, interference with the enzymic assay, reproducibility of the methods, and stability of the extracts. It was concluded that extraction with trichloroacetic acid gave the results most closely reflecting actual levels of adenine nucleotides in intact bacterial cells. A disadvantage with this extraction method is, however, that trichloroacetic acid (TCA) strongly interferes with the assay. Thus extracts have to be extensively diluted or TCA removed by extraction with diethyl ether before the assay (Lundin and Thore, 1975b). Similar studies (with similar results) have been performed on the extraction of adenine nucleotides in algae (Larsson and Olsson, 1979) and in various blood cells (Lundin, unpublished). Considering the analytical interference from TCA it may, at least in applications necessitating the highest possible sensitivity, be worthwhile to try to find the lowest concentration of TCA or other extracting agents giving a reliable extraction of adenine nucleotides. In such studies it is recommended to use extraction with a high concentration of TCA (final concentration $> 10\%$) as a reference. This is also recommended when using commercially available extracting (releasing) agents, since a generally applicable agent for extraction of, e.g., bacterial or somatic cell ATP, is not available. In samples containing a very high concentration of organic material or strongly buffering substances, e.g., certain types of activated sludge, extraction even with a high concentration of TCA may not always be reliable if the sample is not extensively diluted in the TCA solution.

The following parameters are important for comparisons of extraction methods:

1. The time between sampling and the inactivation of nucleotide-converting enzymes should be short because the ATP pool may be completely turned over in a few seconds.
2. The extraction should promptly and irreversibly inactivate all nucleotide-converting enzyme systems. The use of ethylenediaminetetraacetic acid improves this property in several extraction agents.
3. A reproducible yield, reflecting actual levels of adenine nucleotides in the biological material, should be obtained.
4. Extracts should be stable when stored under appropriate conditions.
5. Interference with firefly assay and other enzymatic systems used in the assay should be low.

IV. APPLICATIONS

Many papers have been published on applications of the firefly assay (cf. Introduction). The following section is limited, therefore, to a general description of the types of experiments that may be performed using the firefly assay in a variety of applications. References are given only to papers of general historic or technical interest. Examples of specific applications are given in tabular form (with references). Most of the clinical applications have been compiled from recent review articles on applications of bioluminescence (Gorus and Schram, 1979; Thore, 1979; Whitehead et al., 1979).

A. Biomass Determinations

The use of ATP as an index of biomass (Table 2) is based on two facts. First, ATP can be conveniently determined in low concentrations by the firefly assay. Second, the intracellular concentration of ATP is fairly constant within all organisms. This is due to the central role played by ATP in the metabolism of the cell. As ATP is a universal metabolite, the use of the firefly assay has even been considered a possible tool for the detection of life on other planets (MacLeod et al., 1969).

The number of cells that can be detected by the firefly assay depends on the size of the cell and the detection limit of the assay. With commercially available instruments and reagents, the detection limit of the firefly is approximately 10^{-11} mol/L ATP. As may be calculated from Table 3, this corresponds to approximately 10^{-4} bacterial cells or single HeLa or infusoria cells. That is is actually possible to assay of ATP in single cells has been demonstrated by Wettermark et al. (1975).

General application areas include biomass determinations in marine and fresh water, sediments, soil, wastewater, activated sludge, food, early warning of biological warfare, and for fermentation studies.

Table 2 Application of Biomass Determination

Application area	Sample	References
Environmental studies	Marine water	Hamilton and Holm-Hansen (1967) Holm-Hansen and Booth (1966)
	Fresh water	Jones and Simon (1977), Riemann (1979)
	Marine sediment	Karl and LaRock (1975)
	Lake sediment	Lee et al. (1971)
	Soil	MacLeod et al. (1969)
Drinking water and food industry	Drinking water	Levin et al. (1967)
	Milk	Bossuyt (1978)
	Beer	Hysert et al. (1976)
Sewage treatment plants	Wastewater	Levin et al. (1975)
	Activated sludge	Brezonik and Patterson (1971), Chiu et al. (1973), Kees et al. (1975), Statham and Laugton (1975)
Bacteriuria screening	Urine	Alexander et al. (1976), Conn et al. (1975), Curtis and Johnston (1979), Gutekunst (1975), Johnston and Curtis (1979), Johnston et al. (1976), Thore et al. (1975), Ånsehn et al. (1979).

In clinical bacteriology the firefly assay has been used for quantitation of antibiotic effects (see below and Chap. 10) and for bacteriuria screening. Bacteriuria screening is a good example of how selectivity in biomass determinations can be obtained by sample pretreatment. Human urine contains free ATP, human cell ATP, and bacterial ATP. During the short preincubation of the urine in the presence of the detergent Triton X-100 and the ATP-degrading enzyme apyrase, the human cells are lysed and all nonbacterial ATP is degraded (Thore et al., 1975). The essentially unaffected bacterial ATP is then extracted and determined by the firefly assay. Clinically satisfactory results, with respect to specificity and sensitivity, have

Table 3 Amount of ATP in Different Types of Cells

Cell type	Average ATP/cell (10^{-18} × moles per cell)
Escherichia coli	1.6
Proteus mirabilis	0.9
Staphylococcus aureus	2.3
Candida albicans	26
Red blood cells	60
Mononuclear white blood cells	50
Polymorphonuclear granulocytes	170
HeLa cells	8800
Human embryonic lung fibroblasts	450
Mice macrophages	1000
Infusoria (*Paramecium*)	7900
Dinoflagellates (*Peridinium cinctum*)	4700

Source: Data from Thore et al. (1975) and Wettermark et al. (1975).

been reported by several groups. The use of simplified extraction procedures and improved firefly reagents makes this method an interesting alternative to conventional bacteriuria-screening methods (Ånsehn et al., 1979).

B. Quantitation of Antibiotic and Vitamin Effects in Microorganisms

The firefly assay may be used for the determination of antibiotic and vitamin concentrations, e.g., in serum, as well as for antibiotic susceptibility testing of bacteria isolated from clinical specimens (Table 4). Intracellular as well as extracellular levels of ATP in bacteria and yeast have been used for this purpose.

In methods based on intracellular levels of ATP the firefly assay is used for the determination of ATP as a growth index (biomass determination). Although the amount of ATP per colony-forming unit (CFU) is changed during growth (Harber and Asscher, 1979), these effects can at least partially be attributed to differences in number of cells per CFU (Fig. 5). In vitamin determinations (Harber and Asscher, 1979) the growth stimulation is related to the vitamin concentration. Growth inhibition in the presence of an antibiotic (during 1.5 to 4 hr incubation) is related to the antibiotic concentration and to the antibiotic susceptibility of the organism. Thus, by using an organism with a known susceptibility it is possible to determine antibiotic concentrations and by using known concentrations of antibiotics the antibiotic susceptibility of bacteria in clinical specimens can be ascertained.

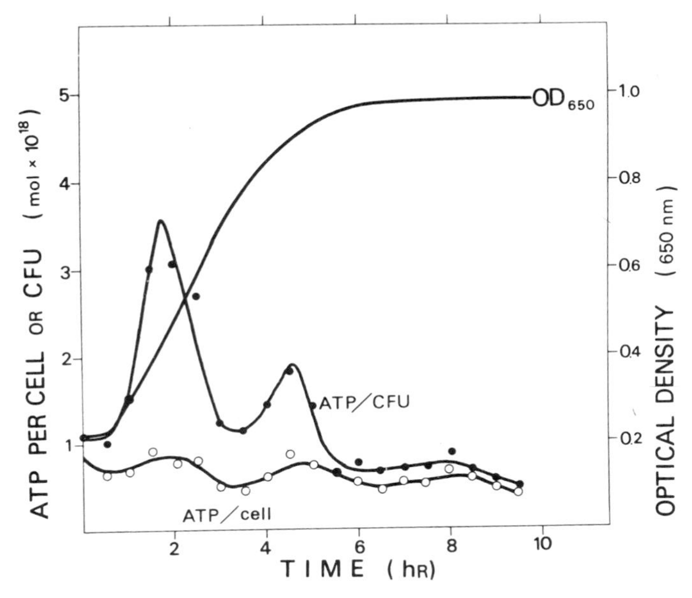

Figure 5 ATP content per cell and per colony forming unit (CFU) in *Escherichia coli* growing in nutrient broth. Cell counts were done microscopically in a Bürker chamber and CFU were determined with the conventional plating technique (Lundin, unpublished).

Aminoglycosides and perhaps also other antibiotics can be determined by an assay of the extracellular ATP which leaks from cells as a result of antibiotic action. This technique for antibiotic determination does not require an extraction step and has been demonstrated for the determination of gentamicin after a 75-min incubation of *E. coli* LU 14 in the presence of 10 μL of serum containing the antibiotic (Nilsson, 1978; Nilsson, 1980).

C. Applications Involving Changes in Cellular ATP Concentrations

ATP can be used as a parameter in the control of fermentation or activated sludge processes. In such applications ATP is both a parameter indicating the metabolic status of the organisms and a growth index (Table 5).

The firefly assay has been extensively used in studies on energy metabolism. One very interesting study indicating the usefulness of firefly assay for rapid identification of, say, bacteria has been published by Klofat et al. (1969). It was shown that mutants of *Bacillus subtilis* deficient in various enzymes could be identified with the

Table 4 Quantification of Antibiotic and Vitamin Effects in Microorganisms

Application area	Principle of test	References
Vitamin determination	Growth stimulation	Harber and Asscher (1979)
Antibiotic determination	Growth inhibition	Harber and Asscher (1977), Harber and Asscher (1979), Höjer et al. (1976), Höjer and Nilsson (1978), Höjer et al. (1979), Nilsson et al (1977), Nilsson et al (1979), Ånsehn et al (1977)
	ATP-converting enzymatic reaction	Daigneault et al. (1979)
	ATP leakage	Nilsson (1978), Nilsson (1980)
Antibiotic susceptibility testing	Growth inhibition	Ånsehn et al. (1977), Lee and Crispen (1977), Thore et al. (1977)

addition of various substrates affecting the ATP level. Effects on the ATP level were often measurable within minutes of the addition of a substrate.

In medicine the firefly assay may be used in the study of viability of cells (erythrocytes, leukocytes, platelets, spermatozoa), erythrocytes in pathological situations, and the effect of ATP and ADP on platelet aggregation. The assay of ATP in blood may be performed using a very simple procedure yielding stable extracts by treatment with Triton X-100 (Table 6).

The lysis of cells can be studied by the firefly assay, since ATP as well as metabolites and enzymes participating in ATP-converting reactions may be determined by the method. The determination of gentamicin concentration by assay of ATP leakage (Nilsson, 1973). In experiments in cell lysis ATP monitoring is a particularly valuable technique. It has been used to monitor the leakage of ATP during immunolysis of erythrocytes (Thore, 1977) and during action of bacterial cytolysins on various mammalian cells (Fehrenbach et al., 1980).

Table 5 Applications Involving Changes in Cellular ATP Concentrations

Application area	Type of cells	References
Energy metabolism	*Bacillus subtilis*	Klofat et al. (1969)
	Escherichia coli	Cole et al. (1967)
	Peptococcus prevotii	Montague and Dawes (1974)
	Selenomonas ruminantium	Hobson and Summers (1972)
	Neurospora crassa	Slayman (1973)
	Pancreatic islets	Andersson et al. (1974)
	Epidermal cells	Hammar (1973), Hammar et al. (1975)
	Spermatozoa	Brookes (1970)
Hematology	Erythrocytes	Féo and Leblond (1974), Wolf (1975), Wolf et al. (1976), Wolf (1977)
	Platelets	David and Herion (1972), Holmsen et al. (1972), Larsson et al. (1977)
Cell lysis	Platelets	Feinman et al. (1977), Detwiler and Feinman (1973)
	Erythrocytes	Thore (1977)
	Ehrlich ascites cells	Nungester et al. (1969)
	Escherichia coli	Nilsson (1978), Nilsson (1980)
	Various mammalian cells	Fehrenbach et al. (1980)

D. Determination of Enzymes and Metabolites

Theoretically, all enzymes and metabolites participating in ATP-converting reactions can be determined by the firefly assay (Strehler and Totter, 1952). The assays can be made in either of two ways:

1. By measuring in aliquots of an ATP-converting reaction mixture, or
2. By ATP monitoring a mixture of an ATP-converting system and the firefly system

The second alternative is generally more convenient. However, the appropriateness of each method for a particular application is dependent upon a number of factors:

1. Inherent properties of the ATP-converting reaction studied (equilibrium constants, turnover numbers of involved enzymes, pH optima, and other properties of analytical importance)
2. Availability of enzymes, cofactors, etc., of adequate purity
3. Analytical interference from components in samples or reagents
4. Time required for the ATP-converting reaction to proceed before a change in the ATP level can be reliably recorded

Metabolites (Table 7) can be assayed either kinetically or by endpoint determination. The choice of method is based upon the time required for each assay and the type of analytical interference. For example, four metabolites (ATP, ADP, AMP, and cAMP) have been determined in a single assay with the stepwise addition of appropriate enzyme systems (Lundin, Rickardsson, and Thore, 1976).

Many enzymes (Table 8) can be (kinetically) determined using the firefly reaction. Some of these determinations have been made by ATP monitoring, e.g., the assays of total creatine kinase and creatine kinase B subunit activities. In the latter assay the sensitivity of the firefly assay is combined with the specificity of an immunoinhibition technique by using an antibody-inhibiting creatine kinase M subunit activity. The assay of creatine kinase B subunit activity in human serum is used in the diagnosis of acute myocardial infarction. The clinical need for the high-sensitivity assays of creatine kinase is presently being evaluated in several hospitals and a routine kit has been developed (Lundin et al., 1982). This will in all probability be the first bioluminescent method to be routinely used in clinical laboratories.

In studies on oxidative phosphorylation and photophosphorylation the firefly assay has been used for the continuous measurement of ATP levels in electron transport linked phosphorylation reactions (Table 9). Lemasters and Hackenbrock (1973) showed that the firefly reaction can be used for continuous measurements of oxidative phosphorylation even at very high ATP concentration, i.e., above the upper limit for ATP monitoring. The effects of product inhibition and the nonlinear relation between light emission and ATP concentration were mathematically compensated for by multiple calibration of the system (addition of known concentrations of ATP). In contrast Lundin et al. (1977b) avoided the product inhibition problem using ATP monitoring of photophosphorylation and ATPase reaction. Suitable equipment for ATP monitoring during illumination of photophosphorylating systems has been described (Lundin et al., 1979a).

Table 6 Extraction of ATP in Human Blood*

Storage	A. TCA extracts	B. Triton X-100 extracts	C. EDTA extracts	D. Bensalkon extracts
0 hr	0.59 mmol/L ATP	0.58 mmol/L ATP	0.56 mmol/L ATP	0.50 mmol/L ATP
2 hr	0.59 mmol/L ATP	0.58 mmol/L ATP	0.54 mmol/L ATP	0.46 mmol/L ATP
2 days	0.58 mmol/L ATP	0.58 mmol/L ATP	0.20 mmol/L ATP	0.13 mmol/L ATP

*Extracts of 20μl heparinized human blood (stored at 0°C) were prepared by addition of 20μl 10% trichloroacetic acid followed by 10 ml 0.1 mol/L tris acetate buffer, pH 7.75, containing 2 mmol/L EDTA (A); 10 ml 0.1% Triton X-100 in 0.1 mol/L tris acetate buffer, containing 2 mmol/L EDTA (B); 10 ml 2 mmol/L EDTA, pH 7.75, (C); and 10 ml 0.1% bensalkon in 0.1 mol/L tris acetate buffer, pH 7.75, containing 2 mmol/L EDTA (D). Extracts were stored for different lengths of time at room temperature and assayed in 20μl aliquots by the firefly method. Results have been compensated for the dilution of the blood (Lundin, unpublished results).

Table 7 Determination of Metabolites

Analyte	Principle	References
ATP, ADP, AMP	Direct reaction with luciferase, formation of ATP by pyruvate kinase and adenylate kinase, respectively	Lundin and Thore (1975b), Lundin et al. (1976)
cAMP	Formation of ATP by phosphodiesterase, adenylate kinase, and pyruvate kinase	Johnson et al. (1970)
Adenosine tetraphosphate	Probably formation of ATP by crude reagent	Manandhar and Van Dyke (1974)
Adenosine phosphosulfate	Formation of ATP by ATP-sulfurylase	Balharry and Nicholas (1971)
GTP, GDP, GMP, CTP, UTP	Probably formation of ATP by crude reagent	Karl (1978), Strehler (1968)

cGMP	Degradation of ATP by guanylate kinase and cGMP phosphodiesterase	Fertel and Weiss (1974)
Coenzyme A	Release of product inhibition in luciferase reaction	Strehler and McElroy (1957)
Glucose	Degradation of ATP by hexokinase	Cavari and Phelps (1977)
Glycerol, triglycerides	Degradation of ATP by glycerol kinase	Hercules and Sheehan (1978)
Creatine phosphate	Formation of ATP by creatine kinase	Jabs et al. (1977)
1,3-Diphosphoglycerate	Formation of ATP by phosphoglycerokinase	Momsen (1977)
Phosphoenolpyruvate	Formation of ATP by pyruvate kinase	McCoy and Doeg (1975)
Pyrophosphate	Formation of ATP by ATP-sulfurylase	Stanley (1974)

Table 8 Determination of Enzymes

Analyte	Principle	References
ATP sulfurylase	Formation of ATP from adenosine sulfate	Balharry and Nicholas (1971)
ATP phosphoribosyl-transferase	Formation of ATP from phosphoribosyl adenosine triphosphate	Kleeman and Parsons (1975)
ATPase	Degradation of ATP	Strehler (1968); Baltscheffsky and Lundin (1979)
Apyrase	Degradation of ATP	Strehler (1968)
Adenylate kinase	Formation of ATP from ADP	Brolin et al. (1979)
Creatine kinase	Formation of ATP from creatine phosphate	Witteveen et al. (1974), Lundin and Styrelius (1978), Lundin (1978), Lundin et al. (1979c), Lundin et al. (1982), Wahlgren and Kjellin (1979)
Hexokinase	Degradation of ATP by glucose	Strehler and Totter (1952)
Pyruvate kinase	Formation of ATP from phosphoenolpyruvate	Lundin et al. (1977a), Hammar et al. (1975)
Nucleotide phosphokinases	Formation of ATP from GTP, UTP, or CTP	Strehler (1968)
cAMP phosphodiesterase	Formation of ATP from cAMP adenylate kinase and pyruvate kinase	Weiss et al. (1972)
cGMP phosphodiesterase	Degradation of ATP by cGMP using guanylate kinase	Fertel and Weiss (1974)

Table 9 Measurement of Electron Transport Linked Phosphorylation

Biological material	Reaction	References
Mitochondria, submitochondrial vesicles, mitoplasts	Oxidative phosphorylation	Lemasters and Hackenbrock (1973, 1976, 1978)
Bacterial chromatophores	Photophosphorylation after single turnover flashes and continuous illumination	Lundin et al. (1977b), Harris and Baltscheffsky (1979), Lundin and Baltscheffsky (1978), Lundin et al. (1979a)
	Coupling factor ATPase activity	Baltscheffsky and Lundin (1979), Lundin et al. (1979b)
Chloroplasts	Photophosphorylation	Strehler and Hendley (1961)

V. CONCLUSION

The introduction of highly purified and standardized reagents with improved kinetic properties has removed a serious obstacle to the widespread routine use of the firefly assay. Beyond use of the assay in the ATP-converting systems described in this chapter, firefly luciferase is also useful in immobilized enzyme systems and immunoluminescence. Thus a rapid growth in the applicability and use of the firefly reaction for analytical purposes can be expected.

ACKNOWLEDGMENTS

The author wishes to thank all collaborators for their continuous contribution throughout the years in the development of the firefly assay. Special thanks to Dr. Anders Thore for helpful information and Ms. Sue Anne Moody for proofreading of this manuscript.

REFERENCES

Alexander, D. N., G. M. Ederer, and J. M. Matsen (1976). Evaluation of an adenosine 5-triphosphate assay as a screening method to detect significant bacteriuria. *J. Clin. Microbiol.* *3*:42-46.

Andersson, A., E. Borglund, and S. Brolin (1974). Effects of glucose on the content of ATP and glycogen and the rate of glucose phosphorylation of isolated pancreatic islets maintained in tissue culture. *Biochem. Biophys. Res. Commun.* *56*:1045-1051.

Ånsehn, S., A. Lundin, L. Nilsson, and A. Thore (1979). Detection of bacteriuria by a simplified luciferase assay of ATP. In *Proceedings of the International Symposium on Analytical Applications of Bioluminescence and Chemiluminescence*, E. Schram and P. Stanley (Eds.). State Printing and Publishing, Westlake Village, pp. 438-445.

Ånsehn, S., L. Nilsson, H. Höjer, and A. Thore (1977). Antibiotic susceptibility testing and determinations of antibiotic concentrations. In *Second Biannual ATP-Methodology Symposium: Proceedings*, G. Borun (Ed.), SAI Technology, San Diego, Cal., pp. 189-203.

Balharry, G. J. E., and Nicholas, D. J. D. (1971). New assay for ATP sulfurylase using the luciferin-luciferase method. *Anal. Biochem.* *40*:1-17.

Baltscheffsky, M., and A. Lundin (1979). Flash-induced increase of ATP-ase activity in *Rhodospirillum rubrum* chromatophores. In *Cation Flux Across Biomembranes*, Y. Mukohaka and L. Packer (Eds.). Academic, New York, pp. 209-218.

Bény, M., and M. Dolivo (1976). Separation of firefly luciferase using an anion exchanger. *FEBS Lett.* *70*:167-170.

Bossuyt, R. (1978). Usefulness of an ATP assay technique in evaluating the somatic cell content in milk. *Milchwissenschaft* *33*:11-13.

Brezonik, P. L., and J. W. Patterson (1971). Activated sludge ATP: Effects of environmental stress. *J. Sanit. Engr. Div. Amer. Soc. Civ. Engrs.* *97*:813-824.

Brolin, S. E., E. Borglund, and A. Ågren (1979). Photokinetic microassay of adenylate kinase using the firefly luciferase reaction. *J. Biochem. Biophys. Meth.* *1*:163-169.

Brooks, D. E. (1970). Observations on the content of ATP and ADP in bull spermatazoa using the firefly luciferase system. *J. Reprod. Fertil.* *23*:525-528.

Carrico, R. J., K-K Yeung, H. R. Schroeder, R. C. Boguslaski, R. T. Buckler and J. E. Christner (1976). Specific protein-binding reactions monitored with ligand-ATP conjugates and firefly luciferase. *Anal. Biochem.* *76*:95-110.

Cavari, B. Z., and G. Phelps (1977). Sensitive enzymatic assay for glucose determination in natural waters. *Appl. Environ. Microbiol.* *33*:1237-1243.

Chiu, S. Y., I. C. Kao, E. E. Erickson, and L. T. Fan (1973). ATP pools in activated sludge, *J. Water Poll. Control Fed.* *45*:1746.

Cole, H. A., J. W. T. Wimpenny, and D. E. Hughes (1967). The ATP pool in *Escherichia coli*. I. Measurement of the pool using a modified luciferase assay. *Biochim. Biophys. Acta* *143*:445-453.

Conn, R. B., P. Charache, and E. W. Chappelle (1975). Limits of applicability of the firefly luminescence ATP assay for the detection of bacteria in clinical specimens. *Amer. J. Clin. Pathol.* *63*:493-501.

Curtis, G. D. W., and H. H. Johnston (1979). A rapid screening test for bacteriuria. In *Proceedings from the International Symposium on Analytical Applications of Bioluminescence and Chemiluminescence*, E. Schram and P. Stanley (Eds.). State Printing and Publishing, Westlake Village, Cal., pp. 448-457.

Daigneault, R., A. Larouche, and G. Thibault (1979). Aminoglycoside antibiotic measurement by bioluminescence with use of plasmid-coded enzymes. *Clin. Chem.* *25*:1639-1643.

David, J. L., and F. Herion (1972). Assay of platelet ATP and ADP by the luciferase method: Some theoretical and practical aspects. *Adv. Exp. Med. Biol.* *34*:341-354.

DeLuca, M. (1976). Firefly luciferase. In *Advances in Enzymology*, Vol. 44, A. Meister (Ed.). Interscience, New York, pp. 37-68.

DeLuca, M., and W. D. McElroy (1974). Kinetics of the firefly luciferase catalyzed reactions. *Biochemistry* *13*:921-925.

Denburg, J. L., and W. D. McElroy (1970). Anion inhibition of firefly luciferase. *Arch. Biochem. Biophys.* *141*:668-675.

Detwiler, T. C., and R. D. Feinman (1973). Kinetics of the thrombin induced release of adenosine triphosphate by platelets. Comparison with release of calcium. *Biochemistry* *12*:2462-2468.

Fehrenbach, F.-J., H. Huser, and C. Jaschinski (1980). Measurement of bacterial cytolysins with a highly sensitive kinetic method. *FEMS Lett.* *7*:285-288.

Feinman, R. D., J. Lubowsky, I. Charo, and M. P. Zabinski (1977). The lumi-aggregometer: A new instrument for simultaneous measurement of secretion and aggregation by platelets. *J. Lab. Clin. Med.* *90*:125-129.

Féo, C. J., and P. F. Leblond (1974). The discocyte-echinocyte transformation: Comparison of normal and ATP-enriched human erythrocytes. *Blood* *44*:639-647.

Fertel, R., and B. Weiss (1974). A microassay for guanosine-3',5'-monophosphate phosphodiesterase activity. *Anal. Biochem.* *59*:386-398.

Gates, B. J., and M. DeLuca (1975). The production of oxyluciferin during the firefly luciferase light reaction. *Arch. Biochem. Biophys.* *169*:616-621.

Gilles, R., A. Pequeux, J. J. Saive, A. C. Spronck, and G. Thome-Lentz (1976). Effect of various ions on ATP determinations using the "luciferine-luciferase" system. *Arch. Int. Phys. Biochim.* *84*:807-817.

Gorus, F., and E. Schram (1979). Applications of bio- and chemiluminescence in the clinical laboratory. *Clin. Chem.* *25*:512-519.

Goto, T., J. Kubota, N. Suzuki and Y. Kishi (1973). In *Chemiluminescence and Bioluminescence*, M. J. Cormier, D. M. Hercules, and J. Lee (Eds.). Plenum, New York, pp. 325-331.

Green, A. A., and W. D. McElroy (1956). Crystalline firefly luciferace. *Biochim. Biophys. Acta* *20*:170-176.

Gutekunst, R. R. (1975). The firefly luciferase assay for adenosine triphosphate: A unique procedure for detecting bacteria in urine. In *ATP-Methodology Seminar*, G. Borun (Ed.). SAI Technology, San Diego, Cal., pp. 358-359.

Hamilton, R. D., and O. Holm-Hansen (1967). Adenosine triphosphate content of marine bacteria. *Limnol. Oceanog.* *12*:319-324.

Hammar, H. (1973). ATP and ADP levels and epidermal replacement rate in the normal human skin and in some papulosquamous diseases of the skin. *Acta Dermatovener (Stockholm)* *53*:251-258.

Hammar, H., G. Wettermark, and W. Wladimiroff (1975). Bioluminescence assay of enzymes obtained from buccal epithelium by superficial scraping. *Scand. J. Dent. Res.* *83*:375-381.

Harber, M. J., and A. W. Asscher (1977). A new method for antibiotic assay based on measurement of bacterial adenosine triphosphate using the firefly bioluminescence system. *J. Antimicrob. Chemother.* *3*:35-41.

Harber, M. J., and A. W. Asscher (1979). Bioluminescence assay for antibiotics and vitamins. In *Proceedings from the International Symposium on Analytical Applications of Bioluminescence and Chemiluminescence*, E. Schram and P. Stanley (Eds.). State Printing and Publishing,, Westlake Village, Cal., pp. 531-542.

Harris, D. A., and M. Baltscheffsky (1979). Bound nucleotides and phosphorylation in *Rhodospirillum rubrum*. *Biochem. Biophys. Res. Commun.* *86*:1248-1255.

Hercules, D. M., and T. L. Sheehan (1978). Chemiluminescent determination of serum glycerol and triglycerides. *Anal. Chem. 50*:22-25.

Hobson, P. N., and R. Summers (1972). ATP pool and growth yield in *Selenomonas ruminantium. J. Gen. Microbiol. 70*:351-360.

Höjer, H., and L. Nilsson (1978). Rapid determination of doxycycline based on luciferase assay of bacterial adenosine triphosphate. *J. Antimicrob. Chemother. 4*:503-508.

Höjer, H., L. Nilsson, S. Ånsehn, and A. Thore (1976). In vitro effect of doxycycline on levels of adenosine triphosphate in bacterial cultures. *Scand. J. Infect. Dis. (Suppl.) 9*:58-61.

Höjer, H., L. Nilsson, S. Ånsehn, and A. Thore (1979). Possible application of luciferase assay of ATP to antibiotic susceptibility testing. In *Proceedings from the International Symposium on Analytical Applications of Bioluminescence and Chemiluminescence*, E. Schram and P. Stanley (Eds.). State Printing and Publishing, Westlake Village, Cal., pp. 523-530.

Holm-Hansen, O., and C. R. Booth (1966). The measurement of adenosine triphosphate in the ocean and its ecological significance. *Limnol. Oceanogr. 11*:510-519.

Holmsen, H., E. Storm, and H. J. Day (1972). Determination of ATP and ADP in blood platelets: A modification of the firefly luciferase assay for plasma. *Anal. Biochem. 46*:489-501.

Hysert, D. W., F. Kovecses, and N. M. Morrison (1976). A firefly bioluminescence ATP assay method for rapid selection and enumeration of brewery microorganisms. *J. Amer. Soc. Brew. Chem. 34*:144-150.

Jabs, C. M., W. J. Ferrell, and H. J. Robb (1977). Microdetermination plasma ATP and creatine phosphate concentrations with a luminescence biometer. *Clin. Chem. 23*:2254.

Johnson, R. A., J-G. Hardman, A. E. Broadus, and E. W. Sutherland (1970). Analysis of adenosine 3',5'-monophosphate with luciferase luminescence. *Anal. Biochem. 35*:91-97.

Johnston, H. H., C. J. Mitchell and G. D. W. Curtis (1976). An automated test for the detection of significant bacteriuria. *Lancet 2*:400-402.

Johnston, H. H., and G. D. W. Curtis (1979). Detection of bacteria by bioluminescence: Problems in removal of non-bacterial ATP. In *Proceedings from the International Symposium on Analytical Applications of Bioluminescence*, E. Schram and P. Stanley (Eds.). State Printing and Publishing, Westlake Village, Cal., pp. 446-447.

Jones, J. G., and B. M. Simon (1977). Increased sensitivity in the measurement of ATP in freshwater samples with a comment on the adverse effect of membrane filtration. *Freshwater Biol. 7*:253-260.

Karl, D. M. (1978). A rapid, sensitive method for the measurement of guanine ribonucleotides in bacterial and environmental extracts. *Anal. Biochem. 89*:581.

Karl, D. M., and P. A. LaRock (1975). Adenosine triphosphate measurement in soil and marine sediments. *J. Fish Res. Bd. Can. 32*: 599-607.

Kees, U., A. Lewenstein and R. Bachofen (1975). ATP pools in activated sludge. *Eur. J. Appl. Microbiol. 2*:59-64.

Kleeman, J., and S. M. Parsons (1975). A sensitive assay for the reverse reaction of the first histidine biosynthetic enzyme. *Anal. Biochem. 68*:236.

Klofat, M., G. Picciolo, E. W. Chappelle, and E. J. Freese (1969). Production of adenosine triphosphate in normal cells and sporulation mutants of *Bacillus subtilis*. *J. Biol. Chem. 244*:3270-3276.

Larsson, C.-M., and T. Olsson (1979). Firefly assay of adenine nucleotides from algae: Comparison of extraction methods. *Plant & Cell Physiol. 20*:145-155.

Larsson, R., A. Rosengren, and P. Olsson (1977). Determination of platelet adhesion to polyethylene and heparinized surfaces with the aid of bioluminescence and ^{51}chromium labelled platelets. *Thromb. Res. 11*:517-530.

Lee, Y. S., and R. G. Crispen (1977). Rapid quantitative measurement of drug susceptibility of mycobacteria. In *Proceedings of the 2nd Bi-annual ATP Methodology Symposium*, G. Borun (Ed.). SAI Technology, San Diego, Cal., pp. 219-235.

Lee, T. R., J. L. Denburg, and W. D. McElroy (1970). Substrate-binding properties of firefly luciferase. II. ATP-binding site. *Arch. Biochem. Biophys. 141*:38-52.

Lee, C. C., R. F. Harris, J. D. H. Williams, D. E. Armstrong, and J. K. Syers (1971). Adenosine triphosphate in lake sediments. I. Determination. *Soil Sci. Soc. Amer. Proc. 35*:82-86.

Lee, Y., I. Jablonski, and M. DeLuca (1977). Immobilization of firefly luciferase on glass rods: Properties of the immobilized enzyme. *Anal. Biochem. 80*:496-501.

Lemasters, J. J., and C. R. Hackenbrock (1973). Adenosine triphosphate: Continuous measurement in mitochondrial suspensions by firefly luciferase luminescence. *Biochem. Biophys. Res. Commun. 55*:1262-1270.

Lemasters, J. J., and C. R. Hackenbrock (1976). Continuous measurement and rapid kinetics of ATP synthesis in rat liver mitochondria, mitoplasts and inner membrane vesicles determined by firefly-luciferase luminescence. *Eur. J. Biochem. 67*:1-10.

Lemasters, J. J., and C. R. Hackenbrock (1978). Firefly luciferase assay for ATP production by mitochondria. In *Methods in Enzymology*, Vol. 57, *Bioluminescence and Chemiluminescence*, M. DeLuca (Ed.). Academic, New York, pp. 36-50.

Levin, G. V., C. S. Chan, and G. Davis (1967). Development of the bioluminescent assay for the rapid quantitative detection of microbial contamination of water. Techn. Rep. TR-67-71, Aerospace Medical Research Laboratories, Wright-Patterson Air Force Base, Dayton, Ohio.

Levin, G. V., J. R. Schrot, and W. C. Hess (1975). Methodology for application of adenosine triphosphate determination in waste water treatment. *Environ. Sci. Technol. 10*:961-965.

Lundin, A. (1978). Determination of creatine kinase isoenzymes in human serum by an immunological method using purified firefly luciferase. In *Methods in Enzymology*, Vol. 57, *Bioluminescence and Chemiluminescence*, M. DeLuca (Ed.). Academic, New York, pp.56-65.

Lundin, A., and M. Baltscheffsky (1978). Measurement of photophosphorylation and ATP-ase using purified firefly luciferase. In *Methods in Enzymology*, Vol. 57, *Bioluminescence and Chemiluminescence*, M. DeLuca (Ed.). Academic, New York, pp. 50-56.

Lundin, A., and I. Styrelius (1978). Sensitive assay of creatine kinase isoenzymes in human serum using M subunit inhibiting antibody and firefly luciferase. *Clin. Chim. Acta 87*:199-209.

Lundin, A., and A. Thore (1975a). Analytical information obtainable by evaluation of the time course of firefly bioluminescence in the assay of ATP. *Anal. Biochem. 66*:47-63.

Lundin, A., and A. Thore (1975b). Comparison of methods for extraction of bacterial adenine nucleotides determined by firefly assay. *Appl. Microbiol. 30*:713-721.

Lundin, A., M. Blatscheffsky and B. Höijer (1979a). Continuous monitoring of ATP in photophosphorylating system by firefly luciferase. In *Proceedings from the International Symposium on Analytical Applications of Bioluminescence*, E. Schram and P. Stanley (Eds.). State Printing and Publishing, Westlake Village, Cal., pp. 339-349.

Lundin, A., B. Jäderlund and T. Löugren (1982). Optimized bioluminescent assay of creatine kinase and creatine kinase B-subunit activity. *Clin. Chem.* (in press).

Lundin, A., U. Karnell Lundin, and M. Baltscheffsky (1979b). Adenylate kinase activity associated with coupling factor ATP-ase in *Rhodospirillum rubrum. Acta Chem. Scand. 333*:608-609.

Lundin, A., A. Myhrman and G. Linfors (1981). Purification of firefly luciferase by ammonium sulphate precipitation and isoelectric focusing. In *Bioluminescence and Chemiluminescence. Basic Chemistry and Analytical Applications*. M. DeLuca and W. D. McElroy (Eds.). Academic, New York, pp. 455-465.

Lundin, A., A. Rickardsson, and A. Thore (1976). Continuous monitoring of ATP converting reactions by purified firefly luciferase. *Anal. Biochem. 75*:611-620.

Lundin, A., A. Rickardsson, and A. Thore (1977a). Substrate and enzyme determinations by continuously monitoring the ATP level by a purified luciferase reagent. In *Proceedings of the 2nd Biannual ATP Methodology Symposium*, G. Borun (Ed.). SAI Technology, San Diego, Cal., pp. 205-218.

Lundin, A., A. Thore and M. Baltscheffsky (1977b). Sensitive measurement of flash induced photophosphorylation in bacterial chromatophores by firefly luciferase, *FEBS Lett. 79*:73-76.

Lundin, A., K. Lindberg, R. Norlander, O. Nyquist, and I. Styrelius (1979c). Determination of creatine kinase B-subunit activity by continuous monitoring of ATP: A new bioluminescent technique applied in clinical chemistry. In *Proceedings from the International Symposium on Analytical Applications of Bioluminescence and Chemiluminescence*, E. Schram and P. Stanley (Eds.). State Printing and Publishing, Westlake Village, Cal., pp. 467-478.

MacLeod, N. H., E. W. Chappelle, and A. M. Crawford (1969). ATP assay of terrestrial soils: A test of an exobiological experiment. *Nature 223*:267-268.

Manandhar, M. S. P., and K. Van Dyke (1974). Adenosine tetraphosphate analysis: Polyethyleneimine (PEI) thin layer purification and firefly-extract liquid scintillation counting. *Anal. Biochem. 58*: 368-375.

McCoy, G. D., and K. A. Doeg (1975). A simplified assay of phosphoenolpyruvate. *Anal. Biochem. 64*:115-120.

McElroy, W. D. (1947). The energy source for bioluminescence in an isolated system. *Proc. Natl. Acad. Sci. USA 33*:342-345.

McElroy, W. D., J. W. Hastings, J. Coulombre, and V. Sonnenfeld (1953). The mechanism of action of pyrophosphate in firefly luminescence. *Arch. Biochem. Biophys. 46*:399-416.

McElroy, W. D., H. H. Seliger, and E. H. White (1969). Mechanism of bioluminescence, chemiluminescence and enzyme function in the oxidation of firefly luciferin. *Photochem. Photobiol. 10*:153-170.

Momsen, G. (1977). Determination of 1,3-diphosphoglycerate and other intermediates in the human red blood cells by firefly luciferase method. *Anal. Biochem. 82*:493-502.

Montague, M. D., and E. A. Dawes (1974). The survival of *Peptococcus prevotii* in relation to the adenylate energy charge. *J. Gen. Microbiol. 80*:291-299.

Myhrman, A., A. Lundin, and A. Thore (1978). The analytical application of ATP monitoring using firefly bioluminescence. *Application Note 314*, LKB-Wallac, Turku.

Nielsen, R., and H. Rasmussen (1968). Fractionation of extracts of firefly tails by gel filtration. *Acta Chem. Scand. 22*:1757-1762.

Nilsson, L. (1978). New rapid bioassay of gentamicin based on luciferase assay of extracellular ATP in bacterial cultures. *Antimicrob. Agents Chemother. 14*:812-816.

Nilsson, L. (1980). Factors affecting gentamicin assay. *Antimicrob. Agents Chemother. 17*:918-921.

Nilsson, L., H. Höjer, S. Ånsehn, and A. Thore (1977). A rapid, semiautomatic assay of gentamicin based on luciferase assay of bacterial adenosine triphosphate. *Scand. Inf. Dis. 9*:232-236.

Nilsson, L., H. Höjer, S. Ånsehn and A. Thore (1979). A simplified luciferase assay of antibiotics in clinical serum specimens. In *Proceedings from the International Symposium on Analytical Applications of Bioluminescence*, E. Schram and P. Stanley (Eds.). State Printing and Publishing, Westlake Village, Cal., pp. 515-522.

Nungester, W. J., L. J. Paradise, and J. A. Adair (1969). Loss of cellular ATP as a measure of cytolytic activity of antiserum. *Proc. Soc. Exp. Biol. Med. 132*:582-586.

Rasmussen, H., and R. Nielsen (1968). An improved analysis of adenosine triphosphate by the luciferase method. *Acta Chem. Scand. 22*:1745-1756.

Riemann, B. (1979). Interference in the quantitative determination of ATP extracted from freshwater microorganisms. In *Proceedings from the International Symposium on Analytical Applications of Bioluminescence and Chemiluminescence*, E. Schram and P. Stanley (Eds.). State Printing and Publishing, Westlake Village, Cal., pp. 316-332.

Seliger, H. H., and W. D. McElroy (1960). Spectral emission and quantum yield of firefly bioluminescence. *Arch. Biochem. Biophys. 88*:136-145.

Slayman, C. L. (1973). Adenine nucleotide levels in *Neurospora*, as influenced by conditions of growth and by metabolic inhibitors. *J. Bacteriol. 114*:752-766.

Stanley, P. E. (1974). Use of bioluminescence procedures and liquid scintillation spectrometers for measuring very small amounts of enzymes and metabolites. In *Liquid Scintillation Counting, Recent Developments*, P. E. Stanley and B. A. Scoggins (Eds.). Academic, New York, pp. 421-430.

Statham, M., and D. Laugton (1975). The use of adenosine triphosphate measurements in the control of the activated sludge process by the solids retention time method. *Process Biochem. (October 1975)*, pp. 25-28.

Strehler, B. L. (1968). Bioluminescence assay: Principles and practice. In *Methods of Biochemical Analysis*, Vol. 16, D. Glick (Ed.). Interscience, New York, pp. 99-181.

Strehler, B. L., and D. D. Hendley (1961). In *A Symposium on Light and Life*, W. D. McElroy and H. B. Glass (Eds.). Johns Hopkins Press, Baltimore, pp. 601-608.

Strehler, B. L., and W. D. McElroy (1957). The assay of adenosine triphosphate. In *Methods in Enzymology*, Vol. 3, S. P. Colowick and N. O. Kaplan (Eds.). Academic, New York, pp. 871-873.

Strehler, B. L., and J. R. Totter (1952). Firefly luminescence in the study of energy transfer mechanisms. I. Substrate and enzyme determination. *Arch. Biochem. Biophys. 40*:28-41.

Thore, A. (1977). An overview of the work of the Swedish bioluminescence group. In *Second Biannual ATP-Methodology Symposium Proceedings*, G. Borun (Ed.). SAI Technology, San Diego, Cal., pp. 171-187.

Thore, A. (1979). Luminescence in clinical analysis. *Ann. Clin. Biochem. 16*:359-369.

Thore, A., S. Ånsehn, A. Lundin and S. Bergman (1975). Detection of bacteriuria by luciferase assay of adenosine triphosphate. *J. Clin. Microbiol. 1*:1-8.

Thore, A., L. Nilsson, H. Höjer, S. Ånsehn and L. Bröte (1977). Effects of ampicillin on intracellular levels of adenosine triphosphate in bacterial cultures related to antibiotic susceptibility. *Acta Pathol. Microbiol. Scand. Sect. B 85*:161-166.

Wahlgren, N. G., and K. G. Kjellin (1979). Cerebrospinal fluid analysis of brain-specific creatine kinase isoenzymes using M subunit inhibiting antibody and firefly luciferase. In *Proceedings from the International Symposium on Analytical Applications of Bioluminescence and Chemiluminescence*, E. Schram and P. Stanley (Eds.). State Printing and Publishing, Westlake Village, Cal., pp. 479-487.

Weiss, B., R. Lehne, and S. Strada (1972). Rapid microassay of adenosine -3',5'-monophosphate phosphodiesterase activity. *Anal. Biochem. 45*:222.

Wettermark, G., H. Stymne, S. E. Brolin, and B. Petersson (1975). Substrate analyses in single cells: I. Determination of ATP. *Anal. Biochem. 63*:293-307.

Whitehead, T. P., L. J. Kricka, T. J. N. Carter, and G. H. G. Thorpe (1979). Analytical luminescence: Its potential in the clinical laboratory. *Clin. Chem. 25*:1531-1546.

Witteveen, S. A., S. E. Sobel, and M. DeLuca (1974). Kinetic properties of the isoenzymes of human creatine phosphokinase. *Proc. Natl. Acad. Sci. USA 71*:1384-1387.

Wolf, P. L. (1975). Decreased ATP and increased calcium in sickle cells. In *Proceedings of ATP Methodology Seminar*, G. Borun (Ed.), SAI Technology, San Diego, Cal., pp. 340-357.

Wolf, P. L. (1977). Hemolytic anemia in hepatic disease with decreased RBC ATP. In *Proceedings of the 2nd Bi-annual ATP Methodology Symposium*, G. Borun (Ed.). SAI Technology, San Diego, Cal., pp. 237-249.

Wolf, P. L., P. Walters, and P. Singh (1976). An investigation of blood ATP and erythrocyte calcium in various diseases. *Fed. Proc. Fed. Amer. Soc. Exp. Biol. 35*:252.

5

ANALYTICAL APPLICATIONS OF BIOLUMINESCENCE: MARINE BACTERIAL SYSTEM

EDWARD G. JABLONSKI* and MARLENE DELUCA University of California at San Diego, La Jolla, California

I. Introduction and General Properties 76

II. Materials 78
- A. Bacteria 78
- B. Commercial Luciferase and NAD(P)H:FMN Oxidoreductase 78
- C. Purification of Luciferase and Oxidoreductase 79
- D. Reagents 79
- E. Instrumentation 79

III. Assays 80
- A. Calibration Curves 80
- B. Quantitation of NAD(P)H 80
- C. Assay of FMN 81
- D. Assay of Various Other Substrates and Enzymes 81
- E. Monitoring Specific Protein-Binding Reactions 82
- F. Protease Assays 82
- G. Immobilized Enzymes 82

IV. Conclusion 84

References 84

*Dr. Jablonski is currently with The Johns Hopkins University, McCollum-Pratt Institute, Baltimore, Maryland.

I. INTRODUCTION AND GENERAL PROPERTIES

The use of bioluminescence for analytical purposes has been well documented and in recent years various luciferases have been used to assay many important biological molecules. The advantage of these assays is that light is produced in the reaction. The availability of modern electronic instruments has made it possible to measure light with great precision and sensitivity. The bacterial luciferase system has been used extensively for assaying a number of reactions involving pyridine nucleotides. Luminescent bacteria contain enzymes which catalyze the following reactions:

$$(1) \quad NAD(P)H + H^+ + FMN \rightarrow NAD(P)^+ + FMNH_2$$

$$(2) \quad FMNH_2 + O_2 + RCHO \rightarrow FMN + RCOOH + H_2O + h\nu$$

Reaction (1) is catalyzed by the NAD(P)H:FMN oxidoreductase and the $FMNH_2$ produced can be utilized by luciferase in the presence of O_2 and a long-chain aldehyde to produce light (Reaction 2).

If the assay conditions are such that NAD(P)H is the limiting component, then the light intensity, quanta per second, is directly proportional to the concentration of NAD(P)H. Any reaction that leads to the production or disappearance of NAD(P)H can be measured by this light-emitting system. Since many clinical assays can be coupled to the appearance or disappearance of reduced pyridine nucleotides, this bioluminescent assay is of great potential importance.

The oxidoreductase and luciferase are soluble enzymes and have been purified from several different bacterial strains. In one species, *Beneckea harveyi (B. harveyi)*, there are two distinct oxidoreductases, one which is specific for NADH and another which utilizes NADPH. In general, the oxidoreductase and the luciferase are obtained from the same bacteria. There has not been a systematic investigation of the coupled assay using luciferase from one species and oxidoreductase from another.

The enzymes may be assayed individually or in the coupled reaction. The oxidoreductases can be measured in a spectrophotometric assay by monitoring the disappearance of NAD(P)H at 340 nm in the presence of FMN. Luciferase is assayed by injecting excess $FMNH_2$ into a solution containing the enzyme, aldehyde, and oxygen. The maximum intensity of the emitted light is a measure of the luciferase concentration. In the coupled assay when all substrates are present in saturating amounts, light intensity is proportional to the product of the concentrations of the two enzymes (Hastings et al., 1965).

Figure 1 shows the time course of light emission when excess $FMNH_2$ is injected into luciferase (A) or when NADH is injected into a solution containing oxidoreductase and luciferase (B). When $FMNH_2$ is injected, the decrease in light intensity is due to the autooxidation of $FMNH_2$ (Reaction 3).

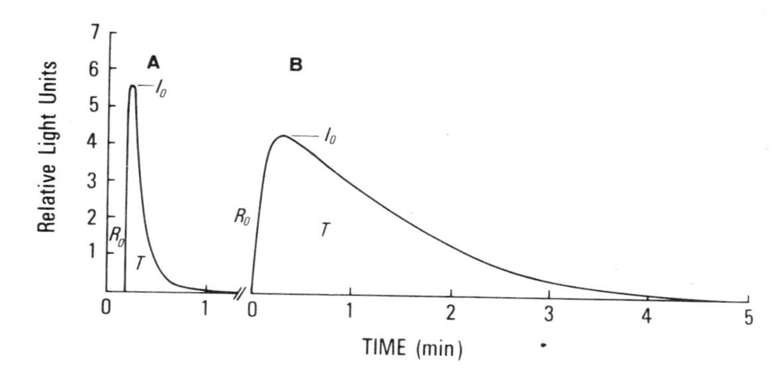

Figure 1 Time course of light emission upon the addition of excess $FMNH_2$ to luciferase (A) and upon the addition of limiting NADH to a coupled reaction system of luciferase and NADH:FMN oxidoreductase (B); I_0, initial maximum light intensity; R_0, initial rate of increase in light emission in relative light units/sec; T, total light emitted for the complete reaction, determined by integration.

$$(3) \quad FMNH_2 + O_2 \rightarrow FMN + H_2O_2$$

The $t_{1/2}$ of $FMNH_2$ oxidation in an air-saturated solution is about 0.1 sec (Gibson and Hastings, 1962). In the coupled assay, the decrease in light intensity is the result of the utilization of NADH.

The use of the bacterial luminescent system for analytical purposes does not require expensive instrumentation. The assays are simple, rapid, and the partially purified enzymes are commercially available. The assay is extremely specific for reduced pyridine nucleotides and is sensitive. The light intensity is linear with NADH concentration in the range of 10^{-12} to 10^{-9} mol (Stanley, 1971). According to one report, as little as 10^{-16} mol of NADH can be detected (Lee et al., 1974). This is much more sensitive than spectrophotometric or fluorometric procedures. The temperature optimum for the coupled assay is 28°C and the pH optima is at 7.0 using 0.1 mol/L phosphate buffer. Higher concentrations of phosphate are inhibitory (Stanley, 1971). The light intensity is dependent on the aldehyde used. Tetradecanal produces the maximal light intensity; however, tetradecanal is a solid at room temperature, and for convenience decanal is often used (Hastings et al., 1963; Lee and Murphy, 1975). The luminescent reaction is unaffected by cyanide, azide, and fluoride at concentrations of 1 mmol/L but it is inhibited by calcium and zinc at concentrations of 10 μmol/L (Stanley, 1971). Both luciferase and the oxidoreductase are inhibited by a variety of sulfhydryl reagents (Nicoli et al., 1974;

Jablonski and DeLuca, 1978) and the presence of 1×10^{-3} mol/L EDTA is advised to prevent heavy-metal inhibition.

The use of the bacterial luminescent assay with other coupled systems has not been explored extensively; therefore, each system consisting of a sequence of reactions should be tested and critically evaluated for errors. There are many reviews dealing with luminescent assays of various metabolites (e.g., Stanley, 1974; Schram, 1974; Lee et al., 1974; Brolin et al., 1977; Stanley, 1976; Gorus and Schram, 1979).

It is the purpose of this chapter to introduce the clinical investigator to the general uses of the bacterial luminescent system in the hope that more assays will be developed or existing ones modified to fit specific needs.

II. MATERIALS

A. Bacteria

The luminous bacteria which contain the enzymes necessary for performing analytical assays are easily maintained and cultured in the laboratory. The most commonly used strains are *Photobacterium fischeri (P. fischeri)* and *B. harveyi.* These bacteria can be grown in regular or artificial seawater (3% NaCl) with ammonium nitrate and a variety of carbon sources. The detailed procedure for handling and growing luminescent bacteria has been reviewed by Nealson (1978). The bacteria should be grown to maximal luminescence in liquid culture, harvested by centrifugation, and stored as a frozen cell paste at -20°C. A procedure for the growth and harvesting of cells from the New Brunswick fermenter has been described by Hastings et al. (1978).

B. Commercial Luciferase and NAD(P)H:FMN Oxidoreductase

Partially purified extracts of *P. fischeri* containing both oxidoreductase and luciferase are available from the Sigma Chemical Company, Lumac, and Worthington Biochemicals. These preparations are adequate for routine analyses of reduced pyridine nucleotides. There are, however, many contaminating enzymes in these preparations which increase the background light emission in the presence of aldehyde and NAD. Specifically, an aldehyde dehydrogenase activity is found in crude preparations of the NAD(P)H:FMN oxidoreductase. Therefore, the commercial samples are of limited use in coupled assays. Brolin and Hjerten (1977) further purified the commercially available luciferase and oxidoreductase and found that the sensitivity is increased and the specificity and reproducibility improved.

C. Purification of Luciferase and Oxidoreductase

Luciferase can be purified from a frozen cell paste. The detailed procedure for the purification from *B. harveyi* and *P. fischeri* was described by Gunsalus-Miguel et al. in 1972 and modified by Baldwin et al. in 1975. A recent description is given by Hastings et al. (1978). Approximately 380 mg of luciferase is obtained from 1 lb (approximately 500 g) of bacteria.

Purified luciferase is stored frozen in 1-ml aliquots containing 15 to 20 mg enzyme per ml in 0.1 mol/L phosphate buffer, pH 7.0, containing 0.1 mmol/L DTT, 0.02% sodium azide. The frozen enzyme is stable for at least a year.

B. harveyi contains two distinct oxidoreductases (Gerlo and Charlier, 1975). One is specific for NADH and the other utilizes NADPH. The procedure for the purification of these enzymes has been published (Jablonski and DeLuca, 1977; Michaliszyn et al., 1977). The yield of NADH and NADPH oxidoreductase is 2.7 and 1.5 mg, respectively, from 1 lb (approximately 500 g) of cells. If a highly purified preparation is not necessary, the enzymes which are eluted from the DEAE cellulose may be used without further purification. *P. fischeri* appears to contain only one oxidoreductase which uses both NADH and NADPH. This enzyme has been partially purified (Duane and Hastings, 1975; Puget and Michelson, 1972).

The oxidoreductase is stored frozen in 0.1 mol/L phosphate buffer with 0.1 mmol/L dithiothreitol (DTT) and 0.2% BSA (bovine serum albumin). It is important to deaerate the buffer before storage. The enzyme slowly loses activity over a period of months under these conditions.

D. Reagents

All of the substrates necessary for the luminescent assays are available commercially and are used without further purification. Commercial dehydrogenases frequently contain traces of the aldehyde dehydrogenase and for extreme sensitivity this contaminant must be removed.

E. Instrumentation

The simplest instrument needed to perform bioluminescent assays consists of a phototube enclosed in a light-tight chamber, with a sample well and a device for measuring and recording the phototube output. The sample to be assayed is then injected through a light-tight gasket into the tube containing the enzymes and other substrates. Either peak height or the total light measured by an appropriate integration circuit may be recorded. Several of the commercially available instruments were evaluated by Picciolo et al. (1978). Gorus and Schram

(1979) recently described all of the commercially available instruments (see also Chap. 12).

Liquid scintillation counters are frequently used for measuring bioluminescent reactions. It is not possible in most cases to record the initial maximum light intensity. Reviews of the applications of liquid scintillation counters for bioluminescence analyses have been written by Schram (1974), Stanley (1974), Lee et al. (1974), and Schram et al. (1976).

III. ASSAYS

A. Calibration Curves

It is important to run a series of standards along with any unknown samples. In addition, an internal standard should be included. This allows calculation of the recovery of the unknown sample and will also indicate if the sample contains inhibitors of either the luciferase or oxidoreductase.

B. Quantitation of NAD(P)H

NADH or NADPH may be assayed by measuring any of three parameters of light emission in the coupled oxidoreductase-luciferase reaction. Maximal peak light intensity, the rate of increase in light intensity, and total light emitted are all proportional to the amount of NAD(P)H when it is the limiting component in the system. In practice it is generally most convenient to measure peak light intensity.

The assay tube contains oxidoreductase and luciferase in 0.5 ml of 0.1 mol/L potassium phosphate buffer pH 7.0; 10 μl of 1.5×10^{-4} mol/L FMN and 5 μl of a saturated aldehyde solution are added, and the unknown or standard NAD(P)H (usually 0.1 ml) is injected and the peak light intensity recorded.

The concentration of enzymes used will depend on the degree of purity, the instrumentation, and the desired sensitivity. Using the purified enzymes from *B. harveyi* and the Aminco Chem-Glow photometer, the lowest amount of NADH or NADPH that can be reliably determined is 1 fmol. In this assay, 14 μg of luciferase and 1.5 μg of oxidoreductase are used (Jablonski and DeLuca, 1979). If higher concentrations of reduced pyridine nucleotides are to be assayed, the amount of enzymes used can be much less. It is possible to measure 1 pmol of NADH using 4 μg of luciferase and 0.05 μg of oxidoreductase.

In order to determine NADH or NADPH in a mixture of the two compounds, it is necessary either to selectively destroy one of these or to use the purified oxidoreductase which is specific for either NADH or NADPH.

C. Assay of FMN

Flavin mononucleotide (FMN) can be measured using the coupled luminescent system. With 0.14 μg luciferase, 0.15 μg oxidoreductase, and 33 μmol NADH, it is possible to measure 5 pmol of FMN.

Wettermark and Brolin (1979) also were able to detect 5 pmol FMN using a commercial source of luciferase from *P. fischeri*.

D. Assay of Various Other Substrates and Enzymes

In principle, any compound or enzyme which can be coupled to the production or disappearance of NADH or NADPH can be measured by the luminescent system. There are two general procedures that have been used. One method consists of quantitatively oxidizing the substrate which produces a stoichiometric amount of NAD(P)H (Reaction 4). The NAD(P)H is then determined in the usual way:

$$(4) \quad XH_2 + NAD(P)^+ = X_{ox} + NAD(P)H + H^+$$

To measure the oxidized form of the substrate, the reverse reaction may be used and the decrease in NAD(P)H is measured. Alternatively, the dehydrogenase reaction may be directly coupled to the luminescent system in the presence of excess NAD(P) and FMN. The rate of production of NAD(P)H is measured by the peak light intensity. The compound to be assayed should be present in concentrations lower than the K_m for the dehydrogenase. The range of linearity for different substrates should be determined experimentally.

Brolin et al. (1971) measured glucose, malate, and NAD using the appropriate dehydrogenases. Agren, Berne, and Brolin (1977) measured 0.5 pmol of pyruvate in 1 to 10 μg of lyophilized tissue. Pyruvate and NADH were first converted to lactate and NAD^+. The NAD^+ was then converted to NADH by a D-3-hydroxybutyrate dehydrogenase. Gawronski and Egghart (1979) assayed TNT in a two-step reaction utilizing a TNT reductase. This system can detect 30 fmol of TNT. Brolin (1977) measured pmol amounts of L-glycerol-3-phosphate and 3-hydroxybutyrate using commercial enzymes. Stanley (1974, 1978) showed that malate, oxaloacetate, and malate dehydrogenase can be determined in one step by measuring the initial rate of increase or decrease in light intensity instead of peak height.

Androsterone and testosterone have been assayed using a commercial hydroxysteroid dehydrogenase preparation. This assay was linear down to 0.1 nmol of steroid. Further purification of the hydroxysteroid dehydrogenase would increase this sensitivity (Jablonski and DeLuca, 1979). The bacterial luminescent system has been used to assay several clinically important enzymes such as alcohol dehydrogenase, lactate dehydrogenase, malate dehydrogenase, and glucose-6-phosphate dehydrogenase (Haggerty et al., 1978; Jablonski and

DeLuca, 1979). Cantarow and Stollar (1976) measured ATP-NMN adenylyltransferase using several enzymes in a coupled reaction. These investigators were able to measure this enzyme at activities 100-fold lower than the standard assay. Berne (1976) measured the activity of 3-hydroxybutyrate dehydrogenase in pancreatic islet cells.

Table 5.1 summarizes the various published procedures for measuring substrates and enzymes.

E. Monitoring Specific Protein-Binding Reactions

Bacterial oxidoreductase-luciferase have been used to monitor specific protein-binding reactions (see also Chap. 9). Various ligand-NADH conjugates have been synthesized which are active as substrates in the bioluminescent reactions. Schroeder et al. (1976) linked biotin and 2,4-dinitrofluorobenzene to nicotinamide 6-(2-aminoethyl) purine dinucleotide (AENAD). After reducing these compounds with alcohol and alcohol dehydrogenase, they could be quantitatively measured by the light production with oxidoreductase and luciferase from *P. fischeri*. If the specific binding proteins, avidin, and antibody to DNP were added to the conjugates, the light production due to free ligand was inhibited. Addition of soluble biotin or DNP reversed the light inhibition. It is possible to assay the amount of soluble ligands in a sample by the amount of ligand-NADH displaced from the binding protein. These assays were linear in the range of 50 to 300 nmol/L of ligand.

F. Protease Assays

Baldwin (1978) described the use of bacterial luciferase to assay proteases. Native bacterial luciferase is very susceptible to inactivation by a wide variety of proteases. The assay involves incubating luciferase with the protease and measuring the rate of activity loss. The pseudo-first-order rate of inactivation is directly proportional to the protease. The assay is simple, rapid, and sensitive.

G. Immobilized Enzymes

Recently luciferase and oxidoreductase from *B. harveyi* have been immobilized on glass rods. These immobilized enzymes are individually active and are also able to produce light in the coupled assay. The light emitted is linear with NADH concentration from 1 pmol to 50 nmol. The immobilized enzymes appear to have properties similar to the soluble forms, but they are much more stable and a single rod has been used for up to 200 consecutive assays (Jablonski and DeLuca, 1976). The immobilized enzymes are useful for assaying low

Table 1 Substrates and Enzymes Which Have Been Detected Using the Bacterial Bioluminescent System

Compound Assayed	Range of Detection	Reference
NADH	10 fmol-100 nmol 0.1 fmol	Stanley (1971) Lee et al. (1974)
NAD^+	1 pmol	Agren et al. (1977b)
NADPH	10 pmol-200 nmol	Jablonski and DeLuca (1976)
AENADH-Biotin	25-200 pmol	Schroeder et al. (1976)
FMN	10 fmol-100 nmol 5 pmol	Stanley (1971) Wettermark and Brolin (1979)
Glucose	0.5-6 pmol	Brolin et al. (1971)
G-6-P	2-1000 pmol	Jablonski and DeLuca (1979)
3-Hydroxybutyrate	50-1000 pmol	Brolin (1977)
Malate	250-1500 pmol 20-250 pmol	Brolin (1977) Stanley (1978)
L-Glycerol-3-phosphate	300-2000 pmol	Brolin (1977)
Oxaloacetate	5-60 pmol	Stanley (1978)
Pyruvate	0.5-4 pmol	Agren et al. (1977a)
EtOH	.003-012%	Haggerty et al. (1978)
Testosterone Androsterone	0.2-5 nmol	Jablonski and DeLuca (1979)
O_2	0.1-100 nmol	Chance and Oshino (1978)
TNT	20 fmol	Gawronski and Egghart (1979)
Myristic acid	10 pmol	Ulitzur and Hastings (1978)
Lactate dehydrogenase	0.001-1 pmol	Haggerty et al. (1978)
Alcohol dehydrogenase	0.01-10 pmol	Haggerty et al. (1978)

Table 1 (Continued)

Compound Assayed	Range of Detection	Reference
Glucose-6-phosphate dehydrogenase	0.001-1 pmol	Haggerty et al. (1978)
Hexokinase	1-10 pmol	Haggerty et al. (1978)
Lipase	–	Ulitzur and Hastings (1978)
Phospholipase	–	
ATP-NMN adenylyl-transferase	–	Cantarow and Stollar (1976)
3-OH butyrate dehydrogenase	–	Berne (1976)

levels of various dehydrogenases. Malate dehydrogenase, lactate dehydrogenases, glucose-6-phosphate dehydrogenase, and alcohol dehydrogenase have all been assayed in the 0.001 to 1.0 pmol range (Haggerty et al., 1978).

Glucose-6-phosphate dehydrogenase, hexokinase, and alcohol dehydrogenase have been coimmobilized with the oxidoreductase and luciferase. This result in rods which are specific for glucose-6-phosphate, glucose, or alcohol, respectively (Jablonski and DeLuca, 1979). As with the soluble coupled assay, the initial light intensity is proportional to the concentration of the limiting substrate, when all other substrates and cofactors are in excess (see also Chap. 9).

IV. CONCLUSION

The number of assays that can be coupled to the bacterial luminescent system is enormous. Only a few of these have thus far been examined in detail. Luminescent assays have the advantage of great sensitivity, specificity, and are inexpensive. With the availability of instruments and reagents it seems as though there will be many future applications of this sytem in both the research and the clinical laboratory.

REFERENCES

Ågren, A., C. Berne and S. E. Brolin (1977a). Photokinetic assay of pyruvate in the islets of Langerhans using bacterial luciferase. *Anal. Biochem.* *78*:229-234.

Ågren, A., S. E. Brolin and S. Hjerten (1977b). Simplified luciferase assay of NAD^+ applied to microsamples from liver, kidney and pancreatic islets. *Biochim. Biophys. Acta 500*:103-108.

Baldwin, T. O. (1978). Bacterial luciferase as a generalized substrate for the assay of proteases. In *Methods in Enzymology*, Vol. 51, M. DeLuca (Ed.). Academic, New York, pp. 198-201.

Baldwin, T. O., M. Nicoli, J. E. Becvar and J. W. Hastings (1975). Bacterial luciferase: Binding of oxidized flavin mononucleotide. *J. Biol. Chem. 250*:2763-2768.

Berne, C. (1976). Determination of D-3-hydroxybutyrate dehydrogenase in mouse pancreatic islets with a photokinetic technique using bacterial luciferase. *Enzyme 21*:127-136.

Brolin, S. E. (1977). Attempts to simplify the analytical performance in microassay of metabolites with bacterial luciferase. *Bioelectrochem. Bioenerg. 4*:257-262.

Brolin, S. E., and S. Hjerten (1977). Microassay with the NADH-induced light reaction, technique improved by means of purified enzymes from *Achromobacter fischeri. Mol. Cell. Biochem. 17*:61-73.

Brolin, S. E., E. Borglund, L. Tegner, and G. Wettermark (1971). Photokinetic microassay based on dehydrogenase reactions and bacterial luciferase. *Anal. Biochem. 42*:124-135.

Brolin, S. E., G. Wettermark, and H. Hammar (1977). Chemiluminescence microanalysis of substrates and enzymes. *Strahlentherapie 153*:124-131.

Cantarow, W., and B. D. Stollar (1976). The use of bacterial luciferase and a liquid scintillation spectrometer to assay the enzymatic synthesis of NAD^+. *Anal. Biochem. 71*:333-340.

Chance, B., and R. Oshino (1978). Luminous bacteria as an indicator. In *Methods in Enzymology*, Vol. 57, M. DeLuca (Ed.). Academic, New York, pp. 223-226.

Duane, W., and J. W. Hastings (1975). Flavin mononucleotide reductase of luminous bacteria. *Mol. Cell. Biochem. 1*:53-64.

Gawronski, T. H., and H. Egghart (1979). Determination of subpicomole quantities of TNT using bacterial luciferase. In *International Symposium on Analytical Applications of Bioluminescence and Chemiluminescence*. State Printing and Publishing, Westlake Village, Cal., pp. 182-192.

Gerlo, E., and J. Charlier (1975). Identification of NADH-specific and NADPH-specific FMN-reductase in *Beneckea harveyi. Eur. J. Biochem. 57*:461-467.

Gibson, Q. H., and J. W. Hastings (1962). The oxidation of reduced flavin mononucleotide by molecular oxygen. *Biochem. J. 83*:368-377.

Gorus, F., and E. Schram (1979). Applications of bio- and chemiluminescence in the clinical laboratory. *Clin. Chem. 25*:512-519.

Gunsalus-Miguel, A., E. A. Meighen, M. Nicoli, K. H. Nealson, and J. W. Hastings (1972). Purification and properties of bacterial luciferase. *J. Biol. Chem. 247*:398-404.

Haggerty, C., E. Jablonski, L. Stav, and M. DeLuca (1978). Continuous monitoring of reactions that produce NADH and NADPH using immobilized luciferase and oxidoreductases from Beneckea harveyi. *Anal. Biochem.* *88*:162-173.

Hastings, J. W., J. A. Spudich, and G. Malnic (1963). The influence of aldehyde chain length on the relative quantum yield of the bioluminescent reaction of *Achromobacter fischeri*. *J. Biol. Chem.* *238*: 3100-3105.

Hastings, J. W., W. H. Riley, and J. Massa (1965). The purification, properties and chemiluminescent quantum yield of bacterial luciferase. *J. Biol. Chem.* *240*:1473-1481.

Hastings, J. W., T. O. Baldwin, and M. Z. Nicoli (1978). Bacterial luciferase: Assay, purification and properties. In *Methods in Enzymology*, Vol. 57, M. DeLuca (Ed.). Academic, New York, pp. 135-152.

Jablonski, E., and M. DeLuca (1976). Immobilization of bacterial luciferase and FMN reductase on glass rods. *Proc. Natl. Acad. Sci. USA* *73*:3848-3851.

Jablonski, E., and M. DeLuca (1977). Purification and properties of the NADH and NADPH specific FMN oxidoreductases from *Beneckea harveyi*. *Biochem.* *16*:2932-2936.

Jablonski, E., and M. DeLuca (1978). Studies of the control of luminescence in *Beneckea harveyi*: Properties of the NADH and NADPH: FMN oxidoreductases. *Biochemistry* *17*:672-678.

Jablonski, E., and M. DeLuca (1979). Properties and uses of immobilized light-emitting enzyme systems from *Beneckea harveyi*. *Clin. Chem.* *25*:1622-1627.

Lee, J., and C. L. Murphy (1975). Bacterial bioluminescence: Equilibrium association measurements, quantum yields, reaction kinetics, and overall reaction scheme. *Biochemistry* *14*:2259-2267.

Lee, J., C. L. Murphy, G. J. Faini, and T. L. Baucom (1974). Bacterial bioluminescence and its applications to analytical procedures. In *Liquid Scintillation Counting: Recent Developments*, P. E. Stanley and G. A. Scoggins (Eds.). Academic, New York, pp. 403-420.

Michaliszyn, G. A., S. S. Wing, and E. A. Meighen (1977). Purification and properties of a NAD(P)H:flavin oxidoreductase from the luminous bacterium *Beneckea harveyi*. *J. Biol. Chem.* *252*:7495-7498.

Nealson, K. H. (1978). Isolation, identification and manipulation of luminous bacteria. In *Methods in Enzymology*, Vol. 57, M. DeLuca (Ed.). Academic, New York, pp. 153-166.

Nicoli, M., E. A. Meighen, and J. W. Hastings (1974). Bacterial luciferase: Chemistry of the reactive sulfhydryl. *J. Biol. Chem.* *249*: 2385-2392.

Picciolo, G. L., J. W. Deming, D. A. Nibley, and E. W. Chappelle (1978). Characteristics of commercial instruments and reagents for luminescent assays. In *Methods in Enzymology*, Vol. 57, M. DeLuca (Ed.). Academic, New York, pp. 550-559.

Puget, K., and A. M. Michelson (1972). Studies in bioluminescence. VII. Bacterial NADH flavin mononucleotide oxidoreductase. *Biochimie* *54*:1197-1204.

Schram, E. (1974). Bioluminescence measurements: Fundamental aspects, analytical applications and prospects. In *Liquid Scintillation Counting: Recent Developments*, P. E. Stanley and B. A. Scoggins (Eds.). Academic, New York, pp. 383-402.

Schram, E., F. Demuylder, J. Derycker, and H. Roosens (1976). On the use of liquid scintillation spectrometers for chemiluminescent assays in biochemistry. In *Liquid Scintillation Science and Technology*, A. A. Noujaim, C. Ediss, and L. Weike (Eds.). Academic, New York, pp. 243-254.

Schroeder, H., R. J. Carrico, R. C. Boguslaski, and J. E. Christner (1976). Specific binding reactions monitored with ligand-cofactor conjugates and bacterial luciferase. *Anal. Biochem.* *72*:283-292.

Stanley, P. E. (1971). Determination of subpicomole levels of NADH and FMN using bacterial luciferase and the liquid scintillation spectrometer. *Anal. Biochem.* *39*:441-453.

Stanley, P. E. (1974). Analytical bioluminescence assays using the liquid scintillation spectrometer: A review. In *Liquid Scintillation Counting*, Vol. 3, M. A. Crook and P. Johnson (Eds.). Heyden & Son, New York, pp. 253-272.

Stanley, P. E. (1976). The use of the liquid scintillation spectrometer in bioluminescence analysis. In *Liquid Scintillation Counting, Science and Technology*, A. A. Noujaim, C. Ediss, and L. Weike (Eds.). Academic, New York, pp. 209-227.

Stanley, P. E. (1978). Quantitation of picomole amounts of NADH, NADPH, and FMN using bacterial luciferase. In *Methods in Enzymology*, Vol. 57, M. DeLuca (Ed.). Academic, New York, pp. 215-222.

Ulitzur, S., and J. W. Hastings (1978). Bioassay for myristic acid and long chain aldehydes. In *Methods in Enzymology*, Vol. 57, M. DeLuca (Ed.). Academic, New York, pp. 189-193.

Wettermark, G., and S. E. Brolin (1979). Electrochemiluminescence measurements of FMN. In *Proceedings of International Symposium on Analytical Applications of Bioluminescence and Chemiluminescence*. State Printing and Publishing, Westlake Village, Cal., pp. 212-220.

6

APPLICATIONS OF COELENTERATE LUMINESCENT PROTEINS

MAURICE B. HALLETT and ANTHONY K. CAMPBELL Welsh National School of Medicine, Heath Park, Cardiff, Wales

I. Coelenterate Luminescence 90
- A. Occurrence 90
- B. Chemistry 92
- C. Cellular Location and Regulation of Coelenterate Luminescence 94

II. Possible Applications of Coelenterate Luminescence 97
- A. Importance of Measuring Ca^{2+} 97
- B. Measurement of PAP and PAPS 99
- C. Energy Transfer 100

III. Preparation of Ca^{2+}-Activated Photoproteins 100

IV. Introduction of Ca^{2+}-Activated Photoproteins into Cells 103
- A. Introduction 103
- B. Pressure Microinjection 105
- C. Problems of Introducing Photoproteins into the Cytoplasm of Large Numbers of Small Cells 108

V. Applications of Membrane Vesicles 111

VI. Quantification of Cytoplasmic Photoprotein 113

VII. Relationship between Free Ca^{2+} and Light Emission 114

VIII. Quantification of Intracellular Ca^{2+} 116
- A. Quantification under Conditions of Negligible Photoprotein Consumption 116
- B. Quantification under Conditions of Significant Photoprotein Consumption 117
- C. Problems of Heterogeneity of Cell Populations 118

IX. Ca^{2+} Distribution within the Cell; Image Intensification 119
X. Problems Associated with the Use of Photoproteins 121
A. Ca^{2+}-Independent Light 121
B. Other Sources of Luminescence 121
C. Release of Photoprotein 121
D. Heterogeneous Intracellular Distribution of Photoprotein 121
E. Heterogeneous Intracellular Distribution of Cytoplasmic Ca^{2+} 122
F. Changes in Photoprotein Affinity for Ca^{2+} 122
G. Toxicity 122
H. Temporal Distortion of Light Signal 122
I. Cell Opacity 122
XI. Conclusion 123
Note Added in Proof 124
References 124

Rasa ligno, parum adeo in tenebris splendet

Forskål, 1775

I. COELENTERATE LUMINESCENCE

A. Occurrence

Anyone who has been out on the beach at night and looked carefully at the seaweed in the splash zone as they walked on it will no doubt have observed, perhaps without realizing, coelenterate luminescence. The quotation above from Forskål's *Fauna Arabica* provides, in fact, the first published description of the luminescence of the jellyfish *Aequorea forskalea* or *Medusa aequorea* as he called it. Much as his observations fascinated him, Forskål would undoubtedly have been amazed by the impact which the luminescence from this organism was to have on cell physiology some 200 years later, for it was the luminescent protein extraced from *Aequorea* which was to provide the first generally applicable method for measuring free Ca^{2+} in living cells (Ridgway and Ashley, 1967; Campbell et al., 1979b).

The phylum Coelenterata consists of about 10,000 species now being regarded as two separate, but evolutionary-related, phyla namely Cnidaria (9500 species) and Ctenophora (200 species). Both phyla abound with luminous species (Table 1) which emit blue or blue-green light when touched. At one time it was thought that no

Table 1 Some Luminescent Coelenterates

Cnidaria		
Hydrozoa	Hydroids	*Stomotoca* spp, *Leuckartiara* spp, *Euphysa* spp, *Obelia geniculata, Obelia dichotoma, Obelia longissima, Obelia commissuralis, Gonothyrea loveni, Clytia edwardii, Cytia johnstonii, Lovenella* spp
	Medusa	*Obelia lucifera, Aequorea forskalea, Aequorea vitrina, Tima bairdii, Mitrocomella polydiademata, Eutoninia indicans, Mitrocoma (Halistaura* or *Thaumantis) cellularia, Phialidium* spp, *Octophialucium* spp, *Colobonema* spp, *Crossota* spp, *Solmissus* spp, *Aeginura* spp
	Siphonophores	*Vogtia* spp, *Hippopodius gleba, Apolemia* spp, *Diphyes* spp, *Abyla* spp, *Praya cyrribiformis, Rosacea* spp, *Maresearsia* spp, *Agalma* spp
Scyphoza	Medusae	*Periphylla* spp, *Atolla* spp, *Pelagia noctiluca, Poralia* spp
Anthozoa	Seapens and pansies	*Veretillum* spp, *Cavernularia* spp, *Cavernularia* spp, *Renilla reniformis, Renilla kolikeri, Funiculina* spp, *Umbellula* spp, *Virgularia* spp, *Acanthoptilum* spp, *Stylatula* spp, *Ptilosarcus guemeyi, Pteroides* spp, *Pennatula* spp
	Zoanthids	*Parazoanthus* spp, *Epizoanthus* spp
Ctenophora		
	Sea combs and sea jellies	*Bolinopsis* spp, *Mnemiopsis leidyi, Leucothea (Eucharis)* spp, *Ocyropsis* spp, *Deiopea* spp, *Eurhamphea* spp, *Beröe albans, Beröe rufescens, Cydippedensa, Pleurobrachia*

Sources: Forskål (1775); Forbes (1848); Hincks (1868); Panceri (1876); Harvey (1952); Russe (1952); Nicol (1958); Morin and Hastings (1971); Anderson and Cormier (1973); Blinks et al. (1976); Herring (1978); Campbell et al. (1979b).

gymnoblastic hydroids (Allman, 1873; Harvey, 1952) were luminous. However, the observations of luminescence in *Leukartiara octona*, *Euphysa*, and *Stomatoca* have invalidated this (Hartlaub, 1914; Russell, 1952; Herring, 1978). Unfortunately, some of the older literature has been confused by different nomenclature as well as misidentification of species.

B. Chemistry

While the components of luminescent reactions extracted from organisms of many phyla conform to the classical "luciferin" + "luciferase" oxygen-requiring system first described by Dubois (1885,1887; see also Harvey, 1952), two particular groups, namely, most of the luminous coelenterates and radiolarians, do not (Macartney, 1810; Harvey, 1952). In fact, it has been known for some time that it is possible to stimulate luminescence from these dried organisms in the absence of O_2 (Harvey, 1926). The puzzle was solved by the elegant work of Shimomura and Johnson, who showed that addition of calcium, in the absence of any cofactors or oxygen, stimulated light emission from the protein aequorin, extracted and purified from the hydrozoan *Aequorea forskalea* (Shimomura et al., 1962,1963). The energy for light emission was shown to result from a chemical reaction within the prosthetic group of the protein (Fig. 1) and not a Ca^{2+}-binding energy. It was subsequently proposed that the term *photoprotein* be applied to precharged, luminescent proteins of this sort (Shimomura and Johnson, 1966,1973). Ca^{2+}-activated photoproteins similar to aequorin have been isolated from several hydrozoa, scyphozoa, and ctenophores (Table 2). However, addition of Ca^{2+} to extracts from luminous scyphozoa such as *Atolla* do not produce an increase in light emission. Furthermore, thanks to the work of Cormier and his group, we now know that the reaction scheme for luminescence in *Renilla* and other luminous anthozoans also does not involve a Ca^{2+}-activated photoprotein (Cormier et al., 1974; Cormier, 1978). However, the structure of the prosthetic group seems to be identical for most, if not all, the luminous coelenterates. It has been given the name coelenterazine and is probably covalently linked to the protein through the carbonyl or group -O-O- groups (see Shimomura and Johnson, 1978). The product of the reaction, the so-called blue fluorescent protein (BFP), contains the reacted prosthetic group coelenteramide. Aequorin is itself nonfluorescent but is converted to a blue fluorescent protein (BFP) which has a fluorescence spectrum virtually identical to the luminescence spectrum.

The coelenteramide can be removed from the BFP after removal of Ca^{2+} and separated from the apoprotein by gel filtration (Shimomura and Johnson, 1973). The apophotoprotein can be recharged by addition of native or synthetic prosthetic group (Fig. 2) in the presence of O_2 and mercaptoethanol to protect the -SH groups on the protein. Although

Table 2 Some Ca^{2+}-Activated Photoproteins

Protein	Organism	k_{sat} (s^{-1})	λ_{max} (nm)	Reference
Cnidaria				
Aequorin	*Aequorea forskalea*	1.5	460	Moisescu et al. (1975)
Obelin (g)	*Obelia geniculata*	4.0	475	Campbell et al. (1979b)
Obelin (1)	*Obelia longissima*	0.6	475	Morin and Hastings (1971)
Clytin	*Clytia edwardsia*	–	–	Anderson and Cormier (1973)
Halistaurin	*Halistaura cellularia*	0.4	–	Morin and Hastings (1971)
Pelagin	*Pelagia noctiluca*	6.0	–	Morin and Hastings (1971)
Ctenophora				
Mnemiopsin	*Mnemiopsis leidyi*	3.5	485	Morin and Hastings (1971)
Berovin	*Beröe* spp	–	–	Ward and Seliger (1974)
Protozoa (Radiolaria)				
Thalassicolin	*Thalassicola*	3	*ca* 440	Campbell and Herring (1979)

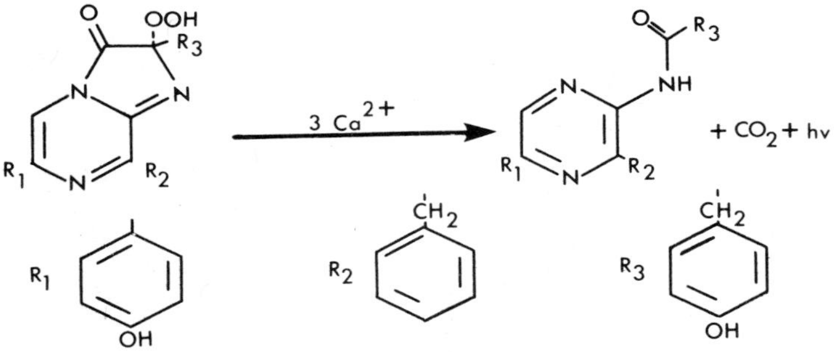

Figure 1 Reaction between prosthetic group of coelenterate photoprotein with calcium ions. (From Campbell et al., 1979b).

the reactivation is slow, the product appears to be identical in its kinetics and luminescence spectrum to the original native photoprotein. One discrepancy exists in that Shimomura and Johnson (1975) claim virtually 100% reactivation, whereas Cormier's group can achieve only 20% (Hart et al., 1979).

It has recently been reported (Herring, 1978) that the luminous radiolarian *Thalassicola* also contains a Ca^{2+}-activated photoprotein. Collection of large quantities of this organism off the Irish coast while on RRS Discovery (Campbell and Herring, 1979) confirmed that thalassicolin is very similar to the Ca^{2+}-activated photoproteins obelin and aequorin and can be reactivated by the same prosthetic group.

C. Cellular Location and Regulation of Coelenterate Luminescence

In spite of the large number of luminous coelenterates which have been identified, the physiological significance of their luminescence is unknown. What is known, however, is that the luminescence can be stimulated by mechanical, electrical, and a variety of chemical stimuli (Bles, 1892; Harvey 1952; Davenport and Nicol, 1955; Morin and Cooke, 1971a,b; Cormier, 1978). Many of these stimuli cause irreversible damage to the organisms, though the effect of high K^+ is reversible (Fig. 3), and mechanical stimulation probably relates to the main cause of luminescence of coelenterates in their natural environment. The stimulation of coelenterate luminescence by KCl is a useful means of finding these organisms on the beach or on rocks at night.

The luminescent cells are found along the marginal canal at the base of the tentacles in hydrozoan medusae such as *Aequorea* and *Obelia*. However, in scyphozoan jelly fish such as *Atolla*, the luminescent cells are also found at other sites, or even on the surface of the bell as in *Pelagia*. In hydroids such as *Obelia* the luminous cells

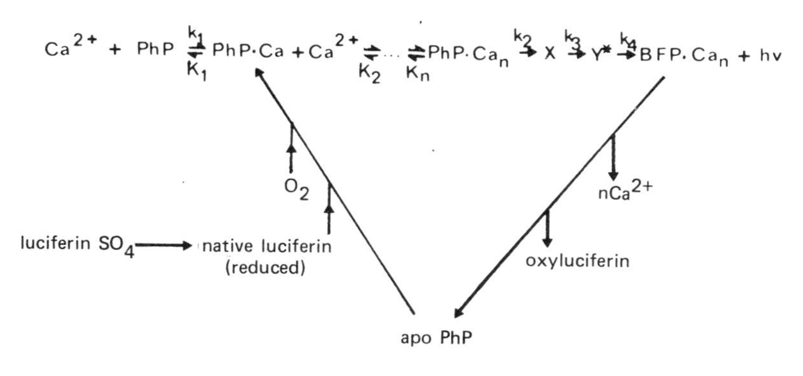

Figure 2 Photoprotein reactions showing regeneration of apophotoprotein by addition of native prosthetic group in the presence of O_2 and mercaptoethanol, protecting -SH groups on the proteins. (From Campbell, Lea, and Ashley, 1979b.)

occur mainly in the stem of the animals, with some in the stolons and virtually none in the hydranths, which are the oral ends of the polyps. Some controversy exists over which cell layer contains the luminous cells. They can only be definitively identified by their luminescence, which requires the aid of an image intensifier (Morin and Reynolds, 1974). In some coelenterates the luminous cells can also be identified by fluorimetry (Morin and Reynolds, 1974; Campbell, Lea, and Ashley, 1979). Coelenterates are diblastic animals with an essentially acellular mesoglea rather than a true mesoderm. It has been reported that the luminous cells of some hydroids and of the scyphozoan *Pelagia* are to be found in the ectodermal, or outer cell, layer (Panceri, 1876; Dahlgren, 1916). While this may be true for *Pelagia* and some siphonophores (Herring, 1978), more recently hydroids, such as *Obelia*, have been reexamined by fluorescence microscopy and image intensification (Morin and Reynolds, 1974; Campbell et al., 1979b) where the luminous cells seemed to be found in the inner gastrodermal cell layer.

Isolated cells can be prepared from *Aequorea* by shaking the dissected marginal canal (Blinks et al., 1978a) or from the coenosarc of *Obelia* hydroids by enzymatic digestion with pronase (Campbell and Hallett, 1979). High K^+ but not normal Na^+ saline stimulates cell luminescence, which can be abolished by removal of extracellular Ca^{2+} or by the presence of the Ca^{2+} channel blocker D600 or La^{3+} (Campbell et al., 1979b). Luminescence is also stimulated by zero Na^+- (Li^+-replaced) saline, but to a lesser extent than high K^+. These

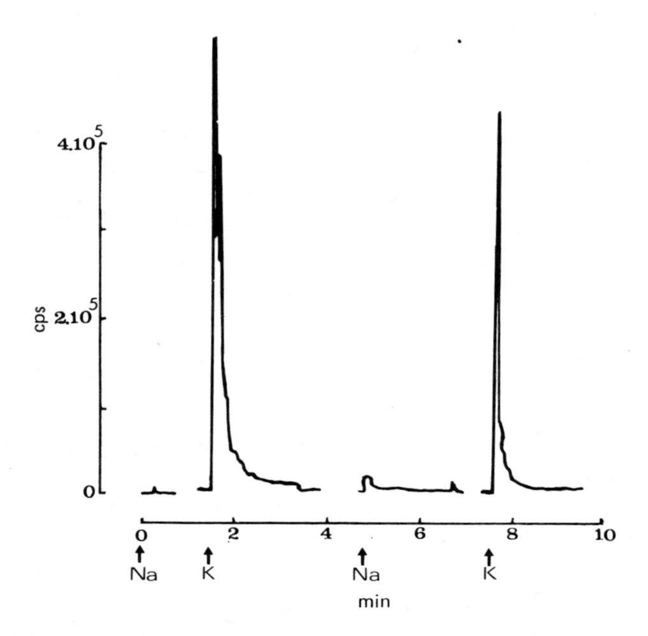

Figure 3 Effect of high K^+ medium on luminescence from chopped *Obelia* hydroids. The ordinate shows luminescence intensity in counts per sec (cps). At points marked Na, artificial seawater was added to the hydroids. When this medium was replaced by artificial seawater in which sodium had been replaced by potassium, at points marked K, luminescence was stimulated. This response could be evoked a number of times from the same chopped *Obelia* hydroids.

observations suggest that the luminous cells from coelenterates could provide a useful model for investigating the effects of various substances on intracellular Ca^{2+}, particularly if a naturally occurring transmitter substance can be identified. More recently (Campbell and Bassot, unpublished) it has been observed that the luminous cells from *Obelia* are extremely unstable in normal seawater, pH 8.0, where they seem to break up into smaller membrane-bound particles. A considerable improvement (approximately 10-fold) in yield and stability of cells was achieved by first narcotizing the hybrids in 0.3 mol/L $MgCl_2$, pH 8.0, before squeezing out the coenosarc and preparing and storing the cells in sea water containing 1 to 10 g bovine serum albumin/L, pH 6.0. This presumably relates more closely to the environment within the coelenteron to which the cells are exposed in situ.

In the hydrozoans, scyphozoans, and ctenophores, Ca^{2+} is required to stimulate luminescence from the photoprotein itself. In *Renilla*, Ca^{2+} regulates luminescence by binding to the protein containing the luciferin, allowing it to react with the luciferase. It is also of interest that a "calmodulin" has recently been isolated from *Renilla* (Jones et al., 1979.)

II. POSSIBLE APPLICATIONS OF COELENTERATE LUMINESCENCE

If a substance is to be measured using a luminescent reaction, then it must either act as or be converted to a substrate, cofactor, or catalyst of that reaction, or be able to interact with one of the components of the reaction in such a way as to affect light emission (Campbell and Simpson, 1979). At present the two most important applications of coelenterate luminescent proteins are in the measurement of Ca^{2+} and the metabolites 3'5'-phosphoadenosine phosphate (PAP) and 3'-phosphoadenosine-5'-phosphosulfate (PAPS) (Table 3).

A. Importance of Measuring Ca^{2+}

Photoproteins such as aequorin and obelin are most suitable for measuring free Ca^{2+} concentrations in the range 10 nmol/L to 100 μmol/L. They could in theory be used to measure the total Ca^{2+} extracted from as little as 10 μg of tissue. However, there have been no reports of such an assay being of practical use compared with techniques such as atomic absorption spectrometry. It has been suggested that aequorin could be used to measure serum Ca^{2+} (Izutsu and Felton, 1972). Coelenterate photoproteins are more than 95% saturated with Ca^{2+} at the concentrations of free Ca^{2+} found in serum and thus only a relatively small proportion of binding sites are available. Furthermore, since the apparent affinity of the Ca^{2+}-activated photproteins for Ca^{2+} is reduced by Mg^{2+} and monovalent cations, it is essential to standardize ionic conditions if accurate values for Ca^{2+} are to be obtained. Thus it is unlikely that this method will be of any real value for the analysis of serum Ca^{2+} in clinical chemistry.

By far the most important application of coelenterate photoproteins is their use in the measurement and study of the distribution of intracellular free Ca^{2+}. During the last 15 years it has been proposed that changes in intracellular Ca^{2+} mediate the effects of many other primary stimuli, such as neurotransmitters and hormones, to activate cell movement, secretion, glycogen breakdown, vision, cell fertilization, and cell-cell communication (Bygrave, 1967; Douglas, 1968; Rasmussen, 1970; Berridge, 1976; Rasmussen and Goodman, 1977). A major problem in the investigation of this hypothesis was the lack of a suitable method for measuring the concentration of free Ca^{2+} inside living cells (Hales et al., 1977; Campbell et al., 1979d; Ashley and Campbell, 1979).

Early attempts to detect changes in intracellular free Ca^{2+} used the red dye alizarin sulfonate (Pollack, 1928) or murexide (Jobsis and

Table 3 Main Application of Coelenterate Luminescent Proteins

Luminescent protein	Substance to be measured	Reference
Aequorin or obelin	Intracellular free Ca^{2+}	Ashley and Campbell (1979)
Aequorin or obelin	Release or uptake of Ca^{2+} by organelles	Azzi and Chance (1969); Yates and Campbell (1979)
Aequorin or obelin	Total Ca^{2+} in tissue extracts	Shimomura and Johnson (1969) Campbell (1974)
Renillin	PAP or PAPS	Burnell and Anderson (1973) Cormier (1978)
Aequorin, obelin, or renillin	Coelenterazine	Shimomura and Johnson (1979)

O'Connor, 1966). However, the injection of the photoprotein aequorin (Ridgway and Ashley, 1967) into a cell provided, for the first time, a method which was generally applicable to many cell types. Ideally, a method for measuring intracellular Ca^{2+} should satisfy seven criteria (Ashley and Campbell, 1979):

1. *Sensitivity and selectivity*: The method must be sensitive to free Ca^{2+} in the range 10 nmol/L to 100 μmol/L and, since the free Mg^{2+} concentration in the cytoplasm is 1 to 5 mmol/L, the method must be highly selective for Ca^{2+} over Mg^{2+}.
2. *Quantification*: It should be relatively easy to determine absolute values for free Ca^{2+} concentration, even though it may be necessary to correct for interference by other ions.
3. *Response time*: The method must be fast enough to detect changes in free Ca^{2+} in cells with response times of the order of msec, e.g., some muscle cells, without distorting the signal.
4. *Incorporation into the cell*: It must be possible to insert the indicator into the living cell.
5. *Cell viability*: Neither the incorporation procedure nor the indicator itself must cause any impairment in cell structure or function, nor must it significantly disturb the Ca^{2+} balance in the cell.
6. *Ca^{2+} distribution*: It should be possible to measure changes in the distribution of free Ca^{2+} within the cell.
7. *Availability*: The reagents should be easily obtained and relatively inexpensive.

One of the aims of the present chapter is to examine critically these parameters in order to see how well they are satisfied by coelenterate photoproteins.

B. Measurement of PAP and PAPS

The observation that acidification of an extract of *Renilla* produced a large increase in the quantity of luciferin led to the discovery of luciferyl sulfate as the storage form of the luciferin (Cormier, 1962). The reversible reaction of luciferyl sulfate with 3',5'-phosphoadenosine phosphate (PAP) catalyzed by a luciferin sulfokinase extractable from *Renilla* has led to the development of a highly sensitive assay for both PAP and the product of the reaction, 3'-phosphoadenosine-5'-phosphosulfate (PAPS).

$$\text{Luciferyl sulfate} + \text{PAP} \underset{\text{sulfokinase}}{\overset{\text{luciferin}}{\rightleftharpoons}} \text{luciferin} + \text{PAPS}$$

$$\text{Luciferin} + O_2 \overset{\text{renillin}}{\rightleftharpoons} \text{oxyluciferin} + CO_2 + h\nu$$

PAP and PAPS occur in plants and microorganisms and have also been measured in mammalian liver, lung, heart, and kidney (Gregory and Lipmann, 1957; Cormier, 1978). Tissue concentrations range from 0.25 nmol/g fresh tissue in spinach leaves to 18 nmol/g fresh tissue in rabbit liver. This puts these metabolites in the same range of concentrations as cyclic AMP in tissues. They are synthesized by the following reactions:

$$\text{ATP} + SO_4^{2-} \xrightarrow{\text{ATP sulfurylase}} \text{adenosine phosphosulfate (APS)} + PP_1$$

$$\text{APS} + \text{ATP} \xrightarrow{\text{APS kinase}} \text{PAPS} + \text{ADP}$$

$$\text{PAPS} \xrightarrow{\text{PAPS sulfatase}} \text{PAP} + SO_4^{2-}$$

The original assays based on this method were able to measure 10 to 100 pmol PAP or PAPS (Burnell and Anderson, 1973; Stanley et al., 1975). More recently, the sensitivity has been increased enabling 0.1 pmol PAP to be detected (Cormier, 1978). The availability of a luminescence assay for PAP and PAPS which is sensitive enough to detect the low concentrations of these metabolites in tissues is likely to lead to an increased understanding of their physiological function.

It has been suggested that renillin might also be used to measure the enzyme aryl sulfatase (Cormier, 1978).

$$\text{Benzyl luciferyl disulfate} \xrightarrow{\text{aryl sulfatase}} \text{benzyl luciferyl sulfate} + SO_4^{2-}$$

$$\text{Benzyl luciferyl sulfate + PAP} \xrightarrow{\text{luciferin sulfokinase}} \text{benzyl luciferin + PAPS}$$

$$\text{Benzyl luciferin} + O_2 \xrightarrow{\text{renillin}} \text{oxyluciferin} + CO_2 + h\nu$$

C. Energy Transfer

Many examples of the transfer of energy between electronically excited molecules are known in photochemistry and biology (Berlman, 1973; Stryer, 1978). One particularly fascinating example seems to exist in many luminous coelenterates. In those organisms exhibiting the phenomenon, a green fluorescent protein (GFP) (Table 4) containing a chromophore different from the luminescent protein (Shimomura, 1979) interacts with the chromophore in the luminescent protein such that the fluorophore in the GFP becomes the actual photon emitter.

The presence of GFP causes the light emitted by the organism to be greener than for the pure luminescent protein, with a sharper luminescence spectrum (Morin and Hastings, 1971; Campbell and Hallett, 1979). However, only in *Renilla* has energy transfer been convincingly demonstrated in vitro (Ward and Cormier, 1978) when a threefold increase in quantum yield occurred. Attempts to demonstrate energy transfer in vitro between aequorin, obelin, and their respective green fluorescent proteins have so far proved unsuccessful. It has been proposed (Campbell et al., 1979c; Campbell et al., 1980b) that a similar type of energy transfer between a chromophore on an antigen and another on an antibody could provide the basis for a highly sensitive homogeneous immunoassay. Antibody-antigen binding would be detected by a change in quantum yield, kinetics, or a wavelength shift. The study of naturally occurring examples of this phenomenon may therefore provide important insights into the conditions necessary to optimize this in immunoassay systems.

III. PREPARATION OF CA^{2+}-ACTIVATED PHOTOPROTEINS

A large number of photoproteins have been extracted from a variety of coelenterate sources, but in practice only two photoproteins have so far been used as intracellular Ca^{2+} indicators in physiological experiments, aequorin and obelin. The proteins have different saturating rate constants and maximum emission wavelengths (Table 2). *Aequorea* is harvested in great numbers in Friday Harbor, Washington between July and September, whereas it is a rare visitor to the British coasts. However, *Obelia geniculata* is found in large quantities growing on

Table 4 Luminous Coelenterates with and without a Green Fluorescent Protein

Coelenterate	Reference
A. *Green emitters with GFP*	
Hydrozoa	
Obelia geniculata	Morin and Hastings (1971) Campbell et al., (1979b)
Aequorea forskalea	Morin and Hastings (1971)
Clytia spp	Herring (1978)
Siphonophores	
Anthozoa (sea pens and sea pansies)	
Renilla reniformis, kollikeri, and *mulleri*	Anderson and Cormier (1973)
Stylatula elonga	
Acanthoptilum gracile	Cormier (1978)
B. *Without GFP*	
Siphonophores	
Hippopodius	Herring (1978)
Scyphozoa	
Pelagia noctiluca	Morin and Hastings (1971)
Atolla	Herring (1978)
Periphylla	Herring (1978)
Ctenophora	
Mnemiopsis spp	Morin and Hastings (1971)
Beroe spp	

Laminaria in Plymouth Sound, United Kingdom, during May to September (Campbell, 1974). Both photoproteins can be purified from the medusa or hydroid extracts by essentially similar means, based on the method of Shimomura et al. (1962) (Campbell et al., 1979b) (Table 5).

Table 5 Extraction and Purification of Aequorin and Obelin

(i) *Extraction and Purification of Aequorin*

1 *Aequorea* netted in Friday Harbour, Washington, USA, washed briefly in isotonic NaCl plus 10 mmol/1 EDTA pH 6.0.

2 Marginal rings removed individually (Blinks et al., 1978a) and placed in cold saturated $(NH_4)_2SO_4$ (∿100 rings/100 ml solution), shaken and resaturated. Rings removed and "squeezed" through cheese-cloth into 50 ml cold saturated $(NH_4)SO_4$ solution and solution resaturated and stirred.

3 To supernatants from step 2, acid-washed celite (John Manville analytical filter aid) is added (∿30% total volume), the resultant "slurry" filtered with fine grade filter paper and the "cake" allowed to dry.

4 "Cake" washed with 3 × 50 ml of 50 mmol/L EDTA (pH 6.0) on the filter paper, and the washings added directly into solid $(NH_4)_2SO_4$ at 0 to 5°C. Precipitated aequorin centrifuged at 50-100,000g for 30 min, stored in this form below -20°C overnight.

5 Aequorin redissolved in 10 mmol/L EDTA pH 6.0 and chromatographed on G-25 Sephadex column.

6 Eluent from several columns loaded on to a DEAE-cellulose column (40 × 3 cm) and eluted with a linear salt gradient from the 10 mmol/L sodium acetate, 10 mmol/L EDTA, pH 6.0, to 1 mmol/L sodium acetate, 10 mol/L EDTA pH 6.0. Fractions collected and assayed for protein concentration, conductivity, and Ca^{2+} stimulated light emission.

7 Fractions containing highest aequorin concentrations pooled and run on a G-25 Sephadex column pre-washed with ammonium formate or acetate.

8 The eluent from the column was lyophilised and stored at -25°C. An alternative method used by Blinks et al. (1978a) starts by shaking the marginal canal to produce isolated cells and may provide a purer final preparation.

(ii) *Extraction and Purification of Obelin*

1 *Obelia geniculata* collected growing on seaweed *Laminaria* by skin diver.

2 Hydroids scraped off seaweed in sea water, and the free hydroids soaked in Ca^{2+} free artificial sea water at 0°C for 15 min.

3 Hydroids homogenised in a blender in 1 L 50 mmol/L EDTA, 100 mmol/L Tris buffer, pH 7.0, 0°C and after soaking, the suspension was squeezed through muslin and the filtrate centrifuged at 10,000 g for 10-20 min.

Table 5 (Continued)

4 The supernatant pH adjusted to 6.0, saturated with $(NH_4)_2SO_4$ and left overnight at 0°C to allow the precipitate to flocculate.

5 The precipitate was filtered and scraped off into saturated $(NH_4)_2SO_4$ and 50 mmol/L EDTA, pH 7.0 (this suspension could be stored for many months at 0°C).

6 A sample of this suspension was filtered and redissolved in 1 mmol/L EDTA, 10 mmol/L Tes buffer, pH 7.0, so that the contaminating pink protein remained in the precipitate, and centrifuged at 100,000g for 30 min.

7 The supernatant was desalted on a G-75 (70 × 3 cm) or G-25 (40 × 3 cm) Sephadex column equilibrated in 1 mmol/L EDTA, 10 mmol/L Tes buffer, pH 7.0, and the active obelin fractions pooled. The conductivity was adjusted to ∿4 mmho.

8 The obelin was loaded on to a DE 52 column equilibrated in 0.1 mmol/L EDTA, 10 mmol/L Tes pH 7.0, and eluted with a linear NaCl gradient (50-500 mmol/L).

9 The eluent was desalted on G-25 Sephadex column equilibrated in 10 mmol/L EDTA, 10 mmol/L Tes buffer pH 7.0.

10 The eluent was aliquoted and gelatin added to give a final gelatin concentration of 0.1 g/L. Aliquots were freeze-dried and stored at -20°C or -70°C.

Source: Campbell et al. (1979b).

IV. INTRODUCTION OF CA^{2+}-ACTIVATED PHOTOPROTEIN INTO CELLS

A. Introduction

The usefulness of Ca^{2+}-activated photoproteins as indicators of cytoplasmic calcium depends upon the ability to introduce the photoprotein into the cell without cell injury.

This can be achieved in giant cells by microinjection (Fig. 4). Cells as large as the barnacle muscle (*Balanus nubilus*) or squid giant axon (*Loligo pealii*) can be microinjected with relatively large volumes (>0.45 μl) of aequorin solution using a syringe with a tip tapering to a diameter of 150 μm (Ashley and Ridgway, 1970; Baker et al., 1971). With smaller cells, microinjection of photoprotein is considerably more difficult. It is not possible to eject the aequorin from the micropipette into cell cytoplasm using iontophoresis. Although the photoproteins

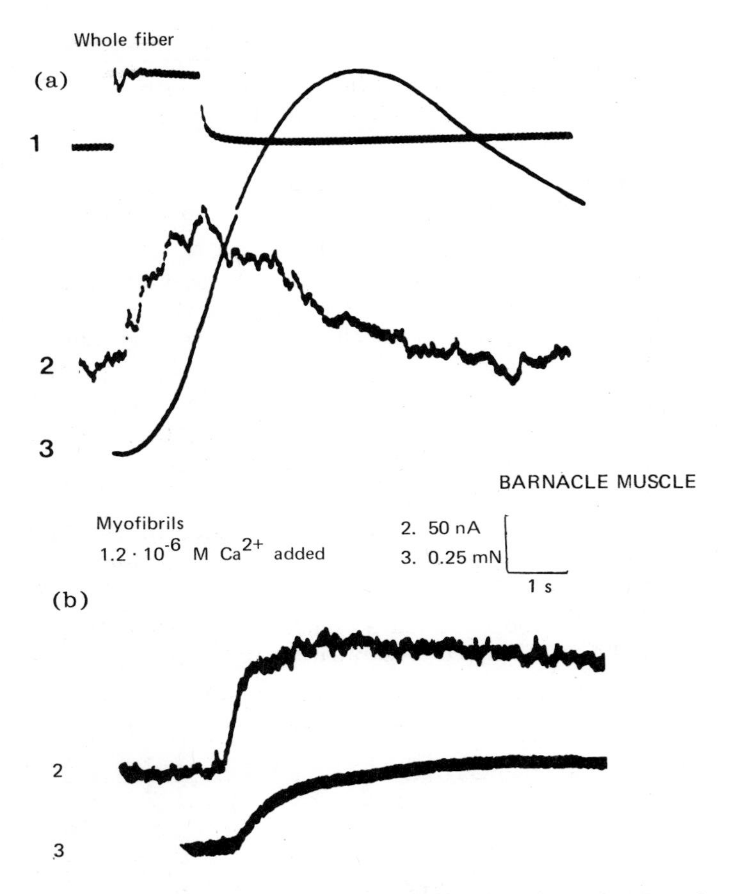

Figure 4 Response of obelin-injected barnacle muscle fiber to electrical stimulation. Traces: 1 shows stimulus marker, 2 obelin luminescence, 3 fiber tension; (a) whole fiber. (From Ashley et al., 1976.) (b) myofibrils. (From Ashley et al., 1975.)

bear a negative charge at physiological pH, their mobility is relatively low compared with small ions. On applying the injection voltage, positive ions, including calcium, enter the micropipette without ejection of photoprotein. Pressure injection, employing either a hydraulic or a gas pressure pulse system, is therefore the only method to be performed successfully.

B. Pressure Microinjection

The glass micropipette must be thoroughly cleaned, minimizing calcium contamination and avoiding particles or glass chips which may block the tip. Calcium contamination can be reduced by acid washing and glass chips removed by fire polishing (Blinks et al., 1976). In order to minimize the damage inflicted upon the cell during impalement, the micropipette should be as small as possible. It is possible to use micropipettes with tips small enough to have resistance greater than 10 megohms in 3 mol/L KCl (Blinks et al., 1978a). Beveling of the tip may lead to easier insertion of the micropipette and several techniques have been reported for achieving this (Barrett and Whitlock, 1973; Brown and Flaming, 1974; Chang, 1975).

Monitoring of the potential at the microelectrode tip provides a method of verifying penetration of the cell. This can be achieved by dissolving aequorin in a conducting solution, normally isotonic KCl, and platinum wire used rather than silver wire, as silver stimulates the aequorin to luminesce (Blinks et al., 1976). The concentration of aequorin in the micropipette cannot be near the maximal solubility of the protein (1.4 mmol/L or 30 g/l (Shimomura and Johnson, 1979) because the high viscosity of concentrated solutions prevents injection. Also there is a tendency at high concentrations for the formation of protein aggregates causing clogging of the micropipette tip. The pipettes can be loaded from either end by capillary action and ejected into the cell using either hydraulic pressure (Llinas et al., 1972) or gas pressure (Stinnakre and Tauc, 1973; Brown and Blinks, 1974; Chang et al., 1974; Kusano et al., 1975; Taylor et al., 1975b). Pulsed gas pressure (between 0.5 and 10×10^5 Pa) can be supplied to the micropipette by the use of solenoid-operated valves (Stinnakre, 1979; Eckert and Tillotson, 1979; Eckert et al., 1977), providing good control during the injection procedure. If the cell undergoing microinjection is large, a second micropipette containing $CaCl_2$ can be inserted into the cell and the progress of aequorin injection monitored by electrophoretic injection of Ca^{2+} through the second pipette (Kusano et al., 1975). Provided the aequorin-filled pipette is blackened to protect the photomultiplier tube from spontaneous luminescence, light emission is observed in response to Ca^{2+} pulses. A plateau of light response to pulses of Ca^{2+} was reached after a few injections of aequorin into *Aplysia* neuron (Stinnakre, 1979). Damage to the cell is often detectable by a large increase in resting glow. The micropipette must be removed from the cell before attempting to measure resting cell light emission as the glass electrode will contain sufficient contaminating calcium to cause a significant emission of light.

Using techniques similar to those outlined above, a variety of cells have been microinjected with aequorin (Table 6). Most are nonmammalian and the smallest cells to be injected are smooth-muscle cells

Table 6 Cells Injected with Ca^{2+}-Activated Photoprotein

Cell injected	Dimension	Cell volume	Resting free cytoplasmic Ca^{2+} (mol/L)	Reference
Giant muscle fiber, barnacle	20 mm long 1 mm diam	>12 μl	$0.8 - 2 \times 10^{-7}$	Ridgway and Ashley (1967) Ashley and Ridgway (1970)
Giant muscle fiber, crab	1.2 mm long 0.5-1 mm diam	∿0.5 μl	—	Ashley (1969)
Protozoa ciliate		30 nl	—	Ettienne (1970)
Giant axon, squid	>1 cm long 1 mm diam	>60 μl	$<3 \times 10^{-7}$	Baker et al. (1971) Hallett and Carbone (1972)
Giant synapse, squid presynaptic, postsynaptic	100-200 μm diam	∿4 nl	—	Llinas et al. (1972) Kusano et al. (1975)
Amphibian egg		>1 μl >1 μl	4.9×10^{-7}	Baker and Warner (1972) Moreau et al. (1980)
Aplysia neuron	>1 cm long 400 μm diam	>1.2 μl	—	Stinnakre and Tauc (1973)
Amphibian muscle fiber	7 mm long 100 μm diam	∿70 nl	ND	Rudel and Taylor (1973)

Photoreceptor, crab + barnacle	100 μm long 50 μm diam	∿0.2 μl	—	Brown and Blinks (1974)
Salivary gland cell, chironomus	100 μm diam	∿0.5 μl	ND	Rose and Lowenstein (1975)
Giant ameba	700 μm diam	20 nl	$0.5 - 1 \times 10^{-7}$	Taylor et al. (1975a) Cobbold (1979)
Plasmodium of slime mold		3-5 μl	—	Ridgway and Durham (1976)
Sea urchin eggs	100 μm diam	∿0.5 μl	—	Steinhardt et al. (1977)
Medaka fish eggs	1 mm diam	0.4 μl	— —	Ridgway et al. (1977) Gilkey et al. (1978)
Neuron, helix	200 μm diam >1 cm long	>0.5 μl	—	Eckert and Tillotson (1979)
Cardiac muscle, frog	250 μm long 10 μm diam	5-10 pl	—	Allen and Blinks (1978)
Starfish oocyte	160 m diam	1.8 nl	5×10^{-7}	Moreau et al. (1978)
Smooth-muscle cell, frog	250 μm long 4.5 μm diam	∿4 pl	ND	Fay et al. (1979)
Rat and human skeletal muscle	∿150 μm long 10-30 μm diam	∿45 pl	—	Eusebi et al. (1980)

(Fay et al., 1979) and cardiac muscle cells (Allen and Blinks, 1978) from frog with cell volumes of approximately 4 pL and 5 to 10 pL, respectively. With cells of this size, the resting free calcium concentration cannot be measured and a number of cells (up to 100 cells) must be injected to give a measurable signal above background when the cells are stimulated. This illustrates a problem that must inevitably accompany attempts to microinject small mammalian cells with sufficient photoprotein to produce detectable signals. A spherical mammalian cell with a diameter of 10 μm will have a volume of 0.5 pL. If the maximal amount injectable into the cells is 10% of the cell volume and a saturated aequorin solution could be microinjected despite the technical problems associated with the high viscosity, then the aequorin injected would produce a mere 6×10^6 photons (from data of Shimomura and Johnson, 1979). If the counting efficiency of the apparatus was 0.1%, a luminescent rate equivalent to 1 cps above background would be detectable with a rate constant k of reaction (see Sec. VII) of 2×10^{-4} sec^{-1}. Such a rate constant is given by a Ca^{2+} concentration of approximately 5 μmol/L (Baker et al., 1971; Allen and Blinks, 1979). In order to detect the luminescence of a calcium concentration of 0.1 μmol/L ($k \sim 2 \times 10^{-6}$ sec^{-1}), 100 cells must be so injected. Furthermore, the concentration of aequorin within each of these cells will be approximately 150 μmol/L, which is sufficient to cause a significant fraction of intracellular calcium to become bound to the photoprotein (see Sec. X). If the intracellular aequorin concentration were to be reduced to 15 μmol/L, 10^3 cells would have to be injected. It is seen, therefore, that for small mammalian cells microinjection is not the method of choice. A method is required that will "microinject" populations of cells.

C. Problems of Introducing Photoprotein into the Cytoplasm of Large Numbers of Small Cells

One approach may be microinjecting a large number of small cells with photoprotein by fusion of small phospholipid vesicles containing the photoprotein with cell membranes.

Before this approach can be used two criteria must be met. It must be possible to show that the photoprotein is entrapped within the aqueous space of the vesicle and that the vesicle membrane is impermeable to Ca^{2+}. Various membrane vesicles satisfying these criteria are shown in Table 7.

Liposomes have been used to introduce a variety of membrane-impermanent agents into the cytoplasms of a number of cells (Tyrrell et al., 1976; Pagano and Weinstein, 1978). For such a method to be useful, (1) a sufficient quantity of active photoprotein must be transferred to the cytoplasm and (2) it must be possible to quantify the amount of photoprotein within the cytoplasm as opposed to other cellular locations.

Table 7 Possible Ways of Introducing Photoprotein into the Cytoplasm of Large Numbers of Small Cells

Membrane vesicle containing photoprotein	Conditions affecting fusion with cell membrane	Reference
Liposomes	Surface change	Stendahl and Tagesson (1977)
	Albumin present	Blumenthal et al. (1977)
	Temperature	Poste et al. (1976)
	Cholesterol	Papahadjopoulos et al. (1973)
	Sonication	Poste et al. (1976)
	Fusogens	–
Erythrocyte ghosts	Polyethylene glycol 6000	Ahkong et al. (1975)
	Other fusogens	Ahkong et al. (1973)
	Sendai virus	Poste (1972)
Lumisomes	Fusogens?	–
Sonicated cell membrane vesicles	Fusogens?	–
Isolated luminescent cells, e.g., *Obelia* cells	Fusogens?	Campbell and Hallett (1979)
Viral Membranes	–	–

Calcium-impermeable liposomes containing obelin have been prepared (Dormer et al., 1978b). Addition of aqueous medium to dried mixtures of phospholipid (phosphatidyl choline + phosphatidyl serine; molar ratio 8:2) results in swelling of the phospholipids to form multimellar vesicles. Upon sonication, these vesicles break down to form unilamellar liposomes, 20 to 50 nm in diameter, entrapping substances originally present in the suspension medium.

Incubation of liposomes with rat-isolated adipocytes resulted in a time-dependent uptake of the liposomes by the cells, which was shown to be dependent on the presence of intact cell membranes (Dormer et al., 1978a). Further studies revealed that approximately 50% of the liposome uptake could be abolished by an inhibitor of endocytosis, cytochalasin B (1 to 500 μmol/L), and from homogenization of the cells after liposome incubation it was found that a fraction of the cell-associated liposomes was probably within secondary lysosomes. Loose adhesion of liposomes to the cell surface causing a transfer of cell surface antigens to the liposome membrane was also observed (Hallett and Campbell, 1980). A sensitive technique that could detect the presence of cytoplasmic obelin in the presence of obelin in these other cellular locations was required. A suitable method was the use of cell antibody plus complement to induce a selective rise in the permeability to calcium of the cell membrane, as has been described in pigeon erythrocyte "ghosts" (Campbell et al., 1979a). Using this method no cytoplasmic obelin was detected and it was concluded that less than 0.1% of cell-associated liposomal obelin was the result of liposome-cell fusion. One possibility was that the fusion event required a change in calcium permeability of the liposome or cell membrane thereby causing activation and consumption of the obelin (Hallett and Campbell, 1980). It was therefore concluded that as a means of introducing photoprotein into the cytoplasm of populations of small cells, liposomes would only be useful in measuring intracellular calcium if significantly more active photoprotein could be transferred to the cell cytoplasm.

The efficiency of transfer of photoprotein from the vesicle to the cytoplasm can be increased by either (1) increasing the amount of photoprotein transferred per fusion, e.g., by using vesicles of greater volume or using vesicles containing higher concentrations of photoprotein, or (2) increasing the percentage of vesicle-cell fusion events. The first condition can be achieved using unilamellar liposomes with internal volume approximately 10^3 times those of normal liposomes (Szoka and Papahadjopoulos, 1978) or erythrocyte ghosts containing obelin (Campbell and Dormer, 1975; see below). The second condition can be achieved using fusogens, e.g., polyethylene glycol 6000 (40% v/v) and Sendai virus.

However, preliminary experiments show that both fusogens cause a significant increase in cell membrane permeability to calcium and a consequent large consumption of entrapped photoprotein (Fig. 5).

Vesicles containing large quantities of photoproteins can be prepared from luminescent cell homogenates (Anderson and Cormier, 1973; Campbell and Hallett, 1979). Since these vesicles (lumisomes) also contain a green fluorescent protein, it may be possible to use them to monitor fusion. Fusion of the vesicle with a cell will result in dilution of the green fluorescent protein and light emitted from within the cell will be blue, whereas light from intact lumisomes will be blue-green (Campbell and Hallett, 1979).

The use of viral membrane vesicles containing photoprotein provide a prospectively interesting way of introducing photoprotein into the cell cytoplasm. Enveloped viruses depend for their reproduc-

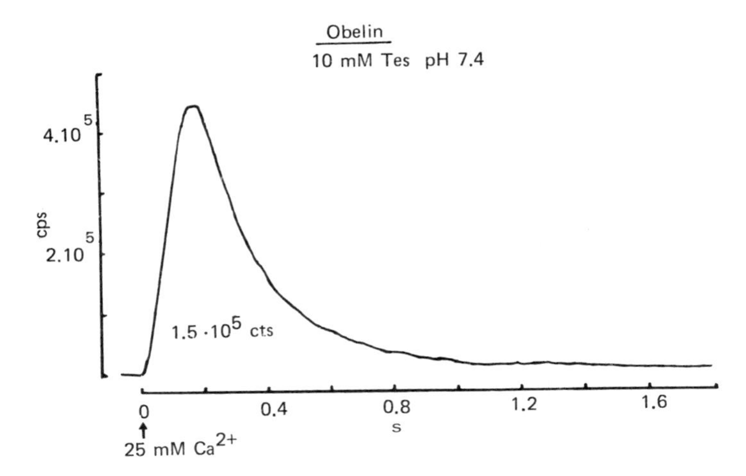

Figure 5 Exponential decay of obelin luminescence following stimulation by saturating calcium concentration.

tion on microinjecting their nucleic acids into the cytoplasm of the host cell and have therefore evolved a highly efficient mechanism of so doing. Replacing the nucleic acids with photoprotein would therefore provide an efficient means of introducing photoprotein into the cytoplasm of nonresistant cells. Also there is the additional advantage that the interaction of Sendai virus with cell membrane can occur in the absence of extracellular Ca^{2+} (Volsky and Loyter, 1978; Impraim et al., 1979; Hallett and Campbell, unpublished observation), thereby allowing the photoprotein to be protected from consumption if any transient increase in membrane permeability to calcium were to occur.

V. APPLICATIONS OF MEMBRANE VESICLES

An alternative means of looking at calcium regulation in small mammalian cells using photoprotein is by producing cell ghosts or cell membrane vesicles containing photoprotein. Pigeon erythrocyte ghosts have been prepared containing obelin (Campbell and Dormer, 1975) which are able to maintain a low intracellular calcium concentration (approximately 0.3 μmol/L) in the presence of 1 mmol/L external calcium (Campbell, Daw, Hallett, and Luzio, 1980). Furthermore, intracellular organelles, nuclei, and mitochondria appear intact when viewed with the electron microscope (Campbell and Dormer, 1978; Campbell et al., 1979b) and cAMP production within the ghosts could be stimulated by extracellular β-adrenergic agonists (Campbell and Dormer, 1978). This model has provided useful information concerning the inhibition of cAMP formation by calcium ions (1 to 10 μmol/L) within the ghosts (Campbell and

Table 7.8 Membrane Vesicles Prepared Containing Ca^{2+}-Activated Photoprotein

Membrane vesicles	Study	Reference
Pigeon erythrocyte ghosts	(i) A23187	Campbell and Dormer (1975)
	(ii) cAMP formation	Campbell and Dormer (1978)
	(iii) Ab + complement	Campbell, et al. (1979a)
Rat erythrocyte ghosts Human erythrocyte ghosts	Ab + complement	Campbell, Hallett, Daw, and Luzio (unpublished)
Sarcoplasmic reticulum	ATP-dependent Ca^{2+} transport	Yates and Campbell (1979)
Adipocyte membrane	Ab + complement	Hallett and Campbell (1980)
Hepatocyte membrane	(i) Hormone action	Westwood (1979)
	(ii) Ab + complement	Campbell et al. (1980b)
Synaptic membrane	Ca^{2+} permeability	Kuroda and Campbell (unpublished)
Rat islet cell membrane	Ab + complement	Luzio, Siddle, and Campbell (unpublished)

Dormer, 1978) and also concerning the effect of antibody plus complement on the intracellular calcium concentration (Campbell et al., 1979).

Other biological membrane vesicles that have been prepared containing obelin for the study of membrane events affecting calcium homeostasis within the cells are shown in Table 8. Each membrane preparation had the ability to maintain low intracellular Ca^{2+} concentrations in the presence of external Ca^{2+} and contained surface membrane proteins.

VI. QUANTIFICATION OF CYTOPLASMIC PHOTOPROTEIN

The amount of photoprotein injected into the cytoplasm can be estimated from a knowledge of the volume and the concentration of the photoprotein solution injected. The concentration of photoprotein in the initial solution can be estimated in two ways: (1) From the measurement of the optical density of the solution the concentration can be calculated using the extinction coefficient for aequorin (Shimomura and Johnson, 1969). (2) The concentration of active photoprotein can be estimated from the number of photons emitted during stimulation of the photoprotein with Ca^{2+}. However, the two estimates may not agree if there is contaminating protein contributing to the optical density or if there is significant fraction of spent photoprotein.

When the volume of injection solution is not known, as with pressure pulse injection (Sec. IV.B), an alternative method of estimating the amount of injected photoprotein is to stimulate rapidly the intracellular photoprotein by injection of Ca^{2+}. With small cells, the addition of the detergent Triton X-100, which solubilizes the membrane, in the presence of saturating Ca^{2+} concentration at the end of the experiment allows stimulation of all photoprotein (Fig. 6), recording light emission with the same geometry as during the experiment (Campbell and Dormer, 1975; Allen and Blinks, 1978; Campbell et al., 1980b).

The time course of Triton solubilization of membranes is relatively slow, but recording total light emitted over 10 or 20 sec provides a measure of the total active photoprotein (Campbell and Dormer, 1975; Campbell et al., 1980b). Alternatively, from the total emitted light, a hypothetical trace peak can be constructed assuming the luminescence decays exponentially at the saturating rate (Allen and Blinks, 1978,1979) and the peak measured. There are three major problems associated with these Triton methods of quantifying total active photoprotein. The first is that Triton X-100 and other agents cause massive shape changes in some cells, e.g., frog cardiac muscle (Allen and Blinks, 1978) or amoeba (Cobbold, 1979), which may alter the optical geometry of the monitoring system. Second, Triton X-100 alone can cause chemiluminescence and can cause luminescence from reacted photoprotein lasting for many minutes. Scintillation grade Triton X-100 must be used to reduce these artifacts (Campbell and Hallett, unpublished). Third, the action of Triton X-100 is unselective and will lyse all membranes. It may therefore stimulate photoprotein in a cellular compartment other than that in which the changes in intracellular Ca^{2+} occur. This last problem can be overcome by lysing the cell membrane specifically using an antibody-complement interaction (Sec. IV.C).

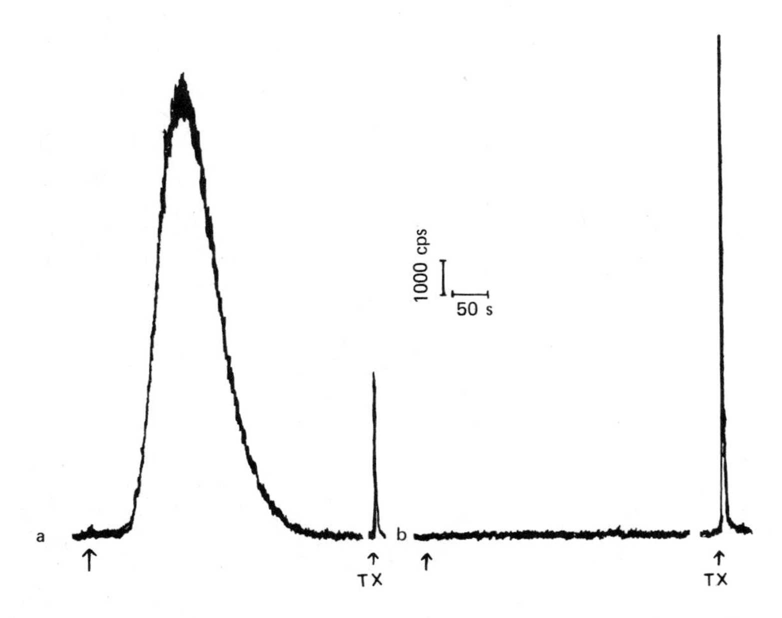

Figure 6 Effect of Sendai virus on luminescence from pigeon erythrocyte ghosts containing the photoprotein obelin. Traces: (a) shows addition of Sendai virus (1000 HAU), (b) shows addition of incubation medium. TX denotes addition of Triton X-100 (3%). The ordinates are calibrated in luminescence counts per second (cps).

VII. RELATIONSHIP BETWEEN FREE CA^{2+} AND LIGHT EMISSION

The reaction of a molecule of photoprotein with Ca^{2+} ions results in the emission of a photon and the inactivation of the photoprotein molecule. The reaction follows first-order kinetics with respect to the amount of photoprotein (X_t) such that

$$-\frac{dX}{dt} = k_t X_t \tag{1}$$

where $-\frac{dX}{dt}$ = rate of photoprotein consumption = I_t/Q

Q = quantum efficiency of reaction

k_t = rate constant at time t

I_t = light intensity at time t

The rate constant (k_t) is dependent upon the fractional saturation by Ca^{2+} of active photoprotein. When the Ca^{2+} concentration is constant, k_t is independent of time and Eq. (1) can be solved, resulting in the usual exponential for first-order reactions:

$$-\frac{dX}{dt} = X_0 ke^{-kt} \qquad (2)$$

The intensity of light recorded (I_t) is therefore given by

$$I_t = QX_0 ke^{-kt} \qquad (3)$$

When the concentration of Ca^{2+} is sufficient to saturate the photoprotein (approximately 1 mmol/L Ca^{2+}), the rate constant is maximal (k_{sat}); $k \cong 1.5\ sec^{-1}$ for aequorin and $k = 4\ sec^{-1}$ for obelin (see Table 2 for comparison of properties of photoproteins). The relationship between k and Ca^{2+} is influenced by pH, Mg^{2+}, monovalent cation concentration, and temperature. Models for the binding of Ca^{2+} to photoprotein have been proposed either as the consecutive binding of three Ca^{2+} ions per photoprotein molecule (Moisescu et al., 1975; Ashley et al., 1976) or as the simultaneous binding of three or four Ca^{2+} ions per photoprotein molecule (Allen et al., 1977). However, in practice the relationship between the rate constant k and Ca^{2+} must be found in the conditions of the experiment, either directly in the cytoplasm or in a medium of the pH, Mg^{2+} concentration, and monovalent ion concentration expected to be found within the cell.

In extruded axoplasm (Baker et al., 1971) or in media containing ions at concentrations thought to exist in the cytoplasm (Blinks et al., 1976), the relationship $Ca^{2+} \propto k^{1/2}$ is approximately followed in the range of Ca^{2+} concentrations 0.1 to 10 μmol/L, although under some conditions the relationship can be described by $Ca^{2+} \propto k^{1/3}$ for both aequorin and obelin (Moisescu et al., 1975). When constructing a standard curve, conditions should be as near as possible to those found within the cell. One way of producing optical geometry similar to that employed during an experiment on squid axon is to introduce a concentrated solution of aequorin into a porous cellulose acetate tube of approximately the same diameter as that of the axon (Baker et al., 1977; Baker, 1979). Molecules up to 1000 daltons can diffuse into the porous tube, whereas aequorin is trapped within. Solutions of EGTA-Ca buffer can be placed in the superfusing medium and light emission monitored. In experiments when a standard curve of Ca^{2+} concentration versus luminescence is determined, it is essential to avoid Ca^{2+} contamination by the use of acid-washed and double-glass distilled-water-rinsed vessels and pipettes. It is possible to achieve accurate Ca^{2+} concentrations down to 0.1 μmol/L Ca^{2+} if these precautions are taken, although some workers prefer to use Ca^{2+}-EGTA buffers. The Tes buffer used in the medium can be passed down a Chelex 100 resin column (Bio-Rad Lab., Richmond, California) to remove contaminating

Ca^{2+} (Campbell and Dormer, 1975; Blinks et al., 1978a), although low Ca^{2+} concentrations can be achieved without this procedure. Unfortunately, Chelex 100 treated Tes may contain contaminating substances that interfere with luminescence (Campbell and Hallett, unpublished). Monitoring the pH of the assay medium should not be done by emersion of a pH electrode into the main body of the medium because contaminating Ag^{+} can seriously affect luminescence (Blinks et al., 1978a). Particular attention must be paid to the concentration of Mg^{2+}, pH, and temperature of the assay medium, and it must be possible to inhibit light production by addition of EGTA. It has been reported that in the absence of Ca^{2+}, there is a low level of Ca^{2+}-independent luminescence from aequorin (Allen et al., 1977), although low levels of luminescence in cells have been reported to be inhibited by injection of EGTA (Baker et al., 1971). It is possible that Ca^{2+}-independent light is an artifact of the preparation or may even be the result of instability of the photoprotein at low protein concentrations, particularly at 37°C where a protein carrier (BSA 0.1 g/L) must be added to stabilize the photoprotein. Whatever its cause, Ca^{2+}-independent light does not significantly interfere with measurements of intracellular free Ca^{2+} in vivo (Campbell et al., 1979a).

VIII. QUANTIFICATION OF INTRACELLULAR CA^{2+}

A. Quantification under Conditions of Negligible Photoprotein Consumption

From Eq. (1), it can be seen that luminescence of the photoprotein follows first-order kinetics. When the rate of photoprotein consumption is very small, X_t can be considered to be constant. Therefore,

$$I_t = QX_0k_t \tag{4}$$

(i.e., at low Ca^{2+} concentrations the intensity of light is virtually constant and no significant decay occurs). These conditions are fulfilled in many cell responses. The low resting concentration of intracellular Ca^{2+} gives rise to rate constants of between 10^{-7} and 10^{-4} sec^{-1}, and if the raised intracellular Ca^{2+} concentration occurs for a short time interval, then the light intensity at time t (I_t) is described approximately by Eq. (4). The rate of light emission in these conditions is proportional to k and approximately proportional to $(Ca^{2+})^2$, $(Ca^{2+})^{2.5}$, or $(Ca^{2+})^3$. Therefore, by plotting the square root of light intensity against time, one has an approximately relative measure of the change in intracellular Ca^{2+} with time (Baker et al., 1971; Eckert and Tillotson, 1979). However, it was found in squid giant axon that increasing the frequency of stimulation caused a linear increase in luminescence in some axons, whereas in others the increase in luminescence was better fitted by a quadratic response (Baker et al., 1971). An explanation for this difference in response was suggested to be due to differences in the increment above the resting level of luminescence from the cells during stimulation. Since the light is proportional to the square of the calcium concentration, this will be given by $(Ca_R + \Delta Ca)^2$, where Ca_R is the resting Ca concentration and ΔCa is the increase in Ca^{2+} concentration during stimulation. This function expands to $Ca_R^2 + \Delta Ca^2 + 2Ca_R\,\Delta Ca$. It follows, therefore, that if $\Delta Ca << Ca_R$, the increment of light emission will approximate to $2Ca_R\,\Delta Ca$; while if $\Delta Ca > Ca_R$, the increment of light emission will approximate to ΔCa^2 (Baker et al., 1971).

In order to convert light emission to absolute Ca^{2+} concentrations in smaller cells it is necessary to relate the rate of luminescence to Ca^{2+} concentration using a standard curve (Sec. V.B). Giant cells allow a different approach (Baker et al., 1971; Campbell et al., 1979a). Microinjections of Ca-EGTA buffer solutions can be made in the giant cell containing aequorin until a suitable Ca-EGTA solution is found that produces no change in resting luminescence. Therefore, this Ca-EGTA solution must contain a free Ca^{2+} concentration close to the resting intracellular free Ca^{2+} concentration. Using this "null method," it may be possible to fix one point absolutely and by extrapolation estimate the intracellular Ca^{2+} concentration during a cell response. However, there are difficulties in determining the free Ca^{2+} concentration in the Ca-EGTA buffer solution once within the cytoplasm, since both the intracellular pH and intracellular Mg^{2+} concentration will affect the equilibrium of the Ca-EGTA buffer system (Portzehl et al., 1964). Although these parameters have been measured in squid giant axon (Caldwell, 1958; Baker and Crawford, 1972), in other giant cells such measurements must be made before this method can be employed.

B. Quantification under Conditions of Significant Photoprotein Consumption

When the rise in intracellular Ca^{2+} is sufficiently high or prolonged, the conditions that allow the approximation in Eq. (4) no longer apply and Eq. (1) must be used. It was shown that when the Ca^{2+} concentration is constant, the decay of luminescent intensity will be exponential. Equation (2) can be converted to a linear equation with respect to time by taking logarithms:

$$\log_e(I_t) = -kt + \log_e(QX_0k) \tag{5}$$

Plotting $\log_e(I_t)$ against t will produce a straight line with gradient (-k) when the Ca^{2+} concentration is constant. Such a semi-log plot is useful in deciding whether the Ca^{2+} concentration is constant because if it is not the resulting curve will be either concave (i.e., Ca^{2+} concentration is falling) or convex (i.e., Ca^{2+} concentration is rising). In order to be more precise about the changes in intracellular calcium, the rate constant k can be calculated at separate time points from traces. The rate constant at time t is equivalent to the fractional utilization and can be defined by rearranging Eq. (1):

$$k_t = \frac{I_t}{X_t \cdot Q} \tag{6}$$

where I_t = light emission at time t (counts sec^{-1})

$Q \cdot X_t$ = photoprotein remaining at time t (counts)

k_t = fractional utilization rate (sec^{-1}) = rate constant

In order to convert traces of light emission against time into rate constant of light emission versus time, the total amount of photoprotein remaining at the end of the experiment must be determined, e.g., by the addition of Triton X-100 (Sec. VI). The rate constant can subsequently be converted to Ca^{2+} concentration by use of a suitable standard curve (Fig. 7).

C. Problems of Heterogeneity of Cell Populations

When quantifying the light response from a number of cells injected with photoprotein, there are additional problems. First, the possibility must exist that the changes of intracellular Ca^{2+} in each cell are different, although this possibility has not been discussed (Allen and Blinks, 1978). If the calcium concentrations within the cells rise to different plateau concentrations with no significant photoprotein consumption, the estimate of k from rate of light emission will be the mean k value ($\bar{k}$). The total light output from the population will be

$$\sum_{n=1}^{n=N} I_n = \sum_{n=1}^{n=N} X_n k_n = X_T \sum \frac{k_n}{N} = X_T \bar{k} \tag{7}$$

where X_n = quantity of photoprotein in the nth cell = X_T/N

I_n = light intensity of the nth cell

N = total number of cells

The subscribed T denotes the paramter pertaining to the whole cell population.

When there is significant consumption of the photoprotein, the situation is more complex. If it is assumed that the cell responses are normally distributed about the mean, the rate of decay of the light response from the cell population will be approximately exponential at constant calcium concentration with a rate constant of k, providing the coefficient of variation is small.

The second problem arises because of the possibility that the cells may not respond simultaneously. Providing the standard deviation of the time of onset of cell responses is small compared with the time course of the total cell response, this can be ignored. However, if this is not the case serious problems of interpretation arise.

IX. CA^{2+} DISTRIBUTION WITHIN THE CELL; IMAGE INTENSIFICATION

The power relationship between the luminescent rate constant k and the calcium concentration (Sec. V.B) leads to an enhancement of contrast of the luminescence from areas within the cell with an elevated Ca^{2+} concentration compared with other areas of the cell. This gives photoproteins a facility that is unique among methods of measuring intracellular Ca^{2+} ions, i.e., the ability to monitor the distribution of Ca^{2+} concentration change within the cell. The method for detecting the distribution of Ca^{2+} changes depends upon image intensification of the microscope image, allowing sufficient amplification of the light signal to allow temporal changes to be recorded on videotape (Reynolds, 1968,1972,1975,1978,1979).

The method has been used to reveal features of cytoplasmic calcium changes not easily observed by other means. In *Chironomus* salivary gland cells, an injection of Ca^{2+} was shown to be rapidly buffered, and only when the raised cytoplasmic Ca^{2+} concentration was in the region of the cell adjacent to the gap junction was electrical conductivity between the cells reduced (Rose and Lowenstein, 1975, 1976). In *Medaka* eggs the technique has revealed that fertilization of the egg results in a wave of raised cytoplasmic Ca^{2+} moving slowly across the cell, with morphological changes following the wave of raised Ca^{2+} (Ridgway et al., 1977; Gilkey et al., 1978). In sea urchin eggs a similar phenomenon was observed (Kiehart et al., 1977). In the giant ameba it was shown that cytoplasmic changes occurred at the anodic end of the cell when the cell was subjected to a weak electrical stimulation, causing psuedopod formation at the cathodic end (Taylor, Reynolds, and Allen, 1975).

However, with the present technology there are problems of sensitivity of detection, and resolution of events spatially and temporally. It has been estimated that in some experiments Ca^{2+} concentrations below 0.5 μmol/L may be undetectable, although it has been reported that the limit may be as low as 35 nmol/L Ca^{2+} (Taylor et al.,

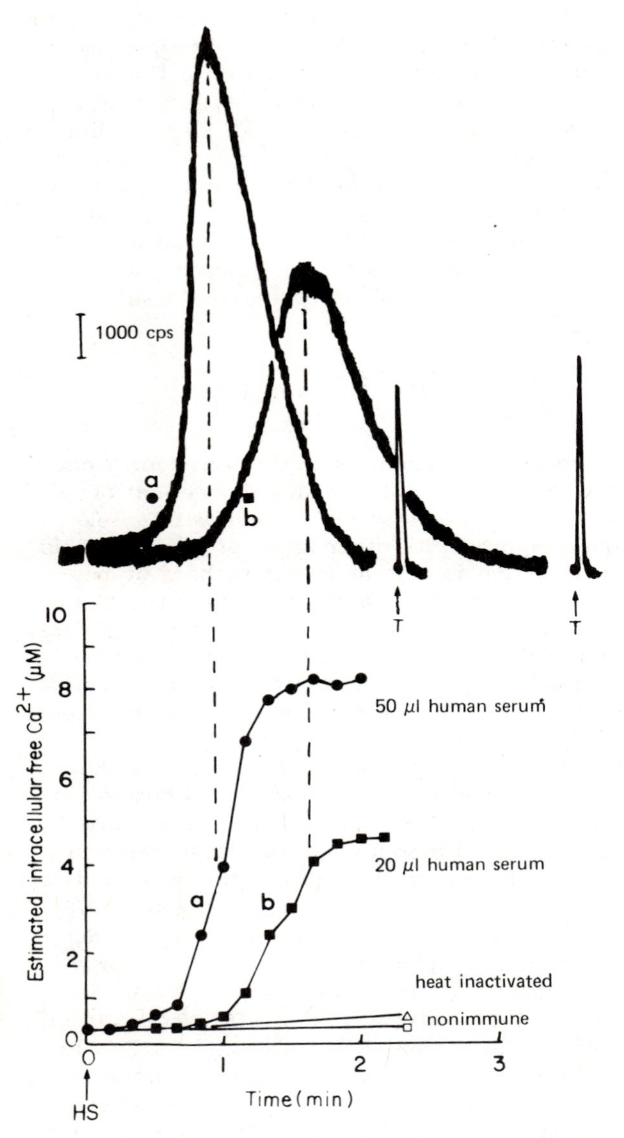

Figure 7 Estimation of intracellular free Ca^{2+} from determination of rate constants at time intervals. Traces a and b show the luminescent responses of obelin-containing pigeon erythrocyte "ghosts" preincubated with antibody, to the addition of human serum as a source of complement. The lower graph shows the intracellular free Ca^{2+} estimated from rate constants calculated from the tracer above. Rate

1975a). The spatial resolution may not be greater than 2 μm (Rose and Lowenstein, 1976). Video recording systems using commercial TV cameras operate at 60 fields/sec, giving a temporal resolution of ∿16 msec, although increased temporal resolution is possible (Reynolds et al., 1978).

X. PROBLEMS ASSOCIATED WITH THE USE OF PHOTOPROTEINS

A. Ca^{2+}-Independent Light

There exists the possibility that a low level of light emission from photoproteins occurs in the absence of calcium (Allen et al., 1977). This can be tested in giant cells by injection of EGTA.

B. Other Sources of Luminescence

A surprising number of reagents give rise to chemiluminescence and all agents should be tested for endogenous luminescence. Endogenous luminescence has also been observed from several cells (Foerder et al., 1978; Trush et al., 1978), often caused by superoxide anion and peroxide moieties.

C. Release of Photoprotein

The possibility that the change in luminescence observed is the result of photoprotein release from the cell into the incubation medium can be tested by stimulating the cell in Ca^{2+}-free medium and subsequently assaying the medium for photoprotein.

D. Heterogeneous Intracellular Distribution of Photoprotein

After microinjection, time must be allowed for diffusion of photoprotein through the cytoplasm, since it is only photoprotein in the region of the cell where Ca^{2+} concentrations changes that responds. It has been shown that following microinjection of aequorin into the middle of squid axon, the luminescent response to stimulation increased as the photoprotein diffused toward the cell periphery, with an apparent diffusion coefficient of 6×10^{-7} cm^2/sec (Baker et al., 1971). In frog muscle the

---

constants were converted to calcium concentration from a calibration curve. It will be noticed that the estimated peak intracellular calcium concentration is reached several seconds after the peak in luminescence. (From Campbell et al., 1980b).

diffusion coefficient was estimated to be 5×10^{-8} cm^2/sec (Blinks et al., 1978b), and in barnacle muscle is 10^{-7} cm^2/sec^{-1} (Ashley, 1978).

E. Heterogeneous Intracellular Distribution of Cytoplasmic Ca^{2+}

If there is an uneven change in Ca^{2+} concentration, difficulty may be experienced in correctly interpreting the total light output from the cell.

F. Changes in Photoprotein Affinity for Ca^{2+}

Caution must be exercised and suitable controls performed to eliminate the possibility that the cell stimulant is causing its effect on luminescence by an apparent increase in photoprotein affinity for Ca^{2+} or a direct stimulation of photoprotein luminescence. Such an artifact has recently been reported, where it was shown that an increase in luminescence occurred when treating squid axon with urethane, a general anesthetic, without stimulating an increased efflux of ^{45}Ca (Baker, 1979; Baker and Schapira, 1980).

G. Toxicity

Although no toxic effects of photoprotein on any of the cells injected with photoprotein have been reported, one consequence of microinjecting the calcium-binding photoprotein into the cytoplasm is the alteration of the equilibrium of free Ca^{2+} and bound calcium in the cell. Providing the concentration of photoprotein is not too high ($<10^{-4}$ mol/L), the other calcium-binding proteins and Ca^{2+}-buffering systems within the cell should prevent significant alteration of the cytoplasmic free Ca^{2+} under study (Blinks, 1978).

H. Temporal Distortion of Light Signal

The response time of aequorin, 10 msec (Hastings et al., 1969), is significantly longer than many physiological responses (e.g., action potential $\sim$1 msec). Distortion of responses will therefore occur (Blinks et al., 1978b) and faster responses may not be recorded at all without a means of slowing down the response (Stinnakre and Tauc, 1973).

I. Cell Opacity

If the cell is opaque, then a decrease in absorption by the cell, occurring, for example, during cell division, may cause an increase in

Table 9 Comparison of the Methods for Measuring Intracellular Free Ca^{2+}

Criterion	Photo-proteins	Dyes	Micro-electrodes
Sensitivity for Ca^{2+}	++	++	++
Selectivity for Ca^{2+}	++	+	++
Quantitation of absolute free Ca^{2+} in vivo	+	-	++
Response time - fast cells	-	+	?
- slow cells	++	++	++
Incorporation into cell - large cell	+	+	+
- small cell	-	-	-
Ca^{2+} distribution in cell cytoplasm	++	?	--
Lack of toxicity	++	-	?
Availability	+	++	+

++ = good; + = adequate; - = problems; -- = bad; ? = not known.
Source: Ashley and Campbell (1979).

the detectable light without a change in photoprotein luminescence (Baker and Warner, 1972).

XI. CONCLUSION

It has been shown in this chapter that coelenterate photoproteins fulfill many of the requirements of an ideal intracellular free Ca^{2+} indicator (see Sec. II.A), providing a sensitive, specific, quantitative, and nontoxic method. It is not surprising, therefore, that many of the advances in the study of intracellular Ca^{2+} regulation over the past 15 years are the result of utilizing coelenterate photoproteins to monitor directly the changes in intracellular free Ca^{2+}. As the technology of photon-detecting instruments advances, so the detection of low-level luminescence will provide a sensitive means of measuring resting intracellular calcium concentrations in small cells and allow the monitoring of intracellular distribution of Ca^{2+} concentrations.

Two other methods are also currently in use by various workers for measurement of free Ca^{2+} concentrations in living cells. As can be seen from Table 9, no one method satisfies all the criteria for an ideal indicator, although the development of fluorescent indicators for free Ca^{2+} offers great promise. While there is general agreement about the need to confirm observations made with one technique by comparison with another, it is still true that nearly all the new observations on intracellular free Ca^{2+} which have been made during the last 15 years have come initially from studies using photoproteins. The most pressing need is for a technique enabling a Ca^{2+} indicator to be incorporated into living mammalian and other small cells without injury.

NOTE ADDED IN PROOF

Since preparing this chapter, two advances in introducing photoproteins into small cells have been reported. (1) A method for microinjecting aequorin into mouse oocytes in sufficient amounts to produce a detectable resting luminescence has been reported (Cuthbertson et al., 1981). (2) The introduction of obelin into the cytoplasm of a cell population by means of fusion with red cell ghosts has also been reported (Hallett and Campbell, 1982).

ACKNOWLEDGMENTS

We are grateful to the director and staff of the Marine Biological Association, Plymouth for facilities to collect *Obelia geniculata*. We thank Drs. C. C. Ashley, R. L. Dormer, J. P. Luzio, and L. H. N. Cooper, F. R. S., and Profs. C. N. Hales and G. H. Elder for useful discussions and the SRC and MRC for financial support.

REFERENCES

Ahkong, Q. F., F. C. Cramp, D. Fisher, J. I. Howell, W. Tampion, M. Verrinder and J. A. Lucy (1973). Chemically-induced and thermally-induced cell fusion: Lipid:lipid interactions. *Nature New Biol. 242*:215-217.

Ahkong, Q.F., J. I. Howell, J. A. Lucy, F. Sufwat, M. R. Davey, and E. C. Cocking (1975). Fusion of hen erythrocytes with yeast protoplasts induced by polyethylene glycol. *Nature 255*:66-69.

Allen, D. G., and J. R. Blinks (1978). Calcium transients in aequorin-injected frog cardiac muscle. *Nature 273*:509-513.

Allen, D. G., and J. R. Blinks (1979). The interpretation of light signals from aequorin-injected skeletal and cardiac muscle cells: A new method of calibration. In *Detection and Measurement of Free Ca^{2+} in Cells*, C. C. Ashley and A. K. Campbell (Eds.). Elsevier-North Holland, Amsterdam, New York, and Oxford, pp. 159-174.

Allen, D. G., J. R. Blinks, and F. G. Prendergast (1977). Aequorin luminescence relation of light emission to calcium concentration: A calcium independent component. *Science 196*:966-998.

Allman, G. J. (1873). *A Monograph of Gymnoblastic or Tubularian Hydroids*. Roy. Soc. Lond., pp. 145-146.

Anderson, J. M., and M. J. Cormier (1973). Lumisomes, the cellular site of bioluminescence in coelenterates. *J. Biol. Chem. 248*:2937-2943.

Ashley, C. C. (1969). Aequorin-monitored calcium transients in single Maia muscle fibres. *J. Physiol. Lond. 203*:32-33P.

Ashley, C. C. (1978). Calcium ion regulation in barnacle muscle fibres and its relationship to force development. *Anal. N.Y. Acad. Sci. 307*:308-329.

Ashley, C. C., and A. K. Campbell (Eds.) (1979). *Detection and Measurement of Free Ca^{2+} in Cells*. Elsevier-North Holland, Amsterdam, New York, Oxford.

Ashley, C. C., and E. B. Ridgway (1970). On the relationship between membrane potential, calcium transients and tension in single barnacle muscle fibre. *J. Physiol. (Lond.) 209*:105-130.

Ashley, C. C., A. K. Campbell, and D. G. Moisescu (1975). A demonstration of some of the properties of obelin; a calcium-sensitive luminescent protein. *J. Physiol. (Lond). 245*:9-10P.

Ashley, C. C., P. C. Caldwell, A. K. Campbell, T. J. Lea, and D. G. Moisescu (1976). Calcium movements in muscle. *Soc. Exp. Biol. Symp. 30*:397-422.

Azzi, A., and B. Chance (1969). The "energized state" of mitochondria: Life time and ATP equivalence. *Biochim. Biophys. Acta 189*:141-151.

Baker, P. F. (1979). The use of aequorin in giant axons. In *Detection and Measurement of Free Ca^{2+} in Cells*, C. C. Ashley and A. K. Campbell (Eds.). Elsevier-North Holland, pp. 175-188.

Baker, P. F., and A. C. Crawford (1972). Mobility of transport of magnesium in squid axons. *J. Physiol. (Lond.) 227*:855-874.

Baker, P. F., and A. F. Warner (1972). Intracellular calcium and cell cleavage in early embryos of Xenopus leavis. *J. Cell. Biol. 53*:579-581.

Baker, P. F., A. L. Hodgkin and E. B. Ridgway (1971). Depolarization and calcium entry in squid giant axons. *J. Physiol. (Lond.) 218*:709-755.

Baker, P. F., and A. H. V. Schapira (1980). Anaesthetics increase light emission from aequorin at constant ionized calcium. *Nature 284*:168-169.

Baker, P. F., D. E. Knight, and R. D. P. Pattni (1977). Porous cellulose acetate tubing provides a suitable support for isolated protoplasm during studies under controlled conditions. *J. Physiol. (Lond.) 266*:6P.

Barrett, J., and D. G. Whitlock (1973). In *Intracellular Staining in Neurobiology*, S. B. Kater and C. Nicholson (Eds.). Springer-Verlag, Berlin and New York, p. 297.

Berlman, I. B. (1973). *Energy Transfer Parameters of Aromatic Compounds*. Academic, New York.

Berridge, M. J. (1976). The interaction of cyclic nucleotides and calcium in the control of cellular activity. *Adv. Cyclic Nucleotide Res. 6*:1-98.

Bles, E. J. (1892). Note on the plankton observed at Plymouth during July, August and September 1892. *J. Marine Biol. Assoc. U.K. 2*: 340-343.

Blinks, J. R. (1978). Measurement of calcium in concentrations with photoproteins. In Calcium Transport and Cell Function, A. Scarpa and E. Carafoli (Eds.). *Ann. N.Y. Acad. Sci. 307*:71-85.

Blinks, J. R., F. G. Prendergast and D. G. Allen (1976). Photoproteins as biological calcium indicators. *Pharmacol. Rev. 28*:1-93.

Blinks, J. R., P. H. Mattingly, B. R. Jewell, M. Van Leeuwen, G. C. Harrer, and D. G. Allen (1978a). Practical aspects of the use of aequorin as a calcium indicator: assay, preparation, microinjection and interpretation of signals. In Methods in Enzymology, Vol. 58, S. P. Colowick and N. O. Kaplan (Eds.). *Bioluminescence and Chemiluminescence*, M. A. De Luca (Ed.) Academic, New York, pp. 292-328.

Blinks, J. R., R. Rudel and S. R. Taylor (1978b). Calcium transients in isolated amphibian skeletal muscle fibres: Detection with aequorin. *J. Physiol. (Lond.) 277*:291-323.

Blumenthal, R., J. N. Weinstein, S. O. Sharrow and P. Henkart (1977). Liposome-lymphocyte interaction: Saturable sites for transfer and intracellular release of liposome contents. *Proc. Natl. Acad. Sci. USA 74*:5603-5607.

Brown, J. E., and J. R. Blinks (1974). Changes in intracellular free calcium concentration during illumination of invertebrate photoreceptors; detection by aequorin. *J. Gen. Physiol. 64*:643-665.

Brown, K. T., and D. G. Flaming (1974). Bevelling of fine micropipette electrodes by a rapid precision method. *Science 185*:693-695.

Burnell, J. N., and J. W. Anderson (1973). Adenosine 5'-sulphatophosphate kinase activity in spinach leaf tissue. *Biochem. J. 134*: 565.

Bygrave, F. L. (1967). The ionic environment and metabolic control. *Nature (Lond.) 214*:667-671.

Caldwell, P. C. (1958). Studies on the internal pH of large nerve and muscle fibres. *J. Physiol. (Lond.) 142*:22-62.

Campbell, A. K. (1974). Extraction, partial purification and properties of obelin, the calcium-activated luminescent protein from the hydroid *Obelia geniculata*. *Biochem. J. 143*:411-418.

Campbell, A. K., and R. L. Dormer (1975). Permeability to calcium of pigeon erythrocyte "ghosts" studied using the calcium-activated luminescent protein, obelin. *Biochem. J. 29*:255-265.

Campbell, A. K., and R. L. Dormer (1978). Inhibition by calcium ions of adenosine cyclic monophosphate formation in sealed pigeon erythrocyte "ghosts". A study using the photoprotein obelin. *Biochem. J. 176*:53-66.

Campbell, A. K., and M. B. Hallett (1979). Luminescence in cells and vesicles isolated from the hydroid *Obelia geniculata*. *J. Physiol. (Lond.) 287*:4-5P.

Campbell, A. K., and P. J. Herring (1979). In Cruise Report No. 82, Institute of Oceanographic Sciences, pp. 18-19.

Campbell, A. K., and J. S. A. Simpson (1979). Chemi- and bioluminescence as an analytical tool in biology. In *Techniques in the Life Sciences: Biochemistry*, Vol. B2/11, Kornberg, H.C., Metcalf, J. C., Northcote, D. H., Pogson, C. I., and Tipton, K. F. (Eds.). Elsevier-North Holland, pp. 1-56.

Campbell, A. K., R. A. Daw and J. P. Luzio (1979a). A rapid increase in intracellular free Ca^{2+} induced by antibody plus complement. *FEBS Lett. 107*:55-60.

Campbell, A. K., T. J. Lea and C. C. Ashley (1979b). Coelenterate photoproteins. In *Detection and Measurement of Free Ca^{2+} in Cells*, C. C. Ashley and A. K. Campbell (Eds.). Elsevier-North Holland, Amsterdam, pp. 13-72.

Campbell, A. K., J. S. A. Simpson, and J. S. Woodhead (1979c). Detecting or quantifying substance using labelling techniques. U.K. Patent application GB 2008247A.

Campbell, A. K., M. B. Hallett, R. A. Daw, J. P. Luzio and K. Siddle (1979d). The importance of measuring intracellular free Ca^{2+}. *Biochem. Soc. Trans. 7*:865-869.

Campbell, A. K., C. J. Davies, R. Hart, F. McCapra, A. Patel, A. Richardson, M. E. T. Ryall, J. S. A. Simpson, and J. S. Woodhead (1980a). Chemi-luminescent labels in immunoassay. *J. Physiol. (Lond.) 306*:3-4P.

Campbell, A. K., R. A. Daw, M. B. Hallett, and J. P. Luzio (1980b). A direct measurement of the increase in intracellular free calcium ion concentration in response to the action of complement. *Biochem. J. 194*:551-560.

Chang, J. J. (1975). A new technique for bevelling the tips of glass capillary micropipettes and microelectrodes. *Comp. Biochem. Physiol. A52*:567-570.

Chang, J. J., A. Gelperin and F. H. Johnson (1974). Intracellular injected aequorin detects transmembrane calcium flux during action potentials in an identified neuron from the terrestrial slug Limax maximus. *Brain Res. 77*:431-442.

Cobbold, P. H. (1979). Measurement of cytoplasmic free calcium in an amoeba and in smaller cells using aequorin. In *Detection and Measurement of Free Calcium Ions in Cells*, C. C. Ashley and A. K. Campbell (Eds.). Elsevier-North Holland, Amsterdam, pp. 245-250.

Cormier, M. J. (1962). Studies on the bioluminescence of *Renilla reniformis*. II. Requirement for 3'5' diphosphoadenosine in the luminescent reaction. *J. Biol. Chem. 237*:2032-2037.

Cormier, M. J. (1978). Comparative biochemistry of animal systems. In *Bioluminescence in Action*, P. J. Herrin (Ed.). Academic, London, pp. 75-108.

Cormier, M. J., K. Hori, and J. M. Anderson (1974). Bioluminescence in coelenterates. *Biochim. Biophys. Acta 346*:137-164.

Cuthbertson, K. S. R., D. G. Whittingham, and P. H. Cobbold (1981). *Nature 294*:754-757.

Dahlgren, U. (1916). The production of light by animals. Cited in Harvey, 1952.

Davenport, D., and J. A. C. Nicol (1955). Luminescence in hydromedusae. *Proc. Roy. Soc. Lond. B. 144*:399-411.

Dormer, R. L., M. B. Hallett, and A. K. Campbell (1978a). The incorporation of the calcium-activated photoprotein obelin into isolated rat fat-cells by liposome-cell fusion. *Biochem. Soc. Trans. 6*:570-572.

Dormer, R. L., G. R. Newman, and A. K. Campbell (1978b). Preparation and characterization of liposomes containing the Ca^{2+}-activated photoprotein, obelin. *Biochim. Biophys. Acta 538*:87-155.

Douglas, W. W. (1968). Stimulus-secretaion coupling: The concept and clues from chromaffin and other cells. *Brit. J. Pharmacol. 34*:451-474.

Dubois, R. (1885). Notes sur la physiologie des pyrophores. *C. R. Soc. Biol. 37*:559-562.

Dubois, R. (1887). Function photogenique chez le Pholas dactylus. *C. R. Soc. Biol. 39*:564-565.

Eckert, R., D. Tillotson, and E. B. Ridgway (1977). Voltage-dependent facilitation of Ca^{2+} entry in voltage-clamped, aequorin-injected molluscan neurons. *Proc. Natl. Acad. Sci. USA 74*:1748-1752.

Eckert, R., and D. Tillotson (1979). Calcium entry and the calcium-activated potassium systems of molluscan neurons: Voltage clamp studies on aequorin-injected cells. In *Detection and Measurement of Free Ca^{2+} in Cells*, C. C. Ashley and A. K. Campbell (Eds.). Elsevier-North Holland, Amsterdam, New York, Oxford, pp. 203-218.

Ettienne, E. M. (1970). Control of contractility in spirostomun by dissociated calcium ions. *J. Gen. Physiol. 56*:168-179.

Eusebi, F., R. Miledi and T. Takashi (1980). Calcium transients in mammalian muscle. *Nature (Lond.) 284*:560-561.

Fay, F. S., H. H. Shlevin, W. C. Granger, and S. R. Taylor (1979). Aequorin luminescence during activation of single isolated smooth muscle cells. *Nature 280*:506-508.

Foerder, C. A., S. J. Klebanoff and B. M. Shapiro (1978). Hydrogen peroxide production, chemiluminescence and the respiratory burst of fertilization: Intracellular events in early sea urchin development. *Proc. Natl. Acad. Sci. USA 75*:3183-3187.

Forbes, E. (1848). *A Monograph of the Naked-Eye Medusae.* Roy. Soc. Lond., pp. 11-14.

Forskål, P. (1775). *Fauna Arabica,* Heineck and Faber, London, pp. 110-111.

Gilkey, J. C., L. F. Jaffe, E. B. Ridgway and G. T. Reynolds (1978). A free calcium wave tranverses the activating egg of the Medaka, oryzias latipes. *J. Cell. Biol. 76*:448-466.

Gregory, J. D., and F. Lipmann (1957). The transfer of sulfate among phenolic compounds with 3',5'-diphosphoadenosine as co-enzyme. *J. Biol. Chem. 229*:1081-1091.

Hales, C. N., A. K. Campbell, J. P. Luzio, and K. Siddle (1977). Calcium as a mediator of hormone action. *Biochem. Soc. Trans. 5*: 866-872.

Hallett, M., and E. Carbone (1972). Studies of calcium influx into squid giant axon with aequorin. *J. Cell Physiol. 80*:219-226.

Hallett, M. B., and A. K. Campbell (1980). Uptake of liposomes by rat isolated fat cells; Adhesion, endocytosis or fusion? *Biochem. J. 192*:587-596.

Hallett, M. B., and A. K. Campbell (1982). *Nature 295*:155-158.

Hart, R. C., J. C. Matthews, K. Hori, and M. J. Cormier (1979). *Renilla reniformis* bioluminescence: Luciferase-catalyzed production of non-radiating excited states from luciferin analogues and elucidation of the excited state species involved in energy transfer to Renilla green fluorescent protein. *Biochemistry 18*:2204-2210.

Hartlaub, C. (1914). Nordisches Plankton Lief 17 XII, *Craspedota medusae* Tiel I Lief 3, pp. 237-363. Cited in Russell, 1952.

Harvey, E. N. (1926). Oxygen and luminescence, with a description of methods for removing oxygen from cells and fluid. *Biol. Bull. (Woods Hole) 51*:89-97.

Harvey, E. N. (1952). *Bioluminescence*. Academic, New York.

Hastings, J. W., G. Mitchell, P. H. Mattingly, J. R. Blinks, and M. Van Leeuwen (1969). Response of aequorin bioluminescence to rapid changes in calcium concentration. *Nature (Lond.) 222*:1047-1050.

Herring, P. J. (1978). *Bioluminescence in Action*. Academic, London.

Hincks, T. (1868). *A History of the British Hydroid Zoophytes*, Vols. 1 and 2. van Voorst, London.

Impraim, C. C., K. J. Micklem and C. A. Pasternak (1979). Calcium, cells and virus: Alterations caused by paramyxovirus. *Biochem. Pharmacol. 28*:1963-1969.

Izutsu, K. T., and S. P. Felton (1972). Plasma calcium assay, with use of the jelly fish protein, aequorin, as a reagent. *Clin. Chem. 18*: 77-79.

Jobsis, F. F., and M. J. O'Connor (1966). Calcium release and reabsorption in the sartorius muscle of the toad. *Biochem. Biophys. Res. Commun. 25*:146-252.

Jones, H. P., J. C. Matthews, and M. J. Cormier (1979). Isolation and characterization of Ca^{2+}-dependent modulator protein from marine invertebrate *Renilla reniformis*. *Biochemistry 18*:55-60.

Kiehart, D. P., G. T. Reynolds, and A. Eisen (1977). Calcium transients during the early development in echinoderms and teleosts. *Biol. Bull. 153*:432.

Kusano, K., R. Miledi and J. Stinnakre (1975). Postsynaptic entry of calcium induced by transmitter action. *Proc. Roy. Soc. Ser. B. Biol. Sci. 189*:49-56.

Llinas, R., J. R. Blinks and C. Nicholson (1972). Calcium transient in pre-synaptic terminal of squid giant synapse: Detection with aequorin. *Science 176*:1127-1129.

Macartney, J. (1810). Observations upon luminous animals. *Phil. Trans. Roy. Soc. Lond. B. 100*:258-293.

Moisescu, D. G., C. C. Ashley and A. K. Campbell (1975). Comparative aspects of the calcium-sensitive photoproteins aequorin and obelin. *Biochim. Biophys. Acta 396*:133-140.

Moreau, M., P. Guerrier, M. Doree and C. C. Ashley (1978). Hormone-induced release of intracellular calcium triggers meiosis in starfish oocytes. *Nature (Lond.) 272*:251-253.

Moreau, M., J. P. Vilain, and P. Guerrier, (1980). Free calcium changes associated with hormone action in amphibian oocytes. *Develop. Biol. 78*:201-214.

Morin, J. G., and I. M. Cooke (1971a). Behaviour physiology of the colonial hydroid Obelia. I. Spontaneous movements and the correlated electrical activity. *J. Exp. Biol. 54*:689-706.

Morin, J. G., and I. M. Cooke (1971b). Behavioural physiology of the colonial hydroid Obelia. II. Stimulus-initiated electrical activity and bioluminescence. *J. Exp. Biol. 54*:707-721.

Morin, J. G., and J. W. Hastings (1971). Biochemistry of the bioluminescence of colonial hydroids and other coelenterates. *J. Cell. Physiol. 77*:305-312.

Morin, J. G., and G. T. Reynolds (1974). The cellular origin of bioluminescence in the colonial hydroid Obelia. *Biol. Bull. 147*:397-410.

Nicol, J. A. (1958). *The Biology of Marine Animals*. Academic, New York.

Pagano, R. E., and J. N. Weinstein (1978). Interaction of liposomes with mammalian cells. *Ann. Rev. Biophys. Bioeng. 7*:435-468.

Panceri, P. (1876). Intomo alla sede del movimento luminos nele *Campannlariae. Atti Della R. Accademia 7*:1-7.

Papahadjopoulos, D., M. Cowden and H. K. Kimelberg (1973). Role of cholesterol in membranes. Effect on phospholipid-protein interactions, membrane permeability and enzymatic activity. *Biochim. Biophys. Acta 330*:10-78.

Pollack, H. (1928). Microsurgical studies in cell physiology: Calcium ions in living protoplasm. *J. Gen. Physiol. 11*:539-545.

Portzehl, H., P. C. Caldwell and J. C. Ruegg (1964). The dependence of contraction and relaxation of muscle fibres from the crab Maia on the internal concentration of free calcium ions. *Biochim. Biophys. Acta 79*:581-591.

Poste, G. (1972). Mechanisms of virus-induced cell fusion. *Int. Rev. Cytol. 33*:157-252.

Poste, G., D. Papahadjopoulos, and W. J. Varl (1976). Lipid vesicles as carriers for introducing biologically active materials into cells. In *Methods in Cell Biology*, Vol. 13, D. M. Prescold (Ed.). Academic, New York, pp. 33-71.

Rasmussen, U. (1970). Cell communication, calcium ion and cyclic adenosine monophosphate. *Science 170*:404-412.

Rasmussen, H., and D. B. P. Goodman (1977). Relationships between calcium and cyclic nucleotides in cell activation. *Physiol. Rev. 57*: 421-509.

Reynolds, G. T. (1968). Image intensification applied to microscope systems. *Adv. Opt. Electron Micros. 2*:1-40.

Reynolds, G. T. (1972). Image intensification applied to biological problems. *Quart. Rev. Biophys. 5*:295-347.

Reynolds, G. T. (1975). Improved photosensitive devices for observing in vivo bioluminescence spectra. *Biol. Bull. 149*:443-444.

Reynolds, G. T. (1978). Application of photosensitive devices to bioluminescence studies. *Photochem. Photobiol. 27*:405-422.

Reynolds, G. T. (1979). Localization of free ionized calcium in cells by means of image intensification. In *Detection and Measurement of Free Calcium Ions in Cells*, C. C. Ashley and A. K. Campbell (Eds.). Elsevier-North Holland, Amsterdam, pp. 227-244.

Reynolds, G. T., J. R. Milch and S. M. Gruner (1978). A high sensitivity image intensifier-TV detector for X-ray diffraction study. *Rev. Sci. Inst. 49*:1241-1249.

Ridgway, E. B., and C. C. Ashley (1967). Calcium transients in single muscle fibres. *Biochem. Biophys. Res. Commun. 29*:229-234.

Ridgway, E. B., and A. C. Durham (1976). Oscillations of calcium ion concentrations in physarum polycephalum. *J. Cell. Biol. 69*:223-226.

Ridgway, E. B., J. G. Gilkey, and L. F. Jaffe (1977). Free calcium increases explosively in activating medakla eggs. *Proc. Natl. Acad. Sci. USA 74*:623-627.

Rose, B., and W. R. Lowenstein (1975). Permeability of cell junction depends upon local cytoplasmic calcium activity. *Nature (Lond.) 254*: 250-252.

Rose, B., and W. R. Lowenstein (1976). Permeability of a cell junction and the local cytoplasmic free ionized calcium concentration: A study with aequorin. *J. Membr. Biol. 28*:87-119.

Rudel, R., and S. R. Taylor (1973). Aequorin luminescence during contraction of amphibian skeletal muscle. *J. Physiol. (Lond.) 233*: 5-6.

Russell, F. S. (1952). *The Medusae of the British Isles*, Vol. 1. University Press, Cambridge.

Shimomura, O. (1979). Structure of the chromophore of aequorea green fluorescent protein. *FEBS Lett. 104*:220-222.

Shimomura, O., and F. H. Johnson (1966). Partial purification and properties of the *Chaetorterus* luminescence system, in *Bioluminescence*-

cence in Progress, F. H. Johnson and Y. Haneda (Eds.). Princeton University Press, pp. 495-521.

Shimomura, O., and F. H. Johnson (1969). Properties of the bioluminescent protein aequorin. *Biochemistry* 8:3991-3997.

Shimomura, O., and F. H. Johnson (1973). Further data on the specificity of aequorin luminescence. *Biochem. Biophys. Res. Commun.* 53:490-494.

Shimomura, O., and F. H. Johnson (1975). Regeneration of the photoprotein aequorin. *Nature (Lond.)* 256:236-238.

Shimomura, O., and F. H. Johnson (1978). Peroxidized coelenterazine, the active group in the photoprotein aequorin. *Proc. Natl. Acad. Sci. USA* 75:2611-2615.

Shimomura, O., and F. H. Johnson (1979). Chemistry of the calcium-sensitive photoprotein aequorin. In *Detection and Measurement of Free Ca^{2+} in Cells*, C. C. Ashley and A. K. Campbell (Eds.). Elsevier-North Holland, Amsterdam, New York, Oxford, pp. 73-84.

Shimomura, O., F. H. Johnson, and Y. Saiga (1962). Extraction, purification and properties of aequorin a bioluminescent protein from the luminous hydromedusan, Aequorea. *J. Cell. Comp. Physiol.* 59:223-239.

Shimomura, O., F. H. Johnson, and Y. Saiga (1963). Extraction and purification of halistanium, a bioluminescent protein from the hydroid Halistanran. *J. Cell. Comp. Physiol.* 62:9-15.

Stanley, P. E., B. C. Kelley, O. H. Tuovinen, and D. J. D. Nicholas (1975). A bioluminescence method for determining adenosine 3'-phosphate 5'-phosphate (PAP) and adenosine 3'-phosphate 5'-sulphatophosphate (PAPS) in biological materials. *Anal. Biochem.* 67:540-551.

Steinhardt, R., R. Zucker and G. Schatten (1977). Intracellular calcium releast at fertilization in the sea urchin egg. *Rev. Biol.* 58: 185-196.

Stendahl, O., and C. Tagesson (1977). Interaction of liposomes with polymorphonuclear leukocytes; studies on the mode of interaction. *Exp. Cell. Res.* 108:167-174.

Stinnakre, J. (1979). Pressure injection of aequorin into molluscan neurones. In *Detection and Measurement of Free Ca^{2+} in Cells*, C. C. Ashley and A. K. Campbell (Eds.). Elsevier-North Holland, Amsterdam, New York, Oxford, pp. 189-202.

Stinnakre, J., and L. Tauc (1973). Calcium influx in active Aplysia neurones detected by injected aequorin. *Nature New Biol.* 242:113-115.

Stryer, L. (1978). Fluorescence energy transfer as a spectroscopic ruler. *Ann. Rev. Biochem.* 47:819-846.

Szoka, F., and D. Papahadjopoulos (1978). Procedure for preparation of liposomes with large internal aqueous space and high capture by reverse-phase evaporation. *Proc. Natl. Acad. Sci. USA* 75:4194-4198.

Taylor, D. L., G. T. Reynolds, and R. D. Allen (1975a). Detection of free calcium ions in amoeba by aequorin luminescence. *Biol. Bull. 149*:Abstr. 448.

Taylor, S. R., R. Rudel, and J. R. Blinks (1975b). Calcium transients in amphibian muscle. *Fed. Proc. 34*:1379-1381.

Trush, M. A., M. E. Wilson, and K. Van Dyke (1978). The generation of chemiluminescence (CL) by phagocytic cells. *Meth. Enzymol. 57*: 462-494.

Tyrrell, D. A., T. P. Heath, C. M. Colley, and B. E. Ryman (1976). New aspects of liposomes. *Biochim. Biophys. Acta 457*:259-302.

Volsky, D. J., and A. Loyter (1978). Role of Ca^{2+} in virus-induced membrane fusion Ca^{2+} accumulation and ultrastructural changes induced by Sendai virus in chicken erythrocytes. *J. Cell. Biol. 78*: 465-479.

Ward, W. W., and M. J. Cormier (1978). Energy transfer in protein-protein interaction in *Renilla* bioluminescence. *Photochem. Photobiol. 27*:389-396.

Ward, H. H., and H. H. Seliger (1974). Extraction and purification of calcium-activated photoproteins from the ctenophores *Mnemiopsis* sp. and *Beroe orata. Biochemistry 13*:1491-1499.

Westwood, S. A. (1979). Studies on liver plasma membrane in relation to endocrine control of metabolism. Ph.D. Thesis, University of Wales.

Yates, D. W., and A. K. Campbell (1979). Calcium transport into sarcoplasmic reticulum vesicles measured by obelin luminescence. In *Detection and Measurement of Free Ca^{2+} in Cells*, C. C. Ashley and A. K. Campbell (Eds.). Elsevier-North Holland, Amsterdam, New York, Oxford, pp. 257-268.

7

ANALYTICAL APPLICATIONS OF CHEMILUMINESCENCE

TIMOTHY J. N. CARTER* Wolfson Research Laboratories, Queen Elizabeth Medical Centre, Edgbaston, Birmingham, England

LARRY J. KRICKA† University of Birmingham, Birmingham, England

I. INTRODUCTION

Chemiluminescent materials have been known for many years, although they have found little application as analytical reagents. This is surprising considering the possible advantages offered by luminescent techniques in terms of sensitivity, specificity, cost, and speed (Seitz and Neary, 1976).

Chemiluminescence can occur in the gas, liquid, or solid phase. While there are several important gas phase reactions that have found

*Dr. Carter is currently with Battelle, Centre de Recherche de Genève, Carouge, Genève, Switzerland.

†Dr. Kricka is currently with the University of California at San Diego, La Jolla, California, on a leave of absence from the University of Birmingham.

application in quantitative analysis, notably those involving ozone (Nederbragt et al., 1965) and nitrogen oxides (Fontijn et al., 1973), these are unlikely to be of importance in clinical or biochemical analysis. Accordingly, only liquid phase chemiluminescence will be discussed here.

For analytical purposes, it is necessary to react the analyte with excess of the luminogenic reagents and thence to record the progress of the reaction by measuring the chemiluminescent intensity as a function of time. With modern instrumentation this can be achieved with great sensitivity. If the kinetics of the reaction are first-order with respect to the analyte, then the rate of change of chemiluminescent intensity (in photons/sec) as analyte is consumed can be related directly to concentration.

The number of individual chemiluminescent reactions is, however, relatively restricted and as a result the number of direct analyses is limited. In addition to the analysis of direct reactants, notably agents which are either catalytic or inhibitory in the chemiluminescent reactions, it is possible to measure oxidizing agents, e.g., hydrogen peroxide, which are essential for luminescence. The range of analytes may be expanded and the specificity of the assay improved by employing oxidase enzymes to generate hydrogen peroxide, which is ultimately determined using chemiluminescent agents.

The chemiluminescent agents most commonly used for analytical purposes are luminol (Albrecht, 1928) and lucigenin (Gleu and Petsch, 1935). Both produce light under alkaline conditions in the presence of an oxidizing agent and a catalyst (White et al., 1964). The optimal pH of light production for luminol in the presence of hydrogen peroxide may be reduced to near-neutral values using a peroxidase-like enzyme to intercede between the oxidizing agent and the chemiluminescent reagent (Harvey, 1940; Cormier and Prichard, 1968).

Several alternative reactions have been proposed to overcome the unfavorable pH required for the luminol and lucigenin reactions, including peroxyoxalate-sensitized chemiluminescence (Williams et al., 1976) and a new range of phenyl acridinium derivatives with defined chemiluminescent properties (McCapra et al., 1974).

II. ACTIVATION AND INHIBITION ASSAYS

Many substances are known to activate* or inhibit the chemiluminescence of luminol and lucigenin, and good reviews of their use in analytical chemistry have been presented by Seitz and Neary (1977) and

*Frequently, the activation is referred to as catalysis, but since in many cases the activator is consumed (Seitz and Neary, 1977), this term is misapplied.

Isacsson and Wettermark (1974). In general, while the reaction mechanisms remain obscure, luminol and lucigenin have found application for the analysis of the various metal ions which catalyze their oxidation, although few of these have been analyzed in biological materials. It has been suggested, however, that for such biological materials, only simple wet-ashing prior to analysis is additionally required (Seitz and Neary, 1977).

Since several metal ions may activate a particular chemiluminescent material under similar conditions, some form of prior treatment is required to gain a degree of analyte selectivity. This lack of specificity has probably hindered attempts to apply chemiluminescent trace metal analysis in the past.

Perhaps the best example of chemiluminescent trace metal analysis applied to biological samples is that of Seitz et al. (1972), in which Cr(III) is determined by adding excess ethylene diaminetetraacetic acid (EDTA) to the sample. EDTA complexation drastically reduces chemiluminescence catalyzed by metal ions. At pH 4.4 and room temperature, however, the Cr(III)-EDTA complex is kinetically slow to form. Therefore, Cr(III) will retain its efficiency as a catalyst even after EDTA is added. In this way, the luminol reaction can be used to quantitate Cr(III) with reasonable specificity. Residual catalysis from other metal-EDTA complexes can be accounted for by means of a blank reaction in which the Cr(III)-EDTA complex formation is accelerated by heating to 100°C. After cooling the sample is again reacted with luminol. Since the heating procedure converts all Cr(III) to Cr(III)-EDTA complex, all that remains is the residual catalysis from other metals complexed by EDTA. This is subtracted from sample-catalyzed chemiluminescence. This method has been used to determine Cr(III) in water, orchard leaves, bovine liver, and blood (Li and Hercules, 1974).

An alternative procedure which may be expected to widen the scope of trace metal analysis quite considerably is the use of chemiluminescence as a detection system after ion exchange chromatography of a mixture of metals (Delumyea and Hartkopf, 1976).

The apparatus employed for chemiluminescent trace metal analysis has generally been of a very simple type. Many Russian workers employ photographic detection systems (Babko and Lukovskaya, 1962), although elsewhere photomultiplier tubes have been found to be more satisfactory. Seitz and co-workers have frequently employed flow systems (e.g., Seitz et al., 1972) incorporating a specially designed flow cell whose contents are mixed either by bubbling through nitrogen (Fig. 1), which also helps to reduce interference from molecular oxygen, or by mechanical means. Alternatively, Hoyt and Ingle (1976) designed a discrete sampling photomultiplier photometer for use in trace Cr(III) determinations essentially as described above. This method has also been applied with considerable success to a centrifugal fast analyzer (Anderson, 1969) using a specially designed parallel mixing transfer disk (Bowling et al., 1975).

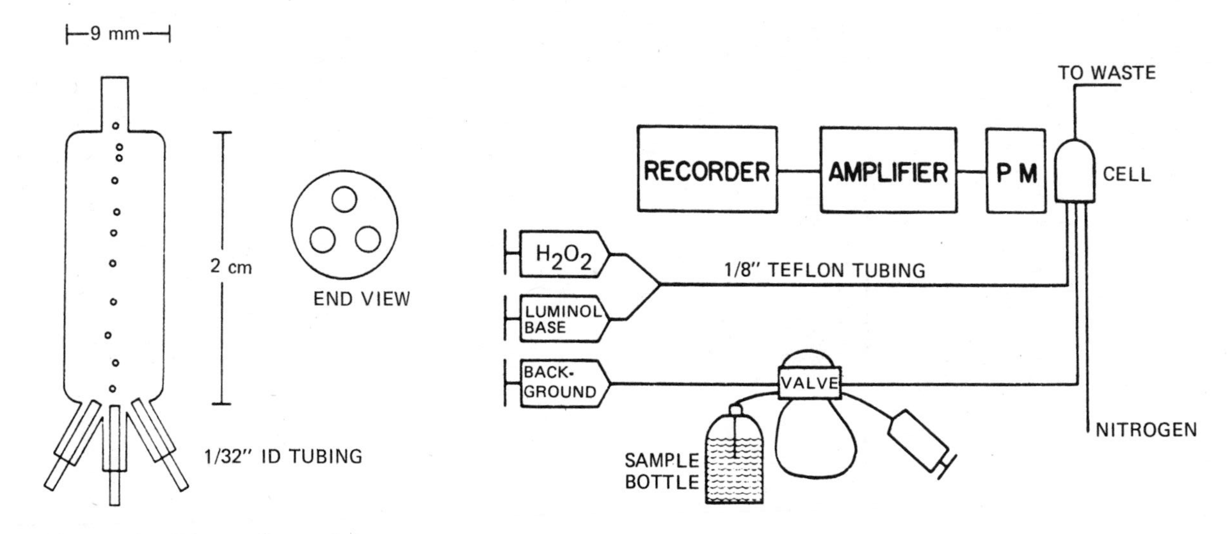

Figure 1 Flow cell used for the chemiluminescent determination of chromium(III) and its associated manifold. (Reprinted with permission from Seitz, W. R., Suydam, W. W., and Hercules, D. M., *Anal. Chem.* *44*:957-963 (1972). Copyright 1972 by the American Chemical Society.)

Various other analytical applications of luminol have been proposed. Since, for example, Co(II) is one of the most efficient catalysts of luminol chemiluminescence at high pH (Seitz et al., 1972; Seitz and Hercules, 1972,1973), this system has been adapted to the analysis of vitamin B_{12} which takes advantage of its cobalt content (Sheehan and Hercules, 1977). This is analogous to the situation for iron in the heme moiety where, although the iron is complexed, the efficiency of its catalysis is scarcely reduced (Isacsson and Wettermark, 1974). The structure of vitamin B_{12} is similar to that of hemin except that the central iron is replaced by Co(III). Co(III) is, however, not a catalyst for luminol chemiluminescence and reduction of the B_{12} prior to the reaction using a Jones reductor column in a flow system was used. It was also found necessary to exclude air totally using a nitrogen-filled enclosure and to purge samples with nitrogen prior to analysis. A detection limit of 2×10^{-9} mol/L was reported.

Various organic substances have been determined either by their catalytic or by their inhibitory effects on the chemiluminescence of luminol or lucigenin. Thus, for example, amino acids have been detected in the eluent of a chromatographic column (Lowery et al., 1977) by their catalytic effect on the copper-luminol-H_2O_2 system. Other substances determined in a similar fashion include nerve gases, insecticides, and isomeric benzene derivatives (Isacsson and Wettermark, 1974).

Perhaps the most important catalytic agent of the luminol reaction with H_2O_2 is the heme moiety because it allows the very sensitive determination of all heme-containing compounds (Neufeld et al., 1965; Ewetz and Thore, 1976). The detection limit for hematin, for example, is less than 10^{-13} mol/L. Substances which have been determined in this way in biological samples include ferritin, cytochrome c, myoglobin, hemoglobin, catalase, and hematin (Neufeld et al., 1965). Moreover, essentially this procedure has been used for many years in forensic science for the detection of blood stains (Sprecht, 1937).

Great interest has been shown recently in the quantitation of bacterial contamination by the use of bioluminescence, notably by the determination of ATP content using firefly luminescence. The reagents for this method are, however, very expensive, and it has been shown that bacterial quantitation may be achieved with adequate sensitivity via their activation of luminol chemiluminescence in alkaline solution (Oleniacz et al., 1968), as a result, presumably, of the presence of heme-containing compounds. While this method is obviously likely to be sensitive to interferences from metal ions, procedures have been presented which overcome these difficulties (Ewetz and Thore, 1976).

III. ENZYMIC PRODUCTION OF HYDROGEN PEROXIDE MONITORED BY CHEMILUMINESCENCE

An alternative to the measurement of catalytic or inhibitory substances is the quantitation of oxidizing agents, notably hydrogen peroxide, in the presence of excess chemiluminescent reagent and catalyst. Since many oxidase enzymes are capable of producing hydrogen peroxide, it is at least theoretically possible to determine these enzymes or their substrates using chemiluminescence. A significant problem with this approach is the high pH required for the chemiluminescence of luminol and lucigenin, necessitating a pH change after hydrogen peroxide production.

Bostick and Hercules (1974,1975) have applied this principle to the analysis of serum and urinary glucose in a flow system similar to that shown in Fig. 1 using immobilized glucose oxidase and the luminol-ferricyanide system at high pH. As an alternative, Auses et al. (1975) employed a discrete system in which preincubation of sample and enzyme was followed by addition of luminogenic reagents buffered at high pH. The resulting chemiluminescence can be related to glucose concentration although a high background light emission, resulting from the reaction of luminol-ferricyanide with oxygen, and interference problems have been encountered (Williams et al., 1976).

While most oxidases generate hydrogen peroxide directly and hence can, in principle, be used as described above, many other oxidoreductases can be coupled to the production of hydrogen peroxide using the procedure shown in Fig. 2 (Williams and Seitz, 1976). NADH reduces oxygen to hydrogen peroxide in the presence of a mediator, methylene blue. The detection limit for NADH using this system is approximately 2×10^{-7} mol/L and it has also been applied to the measurement of lactate dehydrogenase activity.

In an attempt to overcome the disadvantages of the luminol-ferricyanide chemiluminescent system for the determination of enzyme-generated hydrogen peroxide, Williams et al. (1976) investigated the properties of the sensitized chemiluminescent system based on peroxyoxalate with perylene as the fluorescer. The principle of this procedure is shown in Fig. 3 and the detection limit for hydrogen peroxide is 7×10^{-8} mol/L, with a linear range extending up to 10^{-3} mol/L. Although not as sensitive as the luminol system, peroxylate chemiluminescence exhibits a lower background emission and is less sensitive to interfering substances.

McCapra and co-workers (McCapra et al., 1974) described novel acridinium phenyl carboxylates which undergo chemiluminescent oxidation with hydrogen peroxide. The optimal pH of oxidation of these substances depends on the nature of the substituent in the phenyl group. Thus, the acridinium salt may be tailored to fit the pH requirements of, for example, an enzymic hydrogen-peroxide-generating

$$NADH + H^+ + MB_{ox} \rightarrow NAD^+ + MB_{red}$$

$$MB_{red} + O_2 \rightarrow MB_{ox} + H_2O_2$$

$$H_2O_2 + \text{luminol} \xrightarrow[\text{catalyst}]{\text{pH} > 10} \text{light}$$

Figure 2 Reaction sequence for the chemiluminescent determination of NADH (MB = methylene blue). (Modified from Williams and Seitz, 1976.)

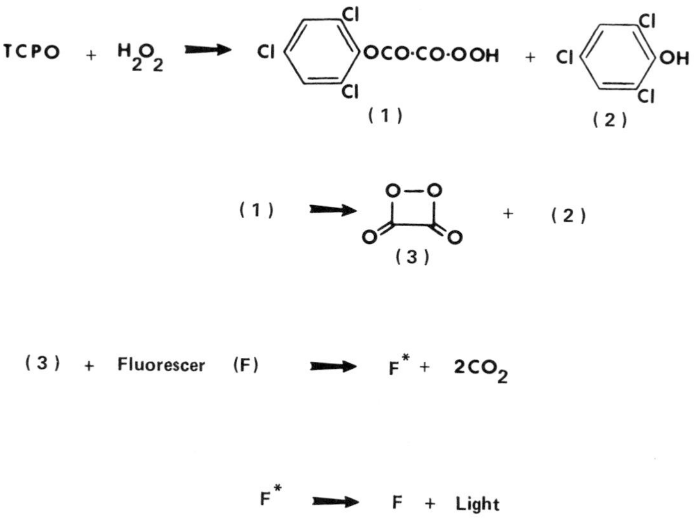

$$\text{TCPO} + H_2O_2 \rightarrow \text{Cl-C}_6\text{H}_2\text{Cl}_2\text{-OCO·CO·OOH (1)} + \text{Cl-C}_6\text{H}_2\text{Cl}_2\text{-OH (2)}$$

$$(1) \rightarrow (3) + (2)$$

$$(3) + \text{Fluorescer (F)} \rightarrow F^* + 2CO_2$$

$$F^* \rightarrow F + \text{Light}$$

Figure 3 Reaction sequence for the determination of H_2O_2 using diaryloxalate-sensitized chemiluminescence. TCPO = bis(2,4,6-trichlorophenyl)oxalate. (Modified from Williams et al., 1976.)

Table 1 Analytical Methods Based on Chemiluminescence

Analyte	Sensitivity or range	Reference
B_{12}	2×10^{-9} mol/L	Sheehan and Hercules (1977)
Catalase	10^{-4} μg	Neufeld et al. (1965)
Chromium(III)	—	Hoyt and Ingle (1976) Li and Hercules (1974)
Cytochrome c	10^{-1} μg	Neufeld et al. (1965)
Ferritin	1 μg	Neufeld et al. (1965)
Glucose	2×10^{-8} mol/L	Bostick and Hercules (1975) Williams et al. (1976) Auses et al. (1975) Bostick and Hercules, (1974)
Hematin	10^{-5} μg	Neufeld et al. (1965)
Hemoglobin	10^{-4} μg	Neufeld et al. (1965)
Hydrogen peroxide	10^{-9} mol/L	Seitz and Neary (1974)
Hypoxanthine	—	Oyamburo et al. (1970) Totter et al. (1960)
Lactate dehydrogenase	—	Williams and Seitz (1976)
Myoglobin	10^{-4} μg	Neufeld et al. (1965)
NADH	2×10^{-7}-10^{-4} mol/L	Williams and Seitz (1976)
Uric acid	10^{-7}-10^{-5} mol/L	Gorus and Schram (1977)

reaction. Moreover, the speed of oxidation can be optimized to the requirements of the analytical equipment employed. These reagents are claimed to be exceedingly stable and have been applied to the analysis of glucose in urine, plasma, serum, and CSF with great sensitivity and linearity, and without significant interference from acetone, ascorbic acid, uric acid, albumin, or globulin (McCapra et al., 1974,1979).

Chemiluminescent methods have also been described for uric acid (Gorus and Schram, 1977) and hypoxanthine (Totter et al., 1960; Oyamburo et al., 1970). A list of published procedures for the chemiluminescent determination of analytes of clinical interest appears in Table 1.

IV. PEROXIDASE-MEDIATED CHEMILUMINESCENCE

The major disadvantage of the determination of hydrogen peroxide produced by oxidase enzymes using luminol is that a high pH is required for light production. A two-stage reaction with a pH change before luminol addition is therefore required. An alternative analytical maneuver, which was first reported by Harvey (1940) and subsequently by Cormier and Pritchard (1968) as a model system for the

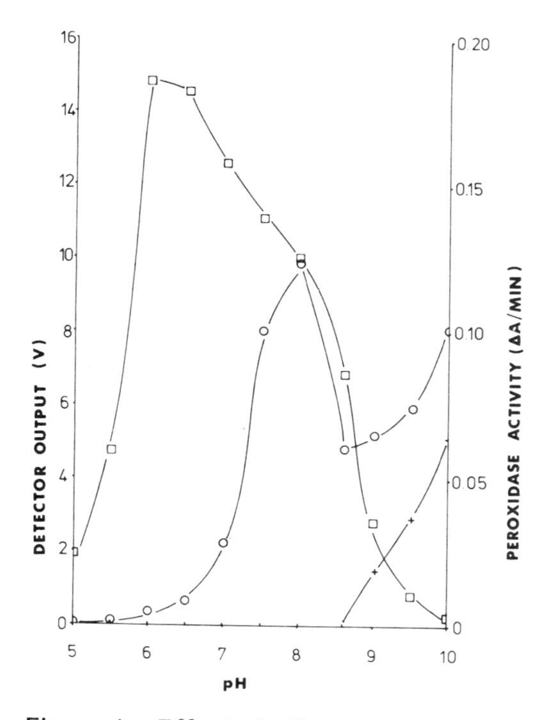

Figure 4 Effect of pH on peroxidase activity, monitored by luminol chemiluminescence (o) and, for comparison, photometrically (□). The light output of blanks in the absence of peroxidase is also shown (+).

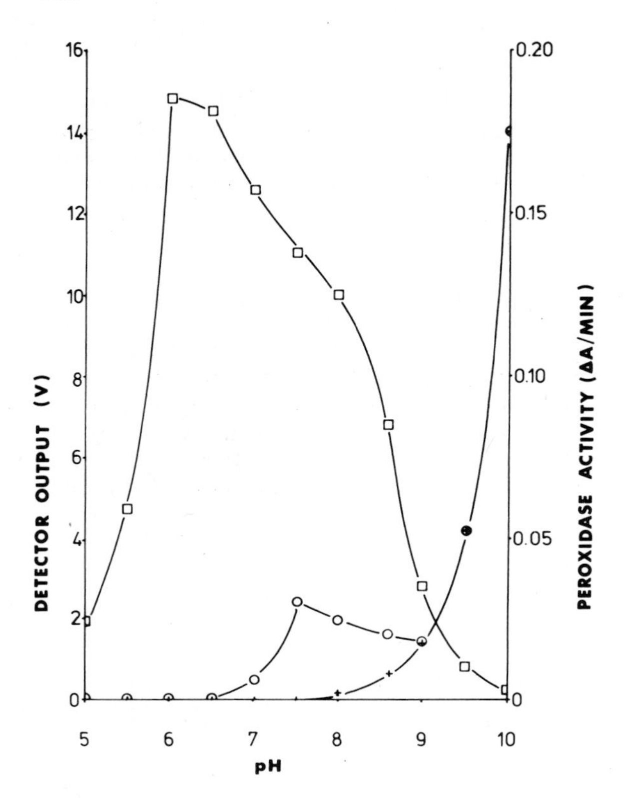

Figure 5 Effect of pH on peroxidase activity monitored by lucigenin chemiluminescence (o) and, for comparison, photometrically (□). The light output of blanks in the absence of peroxidase is also shown (+).

investigation of bioluminescence in *Balanoglossus*, is the use of a peroxidase-like enzyme to intercede between the hydrogen-peroxide-producing oxidase and the luminol indicator reaction. This is analogous to the intercession of peroxidase in the glucose oxidase method now found commonly in clinical laboratories (Trinder, 1969).

The effect of pH on the oxidation of luminol in the presence of horse radish peroxidase (HRP) is shown in Fig. 4 (Carter, 1979). Maximal activity occurs at approximately pH 8 and the reaction exhibits a quantum yield similar to that of the direct reaction. Above pH 8.6, however, direct oxidation predominates. If lucigenin is substituted for luminol, the quantum yield is poor compared with the direct reaction (Fig. 5).

The reaction mechanism remains obscure in spite of the investigations of Cormier and Prichard (1968) as does that of the analogous

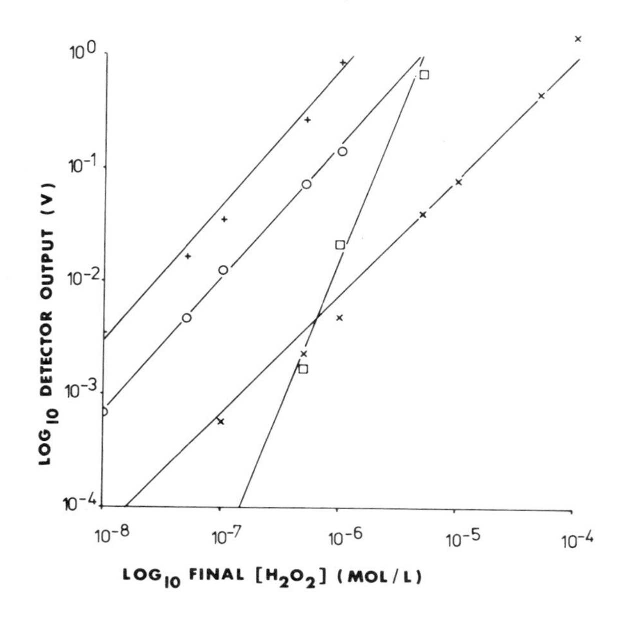

Figure 6 Effect of hydrogen peroxide concentration on the light output of peroxidase-mediated luminol chemiluminescence. The final peroxidase concentration was 0.04 g/L and the final luminol concentration was 0.177 g/L (□); 0.0177 g/L (o); 0.0017 g/L (+); and 0.000177 g/L (×).

reaction of HRP with 2,2'-azino-di-(3-ethylbenzthiazoline-6-sulfonate) (Gallati, 1979), but it is very complex. Investigations in our laboratory have shown that the order of the reaction with respect to hydrogen peroxide and its rate depend primarily on the relative concentrations of HRP and luminol. This may be illustrated by referring to Fig. 6. In addition, the shape of the light emission curve with time varies dramatically (Fig. 7). While the order and hence the linearity of the reaction are not totally independent of hydrogen peroxide concentration, peroxidase and luminol concentrations can be chosen which give reasonable linearity over a range of hydrogen peroxide concentrations of at least 3 orders of magnitude (Fig. 8). The detection limit for hydrogen peroxide using this method is at least 10^{-10} mol/L.

Recently, a novel modification of this system was presented (Freeman and Seitz, 1978) in which HRP is entrapped within a polyacrylamide gel placed at the end of a fiberoptic probe (Fig. 9). When the probe is immersed in solutions containing hydrogen peroxide of various concentrations, chemiluminescence is generated as the hydrogen peroxide diffuses into the HRP-gel and the fiberoptic transmits

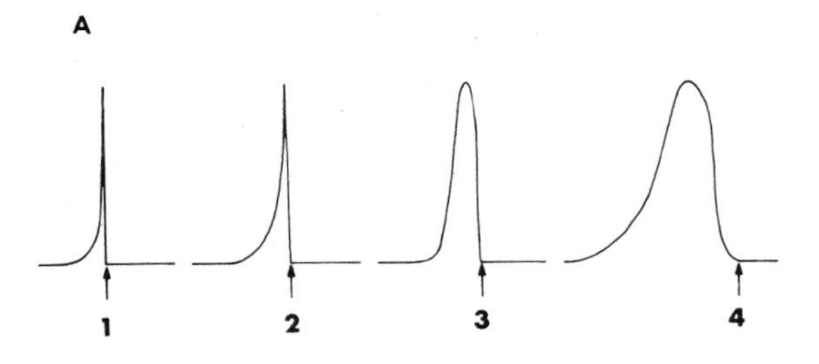

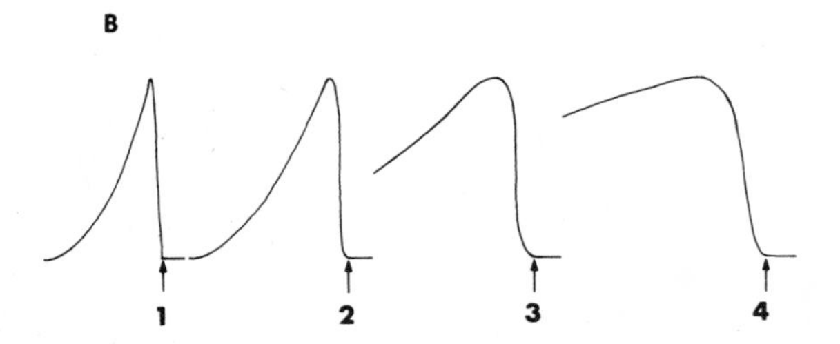

Figure 7 The effect of peroxidase and luminol concentrations on the light output profiles of peroxidase-mediated luminol chemiluminescence. All curves have been normalized for height. Arrows indicate initiation of the reaction. Final hydrogen peroxide concentration: A, 0.04 g/L; B, 0.004 g/L. Luminol concentration: 1, 0.177 g/L; 2, 0.0177 g/L; 3, 0.00177 g/L; 4, 0.000177 g/L.

the light to a detector. The probe shows a response which is second-order with respect to hydrogen peroxide and the detection limit is close to 10^{-6} mol/L. The time taken to reach a steady state depends on the hydrogen peroxide concentration and, generally, the chemiluminescent intensity was shown to be limited by the rate of mass transfer from the solution to the surface of the enzyme phase rather than by the activity of the HRP.

In addition to the measurement of hydrogen peroxide using peroxidase-mediated chemiluminescence, this system now appears to offer an attractive alternative to the use of radiolabels in immunoassay. In the most common embodiment, peroxidase (usually HRP) is employed as the labeling moiety, which is assayed via its catalysis of the luminol reaction. While this aspect of chemiluminescence is more fully considered in Chap. 8, an evaluation of various compounds and

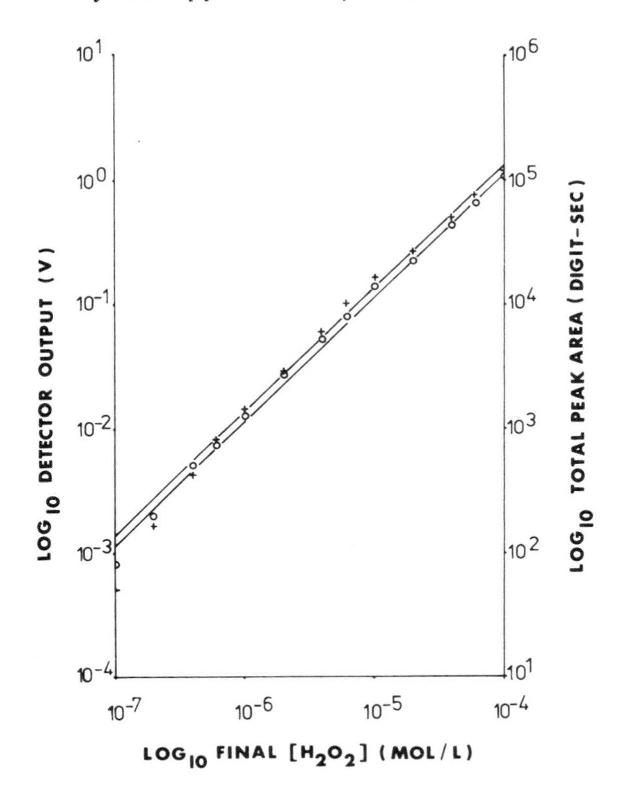

Figure 8 Effect of hydrogen peroxide concentration on light output in peroxidase-mediated luminol chemiluminescence. Light output recorded as maximal peak height (o) and peak area (+).

their effectiveness in different oxidation systems has been presented (Schroeder and Yeager, 1978) for the purpose of immunolabeling. Such data provide valuable assistance in the optimization of hydrogen peroxide detection systems as well as for the detection of label. Enzymes considered for use in various oxidation systems were microperoxidase, catalase, deuterohemin, hematin, lactoperoxidase, and xanthine oxidase.

Evidence has also been presented (Puget et al., 1977) to show that a luciferin isolated from the clam *Pholas dactylus* is a considerably better chemiluminescent substrate for HRP than luminol, although it has only been applied to immunoassay to date, where it appears to offer singular advantages. It remains to be seen whether it is possible to adapt this system to the measurement of hydrogen peroxide.

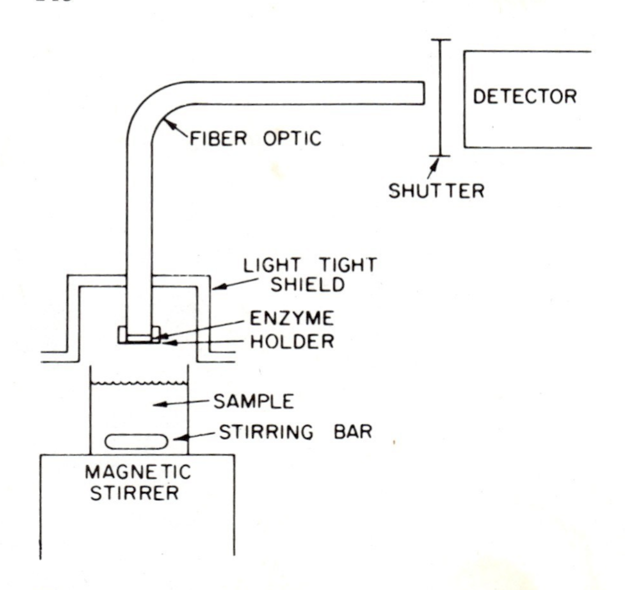

Figure 9 Analytical arrangement for the determination of H_2O_2 using gel-entrapped horse radish peroxidase placed at the end of a fiber-optic probe. (Reprinted with permission from Freeman, T. M., and Seitz, W. R., *Anal. Chem.* *50*:1242-1246 (1978). Copyright 1978 by the American Chemical Society.)

V. CONCLUSION

While chemiluminescence potentially offers many advantages for sensitive analysis, it has yet to be applied widely within the clinical laboratory. It remains to be seen if the procedures detailed in this chapter together with other new analytical strategies may remedy this situation.

REFERENCES

Albrecht, H. O. (1928). Uber die Chemiluminescenz des Aminophthalsaurehydrazids. *Z. Phys. Chem.* *136*:321-330.

Anderson, N. G. (1969). Analytical techniques for cell fractions. XII. A multiple-cuvet rotor for a new microanalytical system. *Anal. Biochem.* *28*:545-562.

Auses, J. P., J. L. Cook, and J. T. Maloy (1975). Chemiluminescent enzyme method for glucose. *Anal. Chem.* *47*:244-249.

Babko, A. K., and N. M. Lukovskaya (1962). Photographic chemiluminescence method for the determination of microquantities of copper and hydrogen peroxide. *Zh. Anal. Khim.* *17*:50-52.

Coleg Meddygaeth Prifysgol Cymru
University of Wales College of Medicine

With the Compliments of
Professor L. E. Hughes

UNIVERSITY DEPARTMENT OF SURGERY,
UNIVERSITY HOSPITAL OF WALES,
HEATH PARK,
CARDIFF CF4 4XN

Tel. 755944
Ext. 2749
2896

Bostick, D. T., and D. M. Hercules (1974). Enzyme-induced chemiluminescence determination of blood glucose using luminol. *Anal. Lett. 7*:347-353.

Bostick, D. T., and D. M. Hercules (1975). Quantitative determination of blood glucose using enzyme induced chemiluminescence of luminol. *Anal. Chem. 47*:447-451.

Bowling, J. L., J. A. Dean, G. Goldstein, and J. M. Dale (1975). Rapid determination of chromium in natural waters by chemiluminescence with a centrifugal fast analyser. *Anal. Chim. Acta 76*:47-55.

Carter, T. J. N. (1979). Luminescence instrumentation and assays in clinical chemistry. Ph.D. Thesis, University of Birmingham.

Cormier, M. J., and P. M. Prichard (1968). An investigation of the mechanism of the luminescent peroxidation of luminol by stopped-flow techniques. *J. Biol. Chem. 243*:4706-4714.

Delumyea, R., and A. V. Hartkopf (1976). Metal catalysis of the luminol reaction in chromatographic solvent systems. *Anal. Chem. 48*:1402-1405.

Ewetz, L., and A. Thore (1976). Factors affecting the specificity of the luminol reaction with haematin compounds. *Anal. Biochem. 71*:564-570.

Fontijn, A., D. Golomb, and J. A. Hodgeson (1973). A review of experimental measurement methods based on gas-phase chemiluminescence. In *Chemiluminescence and Bioluminescence*, M. J. Cormier, D. M. Hercules and J. Lee (Eds.). Plenum, New York, pp. 393-426.

Freeman, T. W., and W. R. Seitz (1978). Chemiluminescence fibre optic probe for hydrogen peroxide based on the luminol reaction. *Anal. Chem. 50*:1242-1246.

Gallati, H. (1979). Peroxidase aus Meerretich:Kinetische Studien sowie Optimierung der Aktivitatsbestimmung mit den Substraten H_2O_2 und ABTS. *J. Clin. Chem. Clin. Biochem. 17*:1-7.

Gleu, K., and W. Petsch (1935). Die Chemiluminescenz der Dimethyldiaridyliumsalze. *Angew. Chem. 48*:57-59.

Gorus, F., and E. Schram (1977). Chemiluminescent assay of uric acid. *Arch. Int. Physiol. Biochim. 85*:981-982.

Harvey, E. N. (1940). *Living Light*. Princeton University Press, Princeton, N.J., pp. 118, 148.

Hoyt, S. D., and J. D. Ingle (1976). A chemiluminescence photometer for trace chromium(III) determinations. *Anal. Chim. Acta 87*:163-175.

Isacsson, U. and G. Wettermark (1974). Chemiluminescence in analytical chemistry. *Anal. Chim. Acta. 68*:339-362.

Li, R. T., and Hercules, D. M. (1974). Determination of chromium in biological samples using chemiluminescence. *Anal. Chem. 46*:916-919.

Lowery, S. N., P. W. Carr, and W. R. Seitz (1977). Determination of L-amino acids and L-amino acid oxidase activity using luminol chemiluminescence. *Anal. Lett. 10*:931-943.

McCapra, F., D. E. Tutt, and R. M. Topping (1974). British Patent No. 1,461,877.

McCapra, F., D. Tutt, and R. M. Topping (1979). The chemiluminescence of acridinium phenyl carboxylates in the assay of glucose and hydrogen peroxide. In *Proceedings of the International Symposium on Analytical Applications of Bio- and Chemiluminescence*, E. Schram and P. Stanley (Eds.). State Printing and Publishing, Westlake Village, Cal., p. 221.

Nederbragt, C. W., A. Van der Horst, and J. Van Duijn (1965). Rapid ozone determination near an accelerator. *Nature 206*:87.

Neufeld, H. A., C. J. Conklin, and K. D. Towner (1965). Chemiluminescence of luminol in the presence of hematin compounds. *Anal. Biochem. 12*:303-309.

Oleniacz, W. S., M. A. Pisano, M. H. Rosenfeld, and R. L. Elgart (1968). Chemiluminescent method for detecting microorganisms in water. *Environ. Sci. Technol. 2*:1030-1033.

Oyamburo, G. M., C. E. Prego, E. Prodanov, and H. Soto (1970). Study of the enzyme-catalysed oxidation of hypoxanthine through the chemiluminescence of luminol. *Biochim. Biophys. Acta 205*:190-195.

Puget, K., A. M. Michelson, and S. Avrameas (1977). Light emission techniques for the microestimation of femtogram levels of peroxidase: Application to peroxidase (and other enzymes)—Coupled antibody cell antigen interaction. *Anal. Biochem. 79*:447-456.

Schroeder, H. R., and F. M. Yeager (1978). Chemiluminescent yields and detection limits of some isoluminol derivatives in various oxidation systems. *Anal. Chem. 50*:1114-1119.

Seitz, W. R., and D. M. Hercules (1972). Determination of trace amounts of iron (II) using chemiluminescence analysis. *Anal. Chem. 44*:2143-2149.

Seitz, W. R., W. W. Suydam, and D. M. Hercules (1972). Determination of trace amounts of chromium (III) using chemiluminescent analysis. *Anal. Chem. 44*:957-963.

Seitz, W. R., and D. M. Hercules (1973). Chemiluminescence analysis for trace metals. In *Chemiluminescence and Bioluminescence*, M. J. Cormier, D. M. Hercules, and J. Lee (Eds.). Plenum, New York, pp. 427-450.

Seitz, W. R., and M. P. Neary (1974). Chemiluminescence and bioluminescence. *Anal. Chem. 46*:188A-202A.

Seitz, W. R., and M. P. Neary (1976). Recent advances in bioluminescence and chemiluminescence assay. *Meth. Biochem. Anal. 23*:161-188.

Seitz, W. R., and M. P. Neary (1977). Chemiluminescence and bioluminescence analysis. *Contemp. Topics Anal. Clin. Chem. 1*:49-125.

Sheehan, T. L., and D. M. Hercules (1977). Analytical study of chemiluminescence from the vitamin B_{12}-luminol system. *Anal. Chem. 49*:446-450.

Sprecht, W. (1937). Die chemiluminescenz des Hamins, ein Hilfsmittel zur Auffinding und Erkennung forensisch wichtiger Blutspuren. *Angew. Chem.* *50*:155-157.

Totter, J. R., V. J. Medina, and J. L. Scoseria (1960). Luminescence during the oxidation of hypoxanthine by xanthine oxidase in the presence of dimethylbiacridylium nitrate. *J. Biol. Chem.* *235*:238-241.

Trinder, P. (1969). Determination of glucose in blood using glucose oxidase with an alternative oxygen acceptor. *Ann. Clin. Biochem.* *6*: 24-27.

White, E. H., O. C. Zafiriou, H. M. Kagi, and J. H. M. Hill (1964). Chemiluminescence of luminol: the chemical reaction. *J. Amer. Chem. Soc.* *86*:940-941.

Williams, D. C., G. F. Huff, and W. R. Seitz (1976). Evaluation of peroxyoxalate chemiluminescence for determination of enzyme generated peroxide. *Anal. Chem.* *48*:1003-1006.

Williams, D. C., and W. R. Seitz (1976). Automated chemiluminescent method for determining the reduced form of NAD coupled to the measurement of LDH activity. *Anal. Chem.* *48*:1478-1481.

8

LUMINESCENT IMMUNOASSAYS

LARRY J. KRICKA* University of Birmingham, Birmingham, England

TIMOTHY J. N. CARTER† Wolfson Research Laboratories, Queen Elizabeth Medical Centre, Edgbaston, Birmingham, England

I. Introduction 154

II. Classification of Luminescent Immunoassays 156

III. Advantages and Disadvantages of Luminescent Immunoassays 157
- A. Sensitivity 157
- B. Toxicity and Stability 159
- C. Cost 159
- D. Equipment 159
- E. Separation Step 159

IV. Luminescent Immunoassay 159
- A. Chemiluminescent Labels 163
- B. Bioluminescent Labels 166
- C. Type of Assay 166

V. Luminescent Enzyme Immunoassay 167

VI. Luminescent Enzyme Multiplied Immunoassay Technique 169

VII. Luminescent Cofactor Immunoassay 170
- A. Nicotinamide Adenine Dinucleotide (NAD) Labels 172
- B. Adenosine-5'-Triphosphate (ATP) Labels 172

---

*Dr. Kricka is currently with the University of California at San Diego, La Jolla, California, on a leave of absence from the University of Birmingham.

†Dr. Carter is currently with Battelle, Centre de Recherche de Genève, Carouge, Genève, Switzerland.

I. INTRODUCTION

Many alternatives to radioactive atoms as labels have been advocated since Yalow and Berson's original description of the application of radioactively labeled antigens in the quantitation of immunological reactions (Yalow and Berson, 1959). Apart from the profit motive, the major stimuli to the search for alternatives to radiolabels have been

1. *Safety*: Radioisotopes constitute a health hazard. Although the hazard is very slight, a safer label would be more acceptable and convenient for use in the clinical laboratory.
2. *Stability*: The relatively short-lived nature of radioisotopes (e.g., the half-life of ^{125}I is approximately 60 days) limits the shelf-life of labeled antigens or antibodies.
3. *Speed*: A radiolabel is not affected by antigen/antibody binding and a separation step must be employed in order to quantitate the extent of binding and hence the concentration of antigen in a specimen (e.g., Scheme 1). This type of assay is termed *hetero-*

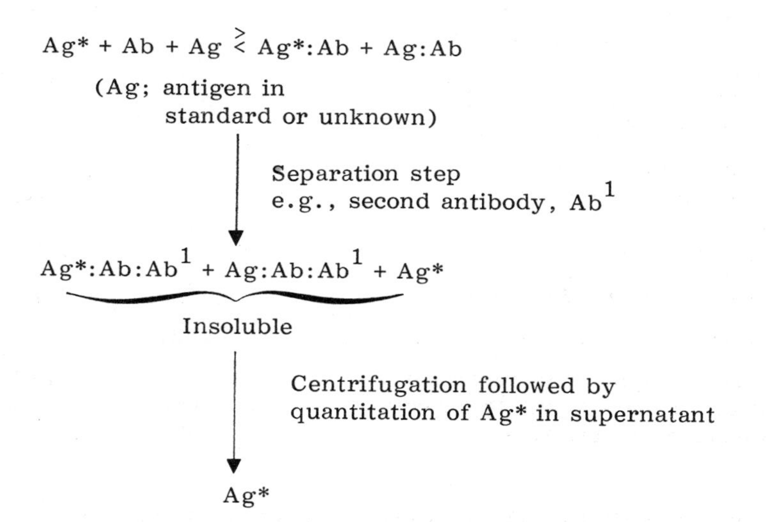

Scheme 1 Heterogeneous immunoassay. The biological properties of the labeled antigen (Ag*) are unaffected by binding to specific antibody (Ab) and thus quantitation of either bound or free Ag* requires a separation step.

Table 1 Labels Employed in Immunoassay

Type of label	Example	Property measured	References
Enzyme	Peroxidase	Enzyme activity	Wisdom (1976) Scharpe et al. (1976)
Fluorescent molecule	Fluorescein	Fluorescence	Dandliker and de Saussure (1970) Soini and Hemmila (1979)
		Fluorescence polarization	Watson et al. (1976)
	Fluorescein	Fluorescence excitation transfer	Ullman et al. (1976)
	Fluorescein/rhodamine	Quenching	Ullman et al. (1976)
Organic free radical	Tetramethyl-piperidinooxyl	Magnetic moment	Leute et al. (1972)
Organometallic or coordination complex	di-π-cyclopen-tadienyliron (ferrocene)	Atomic absorption	Cais et al. (1977)
Bacteriophage	T_2		Makela (1966)
Cofactor	NAD derivative	Cycling reaction with LDH/diaphorase	Carrico et al. (1976a)
Erythrocyte		Hemagglutination inhibition	Adler and Lau (1971)

geneous. The requirement for a separation step makes such assays lengthy. A label attached to an antigen (or antibody) that is sensitive to antigen/antibody binding automatically provides monitoring of the competition between labeled and unlabeled antigen for binding sites on antibodies, and the need for a separation procedure is eliminated. Such an assay is termed homogeneous (Scheme 2).

A listing of some of the alternatives to radioactive labels is presented in Table 1. In theory any substance which is quantitatable

$$Ag^* + Ab + Ag \gtrless Ag^*{:}Ab + Ag{:}Ab$$

Reactive label   Unreactive label

Scheme 2 Homogeneous immunoassay. The biological properties of the labeled antigen are drastically altered when bound to specific antibody and thus quantitation of either bound or unbound Ag* does *not* require a separation step.

either directly by some physical property or indirectly by its involvement in a chemical reaction or process could be used as a label. Most attention, however, has been given to substances with magnetic, fluorescent, or catalytic (enzymic) properties. Commercial fluoro- and enzyme immunoassay kits are now available for a range of analytes (e.g., gentamicin, thyroxine, digoxin, phenytoin, primidone, hepatitis B surface antigen). Although sensitive methods for detection are available for both enzyme and fluorescent substances, they are not as sensitive as detection methods for luminescent substances. This factor has caused luminescent substances to emerge as possible competitors to radio-, enzyme, and fluorescent labels in immunoassays.

This chapter outlines the application of luminescent substances as labels in immunoassay and the indirect application of luminescence to monitor immunological reactions.

II. CLASSIFICATION OF LUMINESCENT IMMUNOASSAYS

Luminescent substances have been used in specific protein binding and immunological assays in two different ways: either directly as labels or indirectly in the luminescent quantitation of enzyme- and cofactor-labeled antigens or antibodies. Based on these applications four different types of luminescent immunoassay have emerged:

1. *Luminescent immunoassay (LIA)*: This type of assay is analogous to radioimmunoassay except that a chemi- or bioluminescent substance is used as the label for the antigen or antibody (Scheme 3).
2. *Luminescent enzyme immunoassay (LEIA).*
3. *Luminescent "enzyme multiplied immunoassay technique" (LEMIT)*: These types of assay employ luminescent detection of bound or free enzyme or, in the case of the homogeneous EMIT type of assay, discrimination between sterically hindered and unhindered enzyme.
4. *Luminescent "cofactor" immunoassay (LCIA)*: In this type of assay an antigen or antibody is labeled with a substance which will participate, directly or indirectly, in a luminescent reaction, e.g.,

$$Ag_{standard} + Ag\text{-}L + Ab \rightleftharpoons Ab{:}Ag + Ab{:}Ag\text{-}L$$

or
unknown

Separation of bound and free Ag-L

$$Ag\text{-}L_{bound} + \text{luminogenic coreactants} \longrightarrow \text{Light}$$

or
free

Scheme 3 Luminescent immunoassay. L = chemi- or bioluminescent molecule; luminogenic coreactants = H_2O_2 when L = luminol or isoluminol, and lauryl aldehyde/$FMNH_2$ when L = bacterial luciferase).

nicotinamide adenine dinucleotide (NAD) or adenosine-5'-triphosphate (ATP). The bound or free fraction is then quantitated by means of a luminescent reaction. In view of the diversity of luminescent reactions, other components of such reactions will doubtless be used as labels in immunoassays.

III. ADVANTAGES AND DISADVANTAGES OF LUMINESCENT IMMUNOASSAYS

Immunoassays based directly or indirectly on luminescence have a number of advantages over their more conventional counterparts. The major advantages are discussed in the following paragraphs.

A. Sensitivity

The sensitivity of an immunoassay depends ultimately upon the avidity of the antibody (or binding protein) employed. Avidity of an antibody can be quantitated by means of the equilibrium constant for the reversible reaction

$$Ag + Ab \gtrless Ag{:}Ab$$

As a simple guide, the minimal detectable concentration of antigen is given by the reciprocal of the equilibrium constant for the reaction of the antigen with the antibody (Yalow, 1978). Any chemical modification or an antigen which reduces its ability to bind with antibody will reduce sensitivity. This problem is negligible for the relatively small radioiodine label but becomes more serious for larger labels, such as enzymes which can sterically hinder antigen/antibody binding. The latter problem can be overcome in part by the introduction of long spacer groups between the antigen and label (Cuatrecasas and Anfinsen, 1971).

A secondary consideration in the sensitivity of an immunoassay is the sensitivity with which the labeled antigen can be detected. The extreme sensitivity of luminescent reactions is the most important advantage of immunoassays based either directly or indirectly on luminescence. For example, the chemiluminescent compound luminol (1) has a detection limit of 1 pmol/L when assayed using hydrogen-peroxide/hematin, while the detection limit for the naphthylhydrazide (3) when assayed with hydrogen-peroxide/microperoxidase is 0.1 pmol/L (Schroeder and Yeager, 1978).

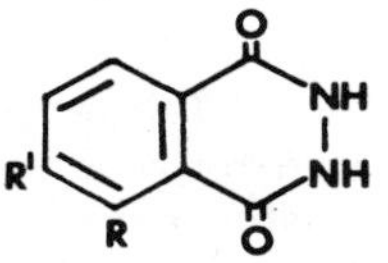

(1) R = NH_2 ; R' = H

(2) R = H ; R' = NH_2

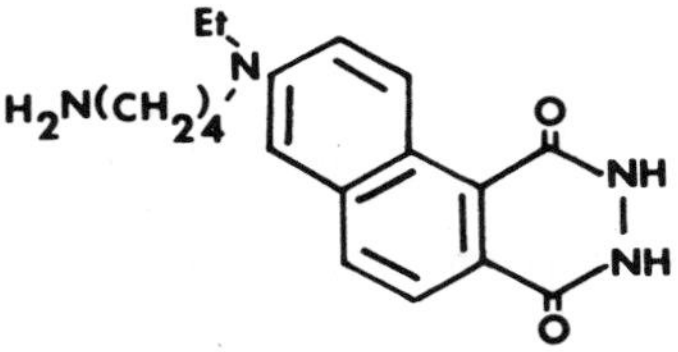

(3)

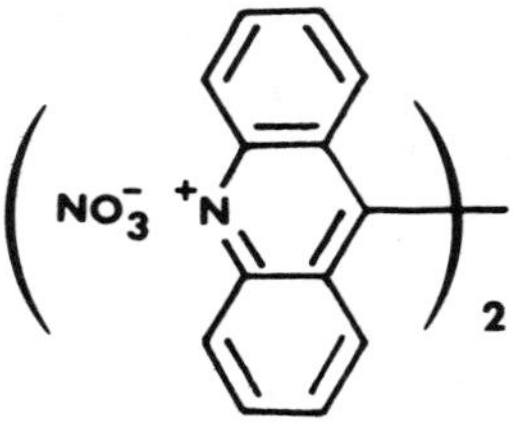

(4)

In the context of enzyme immunoassay, luminescent assays for enzyme-labeled ligands are considerably more sensitive than colorimetric assays. For example, the detection limit for the bioluminescent assay of the enzyme peroxidase using hydrogen peroxide/*Pholas dactylus* lucifern is 5 fg (Puget et al., 1977), whereas the detection limit for a colorimetric assay of this enzyme is 5 pg (Linpisarn, unpublished). Only fluorimetric assays of enzymes surpass the level of sensitivity of luminescent assays, e.g., fluorometric assays for β-galactosidase conjugates have a detection limit of 1 amol/L (Kato et al., 1977).

Quantitation of ligands labeled with dehydrogenases (e.g., malate dehydrogenase) via measurements of NADH formation can be achieved with increased sensitivity using a bioluminescent assay. In comparison with a spectrophotometric assay, a bioluminescent assay employing bacterial luciferase has been estimated to be 25,000 times more sensitive (Stanley, 1974).

B. Toxicity and Stability

Chemi- and bioluminescent substances present no known hazard to the environment. In general, chemiluminescent substances such as luminol, isoluminol (2), and lucigenin (4) are stable compounds. Bioluminescent substances, such as bacterial and firefly luciferase, suffer from the same stability problems as other enzymes.

C. Cost

Generally, chemiluminescent reagents are inexpensive and readily available. In contrast, bioluminescent reagents such as bacterial and firefly luciferase are expensive, less readily available, and the quality of currently available preparations of the former enzyme is prone to considerable batch-to-batch variation.

D. Equipment

The equipment required for the measurement of luminescence is relatively inexpensive. Luminescent immunoassays are amenable to automation although automatic luminometers are not as yet commercially available.

E. Separation Step

Several homogeneous luminescent immunoassays employing isoluminol labels have been described (Kohen et al., 1979, 1980a,b, 1981a,b; Schroeder et al., 1976). It may thus prove possible to develop a range of homogeneous luminescent immunoassays analogous to the EMIT assay.

IV. LUMINESCENT IMMUNOASSAY

Scheme 3 outlines the salient features of a luminescent immunoassay. Both chemiluminescent and bioluminescent labels have been employed in this type of assay. Table 2 summarizes various examples of luminescent immunoassays.

Table 2 Luminescent Immunoassays

Labeled substance	Label	Analyte	Detection system	Sensitivity (range)	Reference
T_4	Isoluminol	T_4	H_2O_2-micro-peroxidase	2-15 μg/dl	Schroeder et al. (1977) Schroeder et al. (1979)
IgG	Luminol	IgG	H_2O_2/hemin	5-100 μg/tube	Hersch et al. (1978)
Insulin	Luminol	Insulin		5-25 ng/ml	Maier (1977,1978)
Digoxin	Luminol	Digoxin			
Biotin	Isoluminol	Biotin	Superoxide or H_2O_2/lacto-peroxidase	50 nmol/L	Schroeder et al. (1976)
IgG	Luminol	Mumps virus antibody	H_2O_2/NaOCl		Konishi et al. (1980)

Anti-mouse IgG	Bacterial luciferase	Cell surface IgG	$FMNH_2$/ lauryl aldehyde	—	Puget et al. (1977)
Testosterone-ovalbumin	Luminol derivative	Testosterone	H_2O_2/cupric acetate	0.1-10 ng	Pratt et al. (1978)
Sheep anti-rabbit IgG	Luminol	Rabbit IgG	H_2O_2/NaOH	∿10 ng/ml	Woodhead et al. (1979); Simpson et al. (1979)
Antipara-thyroid hormone IgG	Luminol	Parathyroid hormone	H_2O_2/NaOH	—	Woodhead et al. (1978)
Antialphafeto-	Luminol	Alphafeto-protein	H_2O_2/NaOH	—	
Progesterone	Isoluminol	Progesterone	H_2O_2/micro-peroxidase	25 pg/tube	Kohen et al. (1979)
Methotrexate	Firefly luciferase	Methotrexate	ATP/luciferin	0.5 pmol	Wannlund et al. (1980)

Table 3 Relative Light Yields of Luminol and Isoluminol Derivatives

	Substituents on the aromatic amino group of isoluminol (2)		Chemiluminescence light yield (%) relative to luminol (100%) a
Isoluminol (2)	H	H	5
	Et	Et	100
	H	$CH_2 \cdot CHOH \cdot CH_2NH_2$	10
	Et	$CH_2 \cdot CHOH \cdot CH_2NH_2$	46
	H	$CH_2 \cdot CHOH \cdot CH_2NHCO\text{-}T_4$	1
	Et	$CH_2 \cdot CHOH \cdot CH_2NHCO\text{-}T_4$	2
	H	$(CH_2)_4NH_2$	14
	Et	$(CH_2)_4NH_2$	84
	Et	$(CH_2)_4NHCO\text{-}T_4$	2
	H	$(CH_2)_6NH_2$	17
	Et	$(CH_2)_6NH_2$	44
	Substituents on the aromatic amino group of luminol (1)		**b**
Luminol (1)	H	H	100
	Et	$(CH_2)_6NH_2$	78
	Et	$(CH_2)_6NHCO\text{-}T_4$	1
	H	$CH_2 \cdot CHOH \cdot CH_2NH_2$	5
	H	$CH_2 \cdot CHOH \cdot CH_2NHCO\text{-biotin}$	5

[a] Isoluminol derivatives assayed with H_2O_2/hematin.
[b] Luminol derivatives assayed with H_2O_2/microperoxidase T_4:thyroxine linked through the carboxyl group.

Sensitivity of the assays is good: for example, the assay for thyroxine has a detection limit comparable with that of radio immunoassay for this analyte.

A. Chemiluminescent Labels

Luminol (1), isoluminol (2), the luminol derivative (3), and lucigenin derivatives have all been employed as labels. Three different types of procedure have been used for the preparation of conjugates of proteins or haptens and luminescent molecules. These are

1. Chemical modification (activation) of the chemiluminescent compound
2. Chemical modification of the protein or hapten
3. Conjugation by means of bifunctional coupling reagents

Modified Label

The luminescent properties of luminol is impaired by substitution of the aromatic amino group, but this group has provided the most convenient site for chemical activation. In contrast, substitution of the aromatic amino group of isoluminol enhances light output (Schroeder et al., 1978). Table 3 illustrates the effect of such substitutions in both luminol and isoluminol. Two examples of the preparation of activated luminol and isoluminol labels and their coupling to proteins or haptens are presented in Schemes 4 and 5. Covalent attachment of thyroxine derivatives leads to a considerable diminution in the light yield of the label (Table 3). This is thought to be a result of internal quenching of luminescence by the iodine atoms in thyroxine.

Modified Protein or Hapten

Only two examples of this approach to the preparation of a luminescent conjugate have been described (Hersch et al., 1978,1979; Maier, 1977,1978). Both involve the generation of reactive aldehyde groups by periodate oxidation of carbohydrate moieties of either a protein or a hapten. Scheme 6 illustrates the use of this reaction for the preparation of a protein-luminol conjugate. A reactive aldehyde group on the protein, generated by periodate oxidation, reacts with the aromatic amino group of luminol to form a Schiff base linkage. This reaction is reversible and in a modification of this procedure, described for the preparation of a digoxin-luminol conjugate, the Schiff base is stabilized by reduction with sodium borohydride (Maier, 1977, 1978). Analogous reactions have been described for the preparation of enzyme-protein conjugates (Nakane and Kawaoi, 1974).

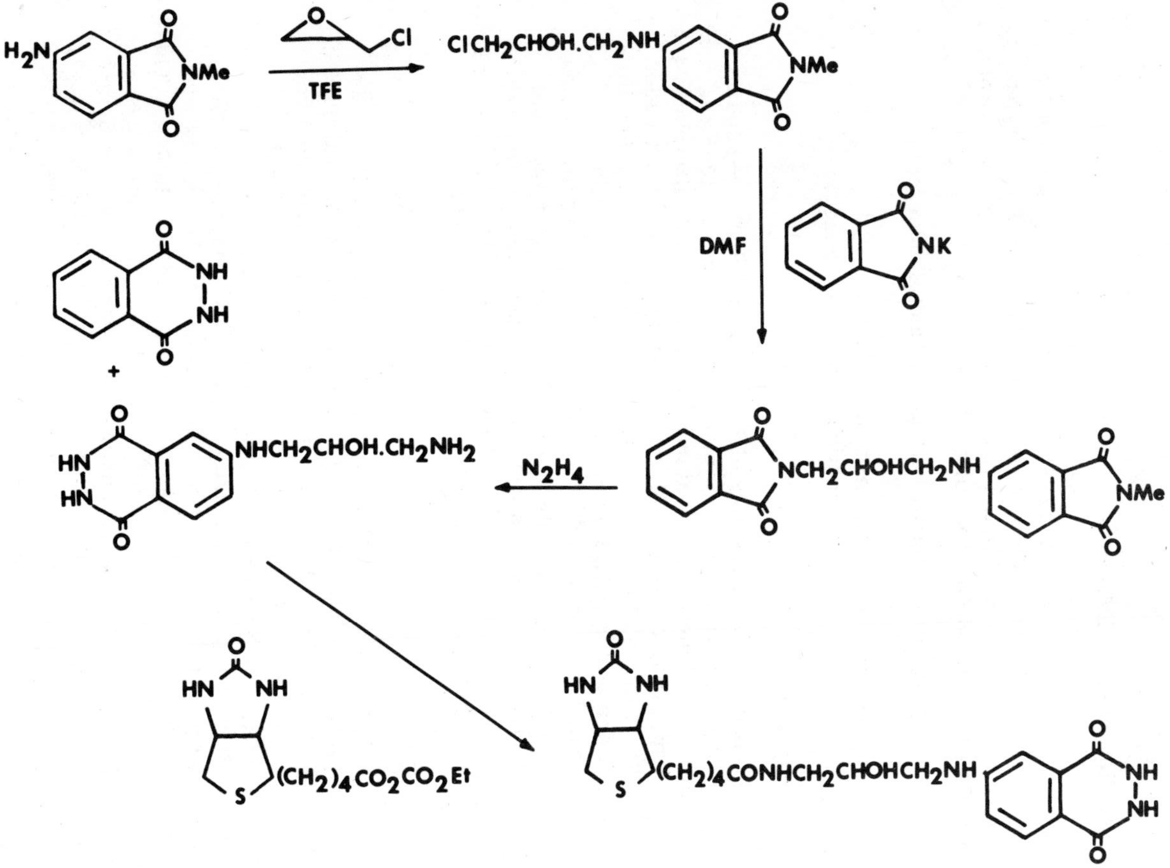

Scheme 4 Sequence of reactions used to prepare a conjugate of isoluminol and biotin. TFE = trifluoroethanol, DMF = dimethylformamide.

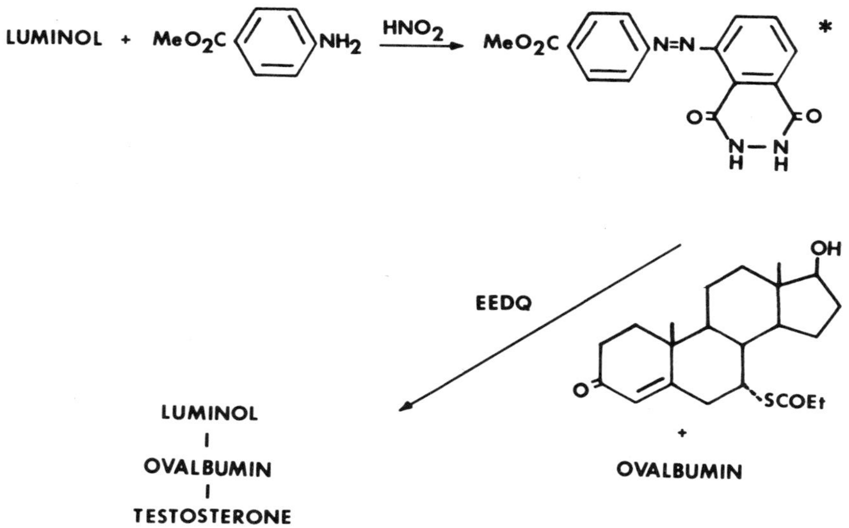

LUMINOL + MeO_2C–C_6H_4–NH_2 $\xrightarrow{HNO_2}$ MeO_2C–C_6H_4–N=N– *

EEDQ

–SCOEt + OVALBUMIN

LUMINOL – OVALBUMIN – TESTOSTERONE

Scheme 5 Sequence of reactions used to prepare a conjugate of luminol and testosterone. E = N-ethoxycarbonyl-2-ethoxyl-1,2-dihydroquinol; * = provisional assignment of structure).

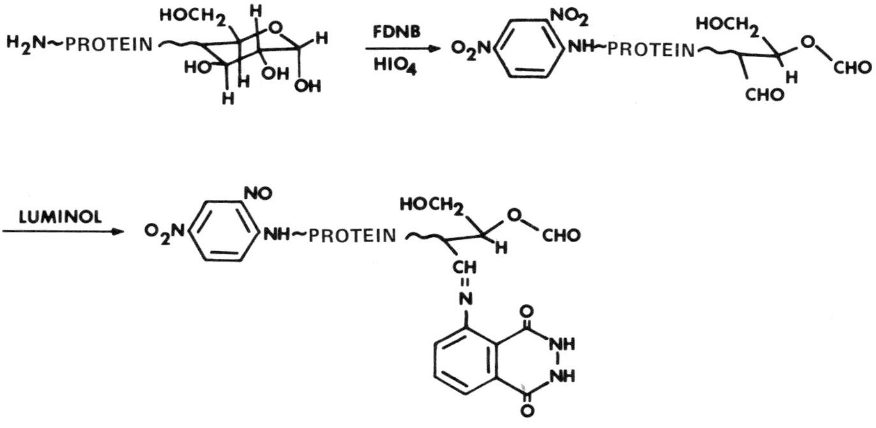

H_2N~PROTEIN $\xrightarrow[HIO_4]{FDNB}$ O_2N–$C_6H_3(NO_2)$–NH~PROTEIN

$\xrightarrow{LUMINOL}$ O_2N–$C_6H_3(NO)$–NH~PROTEIN

Scheme 6 Two-stage periodate coupling procedure for the preparation of a conjugate of luminol and a protein. FDNB = fluorodinitrobenzene.

Bifunctional Coupling Reagents

Only two examples of the use of a bifunctional coupling reagent (a compound with two reactive chemical groups) (Kennedy et al., 1976) for the preparation of a luminescent conjugate has been described. The first involves the reaction of amino groups of insulin and luminol with glutaraldehyde. Glutaraldehyde in solution does not exist exclusively as the monomer $OHC(CH_2)_3CHO$; in fact, a large number of polymeric forms occur (Richards and Knowles, 1968). An advantage of this particular coupling reagent is that it introduces a very large spacer group between the insulin and luminol, hence minimizing steric interactions. The second method involves the use of carbodiimide (Woodhead et al., 1979) to couple luminol or a lucigenin derivative to anti-rabbit IgG.

In the patent literature (Maier, 1977,1978) the possible application of other bifunctional coupling reagents (e.g., carbodiimides, Woodward's reagent K, 2-amino-4,6-dichlorotriazine) for the preparation of luminescent conjugates has also been described.

B. Bioluminescent Labels

Bacterial luciferase has been employed as a label for an antibody in an immunocytochemical assay. Coupling of luciferase to antibody was achieved by means of glutaraldehyde (Avrameas, 1969). p-Benzoquinone has also been advocated for this particular conjugation (Ternynck and Avrameas, 1976,1977). The luciferase conjugate was used to quantitate antigens on the surface of mouse lymphocytes, but results were not very reproducible and linearity was poor (Puget et al., 1977). Recently, firefly luciferase was successfully coupled to methotrexate and the conjugate used in an LIA for this drug. The methotrexate-luciferase conjugate retained 60% of its catalytic activity. A competitive LIA for methotrexate using this conjugate and antimethotrexate immobilized on Sepharose 4B beads was able to detect 2.5 pmol of methotrexate. Sensitivity was increased five-fold by using double-antibody technique (Wannlund and DeLuca, 1981; Wannlund et al., 1980).

C. Type of Assay

The majority of the luminescent immunoassays which have been described are heterogeneous, competitive protein-binding assays. This type of assay requires separation of bound and free labeled fractions, and is therefore time consuming. The time taken to perform an assay is an important consideration, especially in the analysis of clinically urgent analyses, so that homogeneous immunoassays, which do not require separation of bound and free labeled proteins, are preferable. The central feature of an homogeneous immunoassay is that the property of the label, e.g., enzymic activity, is altered when labeled antigen binds to an antibody or specific binding protein.

Two different types of homogeneous luminescent immunoassay have been described, one based on luminescence enhancement and the other on a sensitized chemiluminescent reaction.

Luminescence Enhancement

Luminescence enhancement relies on the enhancement of luminescence when a labeled antigen binds to an antibody or specific binding protein. An example is the luminescent competitive protein-binding assay for biotin (Schroeder et al., 1976). The peak light intensity of an isoluminol-biotin conjugate increased 10-fold when the conjugate was bound to avidin (a binding protein specific for biotin). The cause of the enhanced light output from the bound conjugate is not clear but it has been ascribed to increased chemiluminescence efficiency mediated by avidin.

A similar assay has been described for progesterone in which light production of an isoluminol-progesterone conjugate was enhanced when it was bound to antiprogesterone (Kohen et al., 1979) and, more recently, for other steroids (Kohen et al., 1981a,b; Pazzagli et al., 1981).

Sensitized Chemiluminescence

In the patent literature (Woodhead et al., 1978) an intriguing homogeneous immunoassay for cyclic AMP has been described which involves luminol-labeled anticyclic AMP and fluorescein-labeled cyclic AMP. Unbound luminol-labeled anticyclic AMP emitted light at 460 nm when reacted with peroxidase/hydrogen peroxide; however, when the antibody was bound to fluorescein-labeled cyclic AMP an efficient absorption of light by fluorescein occurred and the wavelength of the emitted light shifted to 540 nm. Thus cyclic AMP could be assayed by competition with fluorescein-labeled cyclic AMP for luminol-labeled anticyclic AMP, and the competition monitored by light emission at 460 or 540 nm.

V. LUMINESCENT ENZYME IMMUNOASSAY

Assay systems involving the use of enzyme-labeled antigens, antibodies, or haptens are termed enzyme immunoassays. Luminescent versions of conventional enzyme immunoassays (Scheme 7) in which the enzyme label is detected by luminescence more sensitively than by colorimetry have been described for assays employing peroxidase and glucose oxidase as labels. Examples of this type of assay are listed in Table 4.

Most of the luminescent enzyme immunoassays described have been based on peroxidase conjugates assayed with luminol/hydrogen

Table 4 Luminescent Enzyme Immunoassays

Labeled substance	Label	Analyte	Detection system	Sensitivity	Reference
Cortisol	Peroxidase	Cortisol	Luminol/H_2O_2	0.02-1 ng	Arakawa et al. (1977)
IgG (anti-SEB)	Peroxidase	Staphylococcal enterotoxin B	Pyrogallol/ H_2O_2	1 ng	Velan and Halmann (1978)
Anti-IgG	Peroxidase Glucose oxidase	Cell surface IgG	Pholad luciferin Glucose/luminol peroxidase	—	Puget et al. (1977)
	Peroxidase	Cortisol insulin		10 pg 1.2 μU/ml	Tsuji et al. (1978)
Goat anti-rabbit IgG	Peroxidase	Anti-HSA HSA	Luminol/H_2O_2	7×10^{-7} diln. of antiserum 10 ng	Olsson et al. (1979)
IgG antibody to the virus	Peroxidase	Sindbis virus	Pyrogallol/ H_2O_2	5×10^7 PFU/ml	Velan et al. (1979)
Goat anti-rabbit IgG	Pyruvate kinase	Goat anti-rabbit IgG	Firefly luciferase/ luciferin	—	Reichard and Miller (1981)

$$Ag + Ag\text{-}E + Ab \rightleftharpoons Ab{:}Ag + Ab{:}Ag\text{-}E$$

Separation of bound and free Ag-E

$$Ag\text{-}E_{\text{bound or free}} + \text{luminogenic coreactants} \longrightarrow \text{light}$$

Scheme 7 Luminescent enzyme immunoassay. E = enzyme, luminogenic coreactants = luminol/H_2O_2 when E = peroxidase.

peroxide, pyrogallol/hydrogen peroxide, *Pholas dactylus* luciferin, or luminol under alkaline conditions.

The latter assay is of particular interest because of the method of estimating the peroxidase. Instead of the enzymatic activity of peroxidase, the ability of the heme contained in peroxidase to catalyze a luminescent reaction is measured (Ewetz and Thore, 1976; Olsson et al., 1979). Thore and his co-workers have drawn attention to the possible use of other heme-containing compounds (e.g., free heme, cytochromes) as labels for luminescent immunoassays.

VI. LUMINESCENT ENZYME MULTIPLIED IMMUNOASSAY TECHNIQUE

The Enzyme Multiplied Immunoassay Technique (EMIT) is an homogeneous immunoassay which relies on the alteration of enzyme activity when an enzyme-labeled antigen binds to its specific antibody (Rubenstein et al., 1972; Bastiani et al., 1973). The principle of the

$$\underset{\text{(enzyme active)}}{\text{Drug-G6PDH}} + \text{Ab} + \text{Drug} \gtrless \text{Drug} \gtrless \underset{\text{(enzyme inactive)}}{\text{Ab:Drug-G6PDH}} + \text{Ab:Drug}$$

$$\text{Substrate} + \text{NAD} \xrightarrow{\text{Drug-G6PDH}} \text{NADH} + \text{product}$$

$$\text{NADH} + \text{FMN} \xrightarrow{\text{NADH:FMN oxidoreductase}} \text{NAD} + \text{FMNH}_2$$

$$\text{FMNH}_2 + \underset{\text{aldehyde}}{\text{long-chain}} \xrightarrow{\text{luciferase}} \text{FMN} + \text{RCO}_2\text{H} + \text{light}$$

Scheme 8 Sequence of events in a luminescent enzyme multiplied immunoassay.

assay is illustrated in Scheme 2. A biological fluid containing the antigen (usually a drug or other small molecule) is added to a specific antibody and binding of the antigen occurs. Enzyme-labeled antigen is then added and this competes with antigen for binding sites on the antibody and enzyme activity is proportionately reduced. The enzyme used as a label is glucose-6-phosphate dehydrogenase and its activity is measured by the rate of production of NADH at 340 nm.

Stanley (1978) described a luminescent version of this type of assay in which bioluminescence replaced the spectrophotometric detection of NADH in an EMIT assay for phenytoin. The bioluminescent reaction utilized bacterial luciferase and is shown in Scheme 8. The greater sensitivity of the bioluminescent indicator reaction for NADH allowed the assay to be scaled down by a factor of 900, thus offering a considerable saving in cost per test.

VII. LUMINESCENT COFACTOR IMMUNOASSAY

Recently, several workers investigated methods for monitoring specific binding reactions using ligand-cofactor conjugates quantitated via a bioluminescent reaction (Table 5). Both NAD and ATP conjugates have been employed and a notable feature of the immunoassays based on such conjugates is that the binding reactions can be monitored without separation of bound and unbound fractions (i.e., a homogeneous assay). A possible disadvantage, however, of luminescent cofactor immunoassays is the complexity of the chemical procedures required to link a cofactor to either an antigen or an antibody.

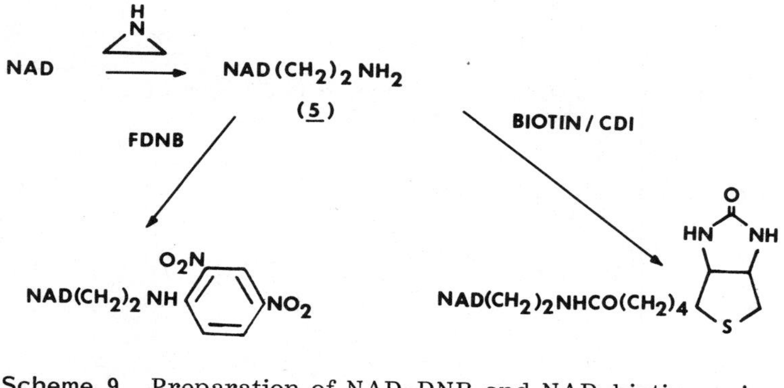

NAD → $NAD(CH_2)_2NH_2$ (5)
FDNB
BIOTIN / CDI
$NAD(CH_2)_2NH$ (O_2N, NO_2)
$NAD(CH_2)_2NHCO(CH_2)_4$
HN NH O S

Scheme 9 Preparation of NAD-DNB and NAD-biotin conjugates. CDI = 1-cyclohexyl-3-(2-morpholinoethyl)carbodiimide metho-p-toluene sulfonate; FDNB = fluorodinitrobenzene.

Table 5 Luminescent Cofactor Immunoassays

Labeled substance	Label	Analyte	Detection system	Sensitivity	Reference
Biotin 2,4-DNFB	NAD derivative	Biotin 2,4-DNB	EtOH/alcohol dehydrogenase bacterial luciferase	—	Schroeder et al. (1976)
2,4-DNFB	ATP derivative	N(2,4-DNP)-β-alanine	Luciferin/firefly luciferase	0.5 nmol/L	Carrico et al. (1976b)

Ag-CF + Ab + Ag $\gtrless$ Ab:Ag-CF + AB:Ag
(active) (cofactor inactive)

Ag-CF + luminogenic coreactants $\longrightarrow$ light

Scheme 10 Luminescent cofactor immunoassay. CF = NAD, luminogenic coreactants = alcohol/alcohol dehydrogenase which reduce Ag-NAD to Ag-NADH, this is then detected by reaction with NADH: FMN oxidoreductase/bacterial luciferase. CF = ATP, luminogenic coreactants = firefly luciferase/luciferin.

A. Nicotinamide Adenine Dinucleotide (NAD) Labels

The first stage in the preparation of a NAD conjugate is modification of NAD by reaction with ethylene imine to produce nicotinamide-6-(2-aminoethylamine)purine dinucleotide (5). Ligands, e.g., biotin or a 2,4-dinitrophenyl (DNB) residue, can then be attached via the reactive 2-amino group (Scheme 9). Generally, yields of conjugates are low (20 to 25%). Conjugates are assayed by means of a two-stage process. First the conjugate is reacted with alcohol/alcohol dehydrogenase, which reduces the NAD label to NADH. The reduced conjugate is then quantitated by means of a bioluminescent reaction involving bacterial luciferase (Scheme 10). The activity of conjugates in the initial reaction with alcohol dehydrogenase is decreased when they are bound to specific binding proteins, thus enabling the unbound fraction to be quantitated in the presence of the bound fraction.

B. Adenosine-5'-Triphosphate (ATP) Labels

Several different chemical procedures have been utilized to prepare conjugates of ATP and the 2,4-dinitrobenzene residue (Carrico et al., 1976b). Two examples of conjugates are shown in structures (6) and

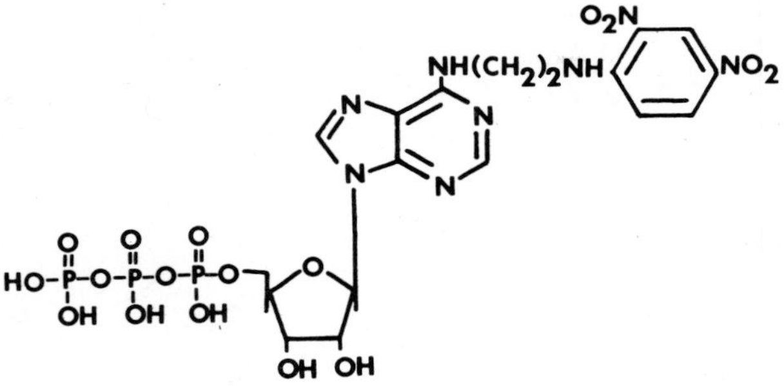

(6)

(7) and the chemical steps in the synthesis of (7) is presented in Scheme 11. A common feature of all three conjugates is the inclusion of a $-CH_2-CH_2-$ spacer group between the DNB residue and its point of attachment to the ATP label. The purpose of this spacer group was to minimize steric inhibition of the binding of the DNB residue to anti-DNB by ATP.

Both conjugates (6) and (7) could be quantitated at the nanomol/L level by the bioluminescent firefly reaction (Scheme 10). Conjugates were inactive in this reaction when bound to antibody and thus unbound conjugate could be quantitated in the presence of bound conjugate. The lower limit of detection in a model homogeneous luminescent cofactor immunoassay for N-(2,4-dinitrophenyl)-β-alanine employing a DNB-ATP conjugate was approximately 10 nmol/L (Carrico et al., 1978).

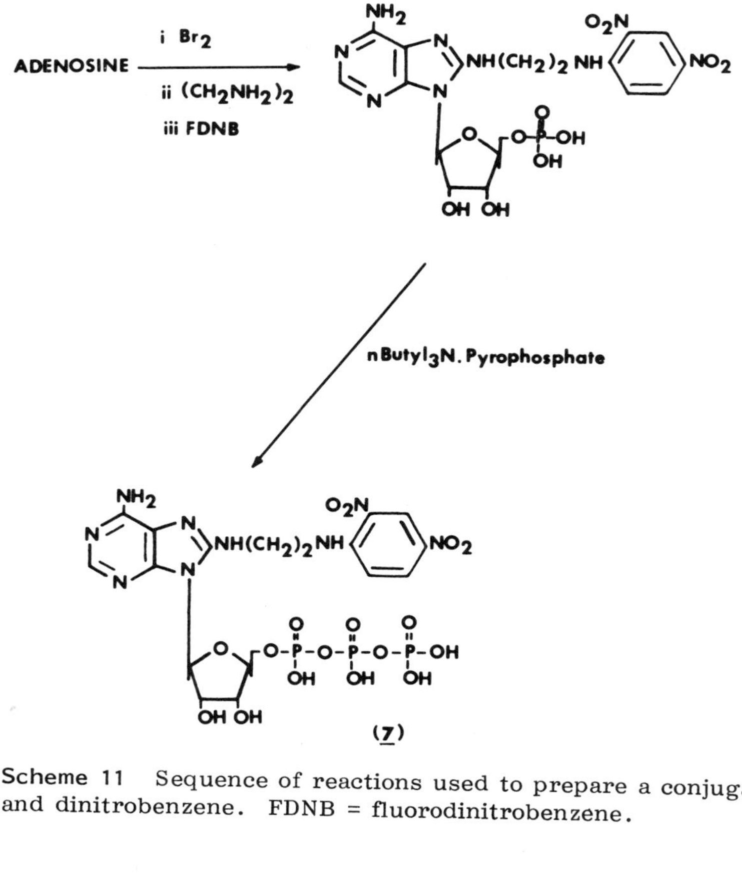
ADENOSINE
i Br2
ii (CH2NH2)2
iii FDNB
NH2
N
NH(CH2)2 NH
O2N
NO2
O-P-OH
OH
OH OH
nButyl3N. Pyrophosphate
O-P-O-P-O-P-OH
(7)

Scheme 11 Sequence of reactions used to prepare a conjugate of ATP and dinitrobenzene. FDNB = fluorodinitrobenzene.

VIII. CONCLUSION

The routine application of luminescent labels and luminescent reactions to the monitoring of immunological reactions is an exciting prospect for the future. It is, however, only one of many possible replacements for radioactive labels. Luminescent immunoassay is still at a very early stage of development and its future success will largely depend upon improvements in the quality of luminescent reagents and labeling techniques.

REFERENCES

Adler, F. L., and C.-T. Lau (1971). Detection of morphine by hemagglutination-inhibition. *J. Immunol.* *106*:1684-1685.

Arakawa, H., M. Maeda, and A. Tsuji (1977). Enzyme immunoassay of cortisol by chemiluminescence reaction of luminol-peroxidase. *Bunseki Kagaku* *26*:322-326.

Avrameas, S. (1969). Coupling of enzymes to proteins with glutaraldehyde. *Immunochemistry* *6*:43-50.

Bastiani, R. J., R. C. Phillips, R. S. Schneider and E. F. Ullman (1973). Homogeneous immunochemical drug assays. *Amer. J. Med. Tech.* *39*:211-216.

Cais, M., S. Dani, Y. Eden, O. Gandolfi, M. Horn, E. E. Isaacs, Y. Josephy, Y. Saar, E. Slivin, and L. Snarsky (1977). Metalloimmunoassay. *Nature* *270*:534-535.

Carrico, R. J., J. E. Christner, R. C. Boguslaski and K. K. Yeung (1976a). A method for monitoring specific binding reactions with cofactor labeled ligands. *Anal. Biochem.* *72*:271-282.

Carrico, R. J., R. D. Johnson, and R. C. Boguslaski (1978). ATP-labelled ligands and firefly luciferase for monitoring specific protein-binding reactions. *Meth. Enzymol.* *57*:113-122.

Carrico, R. J., K. K. Yeung, H. R. Schroeder, R. C. Boguslaski, R. T. Buckler, and J. E. Christner (1976b). Specific protein-binding reactions monitored with ligand-ATP conjugates and firefly luciferase. *Anal. Biochem.* *76*:95-110.

Cuatrecasas, P., and C. B. Anfinsen (1971). Affinity chromatography. *Ann. Rev. Biochem.* *40*:259-278.

Dandliker, W. B., and V. A. de Saussure (1970). Review article of fluorescence polarization in immunochemistry. *Immunochemistry* *7*:799-828.

Ewetz, L., and A. Thore (1976). Factors affecting the specificity of the luminol reaction with haematin compounds. *Anal. Biochem.* *71*:564-570.

Hersch, L. S., W. P. Vann, and S. A. Wilhelm (1978). A luminol-assisted competitive binding immunoassay of human IgG. *Proceedings Abstracts, Xth International Clinical Chemistry Congress*, Mexico.

Hersch, L. S., W. P. Vann, and S. A. Wilhelm (1979). A luminol-assisted competitive-binding immunoassay of human immunoglobulin G. *Anal. Biochem.* *93*:267-271.

Kato, K., Y. Hamaguchi, S. Okawa, E. Ishikawa, K. Kobayashi, and N. Katunuma (1977). Enzyme immunoassay in rapid progress. *Lancet* *1*:40.

Kennedy, J. H., L. J. Kricka, and P. Wilding (1976). Protein-protein coupling reactions and the applications of protein conjugates. *Clin. Chim. Acta* *70*:1-31.

Kohen, F., J. B. Kim, G. Barnard, and H. R. Lindner (1980a). An assay for urinary estriol-16α-glucuronide based on an antibody-enhanced chemiluminescence. *Steroids* *36*:405-419.

Kohen, F., J. B. Kim, and H. R. Lindner (1981a). Assay of gonadal steroids based on antibody-enhanced chemiluminescence. In *Bioluminescence and Chemiluminescence*, M. A. DeLuca and W. D. McElroy (Eds.). Academic Press, New York, pp. 357-364.

Kohen, F., J. B. Kim, H. R. Lindner, and G. Barnard (1981b). An immunoassay for urinary estriol-16α-glucuronide based on antibody-enhanced chemiluminescence. In *Bioluminescence and Chemiluminescence*, M. A. DeLuca and W. D. McElroy (Eds.). Academic Press, New York, pp. 351-356.

Kohen, F., M. Pazzagli, J. B. Kim, and H. R. Lindner (1980b). An immunoassay for plasma cortisol based on chemiluminescence. *Steroids* *36*:421-437.

Kohen, F., M. Pazzagli, J. B. Kim, H. R. Lindner, and R. C. Boguslaski (1979). An assay procedure for plasma progesterone based on antibody-enhanced chemiluminescence. *FEBS. Lett.* *104*:201-205.

Konishi, E., S. Iwasa, K. Kondo and M. Hori (1980). Chemiluminescence-linked immunoassay for detection of mumps virus antibodies. *J. Clin. Microbiol.* *12*:140-143.

Leute, R. L., E. F. Ullman, A. Goldstein, and L. A. Herzenberg (1972). Spin immunoassay technique for determination of morphine. *Nature New Biol.* *236*:93-94.

Maier, C. (1977). Luminescent conjugates for immunological analysis: Comprising a complex formed between a pharmacologically, immunologically or biochemically active ligand and a chemiluminescent compound. Belgian Patent 856,182 (1977).

Maier, C. L. (1978). Procedure for the assay of pharmacologically, immunologically and biochemically active compounds in biological fluids. U.S. Patent 4,104,029.

Makela, O. (1966). Assay of anti-hapten antibody with the aid of hapten-coupled bacteriophage. *Immunology* *10*:81-86.

Nakane, R. K., and A. Kawaoi (1974). Peroxidase labeled antibody, a new method of conjugation. *J. Histochem. Cytochem.* *22*:1084-1091.

Olsson, T., A. Thore, H. E. Carlsson, G. Brunius, and G. Eriksson (1978). Quantitation of immunological reactions by luminescence. In *Proceedings of the International Symposium on Analytical Applications of Bio- and Chemiluminescence*, Brussels, 1978, E. Schram and P. E. Stanley (Eds.), State Printing and Publishing, Westlake Village, Ca.

Olsson, T., G. Brunius, H. E. Carlsson, and A. Thore (1979). Luminescence immunoassay (LIA): A solid-phase immunoassay monitored by chemiluminescence. *J. Immunol. Meth.* *25*:127-135.

Pazzagli, M., F. Kohen, J. B. Kim, and H. R. Lindner (1981). An immunoassay for plasma cortisol based on chemiluminescence. In *Bioluminescence and Chemiluminescence*, M. A. DeLuca and W. D. McElroy (Eds.). Academic Press, New York, pp. 651-657.

Pratt, J. J., M. G. Woldring, and L. Villerius (1978). Chemiluminescence-linked immunoassay. *J. Immunol. Meth.* *21*:179-184.

Puget, K., A. M. Michelson, and S. Avrameas (1977). Light emission techniques for the microestimation of femtogram levels of peroxidase: Application to peroxidase (and other enzymes)—Coupled antibody cell antigen interactions. *Anal. Biochem.* *79*:447-456.

Reichard, D. W., and R. J. Miller, Jr. (1981). Bioluminescent immunoassay: A new enzyme-linked analytical method for the quantitation of antigens. In *Bioluminescence and Chemiluminescence*, M. A. DeLuca and W. D. McElroy (Eds.). Academic Press, New York, pp. 667-672.

Richards, F. M., and J. R. Knowles (1968). Glutaraldehyde as a protein cross-linking reagent. *J. Mol. Biol.* *37*:231-233.

Rubenstein, K. E., R. S. Schneider, and E. F. Ullman (1972). Homogeneous enzyme immunoassay: A new immunochemical technique. *Biochem. Biophys. Res. Commun.* *47*:846-851.

Scharpe, S. L., W. M. Cooreman, W. J. Bloome, and G. M. Laekeman (1976). Quantitative enzyme immunoassay: Current status *Clin. Chem.* *22*:733-738.

Schroeder, H. R., and F. M. Yeager (1978). Chemiluminescence yields and detection limits of some iso-luminol derivatives in various oxidation systems. *Anal. Chem.* *50*:1114-1120.

Schroeder, H. R., R. J. Carrico, R. C. Boguslaski, and J. E. Christner (1976). Specific binding reactions monitored with ligand-cofactor conjugates and bacterial luciferase. *Anal. Biochem.* *72*:283-292.

Schroeder, H. R., P. O. Vogelhut, R. J. Carrico, and R. C. Boguslaski, and R. T. Buckler (1976). Competitive protein binding assay for biotin monitored by chemiluminescence. *Anal. Chem.* *48*:1933-1937.

Schroeder, H. R., F. M. Yeager, R. C. Boguslaski, E. O. Snoke and R. T. Buckler (1977). Immunoassay for thyroxine monitored by chemiluminescence. *Clin. Chem.* *23*:1132 Abstr.

Schroeder, H. R., R. C. Boguslaski, R. J. Carrico and R. T. Buckler (1978). Monitoring specific protein-binding reactions with chemiluminescence *Meth. Enzymol.* *57*:424-445.

Schroeder, H. R., F. M. Yeager, R. C. Boguslaski and P. O. Vogelhut (1979). Immunoassay for serum thyroxine monitored by chemiluminescence. *J. Immunol. Meth.* *25*:275-282.

Simpson, J. S. A., A. K. Campbell, M. E. T. Ryall, and J. S. Woodhead (1979). A stable chemiluminescent-labelled antibody for immunological assays. *Nature (Lond.) 279*:646-647.

Soini, E., and I. Hemmila (1979). Fluoroimmunoassay: Present status and key problems. *Clin. Chem. 25*:353-361.

Stanley, P. E. (1974). Analytical bioluminescent assays using the liquid scintillation spectrometer: A review. In *Liquid Scintillation Counting*, M. A. Crook and P. Johnson (Eds.), Heyden and Son, London, pp. 253-271.

Stanley, P. E. (1978). Applications of analytical bioluminescence to the enzyme multiplied enzyme immunoassay technique. In *Liquid Scintillation Counting*, Vol. 5, P. Johnson and M. Crook (Eds.), Heyden and Son, London, p. 79.

Ternynck, T., and S. Avrameas (1976). A new method using p-benzoquinone for coupling antigens and antibodies to marker substances. *Ann. Immunol. (Inst. Pasteur) 127C*:197-208.

Ternynck, T., and S. Avrameas (1977). Conjugation of p-benzoquinone treated enzymes with antibodies and Fab fragments. *Immunochemistry 14*:767-774.

Tsuji, A., M. Maeda, H. Arakawa, K. Matsuoka, M. Kato, H. Naruse, and M. Irie (1978). Enzyme immunoassay of hormones and drugs by using fluorescence and chemiluminescence reaction. *Proc. International Symposium on Enzyme Labelled Immunoassay of Hormones and Drugs*, S. B. Pal (Ed.). Walter de Gruyter.

Ullman, E. F., M. Schwarzberg, and K. E. Rubenstein (1976). Fluorescent excitation transfer immunoassay. *J. Biol. Chem. 251*:4172-4178.

Velan, B., and M. Halmann (1978). Chemiluminescence immunoassay, a new sensitive method for determination of antigens. *Immunochemistry 15*:331-333.

Velan, B., H. Schupper, T. Sery, and M. Halmann (1978). Solid-phase chemiluminescent immunoassay. In *Proceedings of the International Symposium on Analytical Applications of Bio- and Chemiluminescence*, Brussels, 1978, E. Schram and P. E. Stanley (Eds.). State Printing and Publishing, Westlake Village, Ca.

Wannlund, J., and M. DeLuca (1981). Bioluminescent immunoassays: Use of luciferase-antigen conjugates for determination of methotrexate and DNP. In *Bioluminescence and Chemiluminescence*, M. A. DeLuca and W. D. McElroy (Eds.). Academic Press, New York, pp. 693-696.

Wannlund, J., J. Azari, L. Levine, and M. DeLuca (1980). A bioluminescent immunoassay for methotrexate at the subpicolmole level. *Biochem. Biophys. Res. Commun. 96*:440-446.

Watson, R. A. A., J. Landon, E. J. Shaw, and D. S. Smith (1976). Polarization fluoroimmunoassay of gentamicin. *Clin. Chim. Acta 73*:51-55.

Wisdom, G. B. (1976). Enzyme-immunoassay. *Clin. Chem. 22*:1243-1255.

Woodhead, J. S., J.S.A. Simpson, and A. K. Campbell (1978). Detecting or quantifying substances using labelling techniques. U.K. Patent Application, G.B. 2,008,247A.
Woodhead, J. S., J. S. A. Simpson, M. E. T. Ryall, and A. K. Campbell (1979). *The Preparation and Properties of Chemiluminescent-Labelled Antibodies.* Abstracts Third European Congress of Clinical Chemistry, Brighton, 1979, p. 24.
Yalow, R. S. (1978). Heterogeneity of peptide hormones. Its relevance to clinical radioimmunoassay. *Adv. Clin. Chem. 20*:1-47.
Yalow, R. S., and S. A. Berson (1959). Assay of plasma insulin in human subjects by immunological methods. *Nature 184*:1648-1649.

9

IMMOBILIZED LUMINESCENCE REAGENTS

LARRY J. KRICKA* University of Birmingham, Birmingham, England

TIMOTHY J. N. CARTER† Wolfson Research Laboratories, Queen Elizabeth Medical Centre, Edgbaston, Birmingham, England

I.	Introduction	180
II.	Immobilization Techniques	180
	A. Physical	180
	B. Chemical	181
III.	Solid Supports	182
IV.	Choice of Immobilization Method	182
V.	Immobilized Bacterial Luciferase	183
VI.	Immobilized Firefly Luciferase	183
VII.	Luciferase Coimmobilized with Other Enzymes	186
VIII.	Conclusion	186
	References	186

*Dr. Kricka is currently with the University of California at San Diego, La Jolla, California, on a leave of absence from the University of Birmingham.

†Dr. Carter is currently with Battelle, Centre de Recherche de Genève, Carouge, Genève, Switzerland.

I. INTRODUCTION

Immobilized proteins are widely used in the clinical laboratory, e.g., immobilized antigens and antibodies in immunoassay (Catt and Tregear, 1967; Wide et al., 1967; Line and Becher, 1975) and immobilized enzymes in assays for analytes such as glucose and uric acid (Bowers and Carr, 1976; Gray et al., 1977; Zaborsky, 1973). The main advantages of immobilized proteins over their native counterparts are

1. *Economy*: Immobilized proteins are recoverable and thus reusable.
2. *Convenience*: Immobilization of a protein, e.g., an antibody, facilitates the separation of the protein from other components in solution.
3. *Novelty*: Immobilization of a protein creates opportunities for novel analytical manipulations of a protein and a detection system, e.g., the enzyme electrode (Gough and Andrade, 1973; Kessler et al., 1976), thermal enzyme probes (Weaver et al., 1976), and immunoelectrodes (Boitieux et al., 1979). Luminescent reagents immobilized onto insoluble solid supports have not as yet found extensive application in the clinical laboratory. The success, however, of immobilized enzymes and antibodies as reagents elsewhere suggests that immobilized luminescent reagents, e.g., luciferase, will eventually play a major role in the assay of clinically important analytes.

The purpose of this chapter is to review briefly such immobilization techniques and the progress to date on immobilized luminescent reagents.

II. IMMOBILIZATION TECHNIQUES

Numerous procedures for immobilizing a protein onto an insoluble solid support or matrix have been described (Zaborsky, 1973). These may be classified under the headings of physical and chemical immobilization procedures.

A. Physical

This is the simplest form of immobilization procedure and relies on physical forces (hydrophobic or hydrophilic interactions) between a protein and a solid support. A disadvantage of this procedure is that it may be reversible and desorption of protein due to changes in temperature, pH or ionic strength may occur due to the weak nature of the bonds involved (Barker and Kay, 1975).

Inclusion of proteins within the matrix of a polymer and entrapment of proteins between semipermeable membranes are also forms of physical immobilization.

B. Chemical

Covalent attachment of proteins via amino acid residues to water-insoluble functionalized solid supports is the commonest method of immobilizing proteins. Functional groups on proteins suitable for covalent bond formation with a solid support include terminal amino groups and amino groups of lysine and arginine residues, terminal carboxyl groups and carboxyl groups of aspartate and glutamate residues, the phenolic ring of tyrosine residues, the thiol group of cysteine residues, and the hydroxyl groups of serine and threonine residues.

Attachment may be achieved either directly, by reaction of a reactive chemical group on the protein and a group on the solid support, or indirectly, using a bifunctional reagent (Fig. 1). Table 1 lists some of the more commonly used coupling reagents. The chemistry of bifunctional coupling reagents has been reviewed by Kennedy et al. (1976).

Chemical immobilization of proteins has also been achieved by intermolecular cross-linking of the protein to other proteins using bifunctional coupling reagents.

Table 1 Bifunctional Coupling Reagents

Diisocyanates
toluene-2,4-diisocyanate, m-xylylene diisocyanate
Dialdehydes
glutaraldehyde
Halonitrobenzenes
1,5-difluoro-2,4-dinitrobenzene, 4,4'-difluoro-3,3'-dinitrophenyl-sulfone
Carbodiimides
1-ethyl-3-(3-dimethylaminopropyl)carbodiimide
Bisdiazonium salts
diazonium salts of 3,3'-dimethoxybenzidine, p-phenylene diamine
Isoxazoles
N-ethyl-3-phenylisoxazolium-3'-sulfonate (Woodward's reagent K)
Imidoesters
dimethylsuberimidate

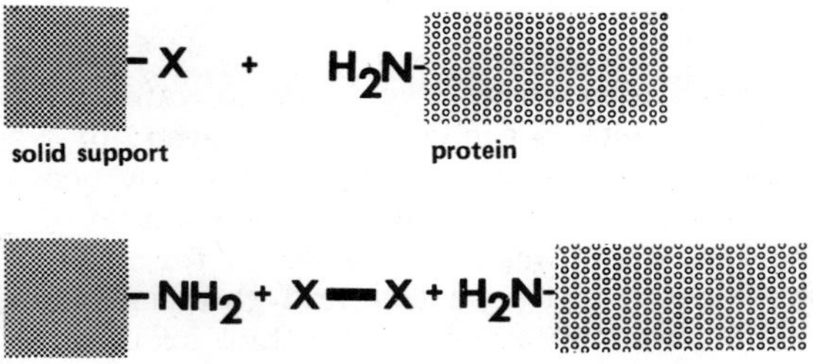

–X + H_2N–

solid support protein

–NH_2 + X–X + H_2N–

Figure 1 Direct (top) and indirect (bottom) chemical immobilization of protein onto a solid support.

III. SOLID SUPPORTS

A multitude of water-insoluble solid supports have been utilized for the immobilization of proteins. Supports can be organic or inorganic, natural (e.g., cellulose, charcoal, silica) or synthetic substances (e.g., polyacrylamide, polystyrene). More detailed information on the range and properties of solid supports is to be found in the monograph by Zaborsky (1973).

IV. CHOICE OF IMMOBILIZATION METHOD

It is most important when immobilizing a protein to preserve as much of its biological activity as possible, usually the enzymic or immunological activity. It follows that a knowledge of the residues important in maintaining the biological activity of a protein will enable methods to be chosen which avoid or minimize loss of such activity. Loss of activity during immobilization can occur in a number of ways:

1. Reaction of the solid support with residues in the biologically active site(s)
2. Secondary conformational changes as a result of reaction with part of the protein distant from the active site
3. Steric hindrance by the solid support and/or adjacent immobilized moieties
4. Microenvironmental effects

Active sites in a protein may be protected during attachment; of course, it is essential that the protective groups can be removed without loss of activity (Goldstein et al., 1970). It is difficult to control attachment such that loss of activity due to secondary conformational changes is minimized. Steric hindrance by the solid support, however, can be minimized by the use of "spacer" groups, e.g., hexamethylene groups (Cuatrecasas, 1971; Cuatrecasas and Anfinsen, 1971). The surface of a solid support provides a new physicochemical environment in the immediate vicinity of the protein ("microenvironment") and this may influence its biological activity. For example, the anionic or cationic nature of the surface of a solid support has been found to cause displacement in the pH optimum of an enzyme by up to 2 pH units and this is usually accompanied by a general broadening of the pH region in which the enzyme can work efficiently (Engasser and Horvath, 1976). In some cases it has proved possible to manipulate pH optima and stability by the introduction of charged groups (Carter et al., 1979) and stabilizing groups (e.g., thiol groups), respectively (Kricka and Carter, 1977).

V. IMMOBILIZED BACTERIAL LUCIFERASE

Covalent attachment of bacterial luciferase to glass has been achieved via a diazotization reaction (Table 2). During this process the enzyme suffers considerable loss of activity (Haggerty et al., 1978; Jablonski and DeLuca, 1976,1978). This may be due in part to interference either by the coupling reagent or the coupling conditions in the association of the luciferase α and β subunits. The pH activity profile and optimal substrate concentration, however, are very similar to those of the native enzyme. The immobilized luciferase is reported to be stable indefinitely when stored at 0°C (see also Chap. 5).

Physical immobilization of bacterial luciferase (*Vibrio fischeri*) has also been attempted using controlled pore ceramics (mesh size 80/120), e.g., silica, titania, and alumina. However, considerable loss of activity of bound enzyme was encountered.

VI. IMMOBILIZED FIREFLY LUCIFERASE

The direct covalent attachment of firefly luciferase onto glass beads via a diazonium-coupling reaction has been reported but the bound enzyme retained only 0.07 to 0.16% of its original activity. The immobilized luciferase had a lower pH optimum than the native enzyme and it emitted light with a major peak at 615 nm, as compared with 562 nm for the native enzyme (Lee et al., 1977). Recently, luciferase immobilized into Sepharose beads and cellophane film has been prepared and shown to retain as much as 20% of its initial activity (Ugarova et al., 1980). On storage the Sepharose 4B preparation lost 40% of its activity after 1 month at 4°C while the luciferase immobilized on cellophane film lost 50% of its initial activity after only 5 days at 4°C. Improvements in the yield and stability of immobilized luciferase can be achieved using cyanogen bromide activated Sepharose 4B and Sepharose CL 6B.

Table 2 Immobilized Luminescent Reagents

Immobilized enzyme(s) (source)	Solid support	Mode of attachment	Analysis	Sensitivity or range of optimal sensitivity	Reference
Luciferase (firefly)	Aminoalkylated glass beads	Glutaraldehyde	ATP	1×10^{-8} -1×10^{-5} mol/L	Lee et al. (1977)
Luciferase (*Photobacterium Fischeri* or *leiognathi*)	Polyacrylic hydrazide	Diazo coupling	$FMNH_2$		Erlanger et al. (1970)
Luciferase/oxidoreductase (*Beneckea harveyi*)	Porous glass	Diazo coupling	NADH	1 pmol-50 nmol	Jablonski and DeLuca (1976)
			NADPH	10 pmol-200 nmol	
Luciferase/oxidoreductase (*Beneckea harveyi*)	Porous glass	Diazo coupling	Malate dehydrogenase	0.007-0.70 pmol	Haggerty et al. (1978)
			Lactate dehydrogenase	0.003-0.70 pmol	
			Alcohol dehydrogenase	0.015-3.0 pmol	
			Glucose-6-phosphate dehydrogenase	0.0015-0.10 pmol	
			Hexokinase	0.1-2.0 pmol	
			Glucose	150-15,000 pmol	

Luciferase/oxido-reductase/alcohol dehydro-genase	Porous glass	Diazo coupling	Ethanol	0.004-0.015%	
Luciferase/oxido-reductase (*Beneckea harveyi*)/glucose-6-phosphate de-hydrogenase	Porous glass	Diazo coupling	Glucose-6-phosphate	1 pmol	Jablonski and DeLuca (1979)
Luciferase/oxido-reductase/glucose-6-phosphate dehy-drogenase/hexo-kinase			Glucose	20 pmol/L	
Luciferase/oxido-reductase/3α and 3β, 17β-hydroxysteroid dehydrogenase	Sepharose 4B	Cyanogen bromide	Testosterone Androsterone	0.8 pmol 0.3 pmol	Ford and DeLuca (1981)
Luciferase (firefly)	Sepharose 4B		ATP ATPase Pyruvate kinase	0.2 fmol	Ugarova et al. (1980)
	Cellophane film		ATP		

VII. LUCIFERASE COIMMOBILIZED WITH OTHER ENZYMES

Bacterial luciferase has been coimmobilized with NAD(P)H/FMN oxidoreductase onto porous arylamine beads glued onto a glass rod, and the immobilized enzymes have been used to assay NADH and NADPH (Jablonski and DeLuca, 1976), malate dehydrogenase, lactate dehydrogenase, alcohol dehydrogenase, glucose-6-phosphate dehydrogenase, and hexokinase (Jablonski and DeLuca, 1978). A further development of this work has been the coimmobilization of this pair of enzymes with alcohol dehydrogenase and the application of such immobilized enzymes to the assay of alcohol (Jablonski and DeLuca, 1978) (see also Chap. 5). This work has now been extended to include the coimmobilization of luciferase/oxidoreductase with glucose-6-phosphate dehydrogenase, hexokinase, and hydroxysteroid dehydrogenase (Jablonski and DeLuca, 1979). Recently, Analytical Luminescence Laboratories of California announced the first commercially available luminescence immobilized enzyme test systems (LIETS). These consist of either bacterial or firefly luciferase coimmobilized with up to four enzymes on a Sepharose bead, and may be used for the assay of clinically important compounds, e.g., alcohol, glucose, and testosterone.

VIII. CONCLUSION

Immobilized luminescence reagents are still in their infancy. The high cost of bacterial and firefly luciferase makes immobilization of these enzymes an attractive proposition, because immobilized luciferase is recoverable and reusable. As yet, few laboratories have investigated the analytical possibilities of immobilized luciferase and luciferase coimmobilized with other enzymes. However, the advent of commerically available immobilized luciferases will facilitate the wider use of this technique and stimulate its application to the analysis of clinically important analytes.

REFERENCES

Barker, S. A., and I. Kay (1975). Principles of immobilized-enzyme technology. In *Handbook of Enzyme Biotechnology*, A Wiseman (Ed.). Halsted, London, pp. 89-110.

Boitieux, J.-L., G. Desmet, and D. Thomas (1979). An antibody electrode, preliminary report on a new approach in enzyme immunoassay. *Clin. Chem.* *25*:318-321.

Bowers, L. D., and P. W. Carr (1976). Immobilized enzymes in analytical chemistry. *Anal. Chem.* *48*:544A-558A.

Carter, T. J. N., L. J. Kricka, and B. R. Stanbridge (1979). Immobilized enzymes as laboratory reagents: Modifications of the pH optimum of immobilized glucose oxidase. *Protides Biol. Fluids 26*:661-664.

Catt, K. J., and G. W. Tregear (1967). Solid-phase radioimmunoassay in antibody-coated tubes. *Science 158*:1570-1572.

Cuatrecasas, P. (1971). Selective adsorbents based on biochemical specificity. In *Biochemical Aspects of Reactions on Solid Supports*, G. R. Stark (Ed.). Academic, New York, pp. 79-109.

Cuatrecasas, P., and C. B. Anfinsen (1971). Affinity chromatography. *Ann. Rev. Biochem. 40*:259-278.

Engasser, J.-M., and C. Horvath (1976). Diffusion and kinetics with immobilized enzymes. *Appl. Biochem. Bioeng. 1*:27-220.

Erlanger, B. F., M. F. Isambert, and A. M. Michelson (1970). Insoluble bacterial luciferases: A new approach to some problems in bioluminescence. *Biochem. Biophys. Res. Commun. 40*:70-76.

Ford, J., and M. DeLuca (1981). A new assay for picomole levers of androsterone and testosterone using coimmobilized luciferase, oxidoreductase, and steroid dehydrogenase. *Anal. Biochem. 110*:43-48.

Goldstein, L., M. Pecht, S. Blumberg, D. Altas, and Y. Levin (1970). Water insoluble enzymes. Synthesis of a new carrier and its utilization for preparation of insoluble derivatives of papain, trypsin, and subtilopeptidase A. *Biochemistry 14*:61-89.

Gough, D. A., and J. D. Andrade (1973). Enzyme electrodes. *Science 180*:380-384.

Gray, D. N., M. H. Keyes, and B. Watson (1977). Immobilized enzymes in analytical chemistry. *Anal. Chem. 49*:1067A-1078A.

Haggerty, C., E. Jablonski, L. Stav, and M. DeLuca (1978). Continuous monitoring of reactions that produce NADH and NADPH using immobilised luciferase and oxidoreductases from *Beneckea harveyi*. *Anal. Biochem. 88*:162-173.

Jablonski, E., and M. DeLuca (1976). Immobilisation of bacterial luciferase and FMN reductase on glass rods. *Proc. Natl. Acad. Sci. USA 73*:3848-3851.

Jablonski, E., and M. DeLuca (1978). Immobilization of bacterial luciferase and oxidoreductase and assays using immobilized enzymes. *Meth. Enzymol. 57*:202-214.

Jablonski, E., and M. DeLuca (1979). Properties and uses of immobilized light-emitting enzyme systems from *Beneckea harveyi Clin. Chem. 25*:1622-1627.

Kennedy, J. H., L. J. Kricka, and P. Wilding (1976). Protein-protein coupling reactions and the applications of protein conjugates. *Clin. Chim. Acta 70*:1-31.

Kessler, M., L. C. Clark, D. W. Lubbers, I. A. Silver and W. Siman (Eds.) (1976). *Ion and Enzyme Electrodes in Biology and Medicine*, Urban and Schwarzenberg, Munchen.

Kricka, L. J., and T. J. N. Carter (1977). A method for enhancing the stability of thiol-containing enzymes immobilised on nylon. *Clin. Chim. Acta 79*:141-147.

Lee, Y., E. Jablonski, and M. DeLuca (1977). Immobilization of firefly luciferase on glass rods: Properties of the immobilised enzyme. *Anal. Biochem. 80*:496-501.

Line, W. F., and M. J. Becher (1975). Solid phase immunoassay. In *Immobilized Enzymes, Antigens, Antibodies and Peptides*, H. H. Weetall (Ed.). Marcel Dekker, New York, pp. 497-566.

Ugarova, N. N., L. Yu. Brovko, and I. V. Berezin (1980). Immobilised firefly luciferase and its use in analysis. *Anal. Lett. 13*:881-892.

Weaver, J. C., C. L. Cooney, S. P. Fulton, P. Schuler, and S. R. Tannenbaum (1976). Experiments and calculations concerning a thermal enzyme probe. *Biochim. Biophys. Acta 52*:285-291.

Wide, L., R. Axen, and J. Porath (1967). Radioimmunosorbent assay for proteins; chemical coupling of antibodies to insoluble dextran. *Immunochemistry 4*:381-386.

Zaborsky, O. R. (1973). *Immobilized Enzymes*, CRC, Cleveland, Ohio.

10

APPLICATIONS OF LUMINESCENCE IN MEDICAL MICROBIOLOGY AND HEMATOLOGY

MICHAEL J. HARBER K.R.U.F. Institute, Royal Infirmary, Welsh National School of Medicine, Cardiff, Wales

I.	Introduction	190
	A. Rationale for Bioluminescence Analysis	190
	B. Absolute Cellular ATP Levels	192
II.	Measurement of Bacterial Adherence	192
	A. Polystyrene Tube Test	192
	B. Viable Mass in Dental Plaque	193
III.	Screening Test for Bacteriuria	195
	A. Principle	195
	B. Problems	195
	C. Comparative Assessment	196
IV.	Antimicrobial Susceptibility Testing	197
	A. Basic Procedure	197
	B. Applicability	198
V.	Antibiotic Assay	199
	A. Principle	199
	B. Basic Assay Procedure	199
	C. Practical Application	201
	D. Comparative Assessment	203
VI.	Vitamin Assay	203
VII.	Luminescence Analyses of Blood Cells	205
	A. Erythrocytes	205
	B. Platelet Viability and Function	206
	C. Leukocyte Function	207

I. INTRODUCTION

A. Rationale for Bioluminescence Analysis

The vast majority of conventional microbiological procedures depend on estimation of bacterial numbers in clinical specimens, assessment of microbial growth in liquid culture, or measurement of inhibition zone diameters in agar diffusion assays. All of the methods in current use are subject to many sources of error and, in addition, generally have severe limitations, such as a requirement for a lengthy incubation period. Viable count procedures, for instance, are based on the assumption that one colony originates from only one bacterial cell, but in practice this ideal is rarely likely to be correct. The results of agar diffusion assays or viable count studies may be influenced by the nature of the specimen under test, the presence of drugs, the nature of any diluent used, and the constituents of the growth medium. Microbial growth in broth culture is commonly assessed turbidimetrically and results of this analysis can be affected by such factors as cell size, shape, or age; clumping and sedimentation of cells; and even by the carbohydrate or lipid content of the medium. Furthermore, metabolic death may go undetected if cell lysis does not occur. Other methods for measuring bacterial growth based on changes in pH or color using redox dyes are generally no more reliable than turbidimetry, and all of these procedures are limited to a sensitivity in the order of 10^7 bacterial cells per ml.

Clearly, more accurate, rapid, and sensitive methods for measuring microbial growth and cell numbers are desirable. A logical line of approach is to use a chemical assay to measure an essential intracellular metabolite which, under defined conditions of culture, will reflect the number of actively metabolizing viable cells. Adenosine triphosphate (ATP) is an obvious candidate for such a system since universally it plays an essential role in cell metabolism, is present in high concentrations compared with other metabolites, is uniformly distributed in the protoplasm of microorganisms from where it may be readily extracted, and—due to its high rate of turnover and rapid loss from dead cells—should provide a good index of cell viability. A number of studies, in fact, have indicated that bacterial levels of ATP may be correlated directly with cell numbers (Hamilton and Holm-Hansen, 1967; Chappelle and Levin, 1968; D'Eustachio and Johnson, 1968).

Spectrophotometric analysis is insufficiently sensitive to be used to measure cellular ATP levels, and enzymatic cycling procedures are too cumbersome for routine applications. However, the sensitivity, simplicity, and rapidity of the firefly bioluminescence system makes this technique ideally suited for measuring low concentrations of ATP and therefore for assessing bacterial growth or cell numbers in microbiological tests. Using modern reagents and commercially available instruments it is possible to detect as few as 1000 bacterial cells per ml using the firefly system, and further improvements in sensitivity are anticipated in the future. Firefly bioluminescence may also be used as a sensitive indicator of the metabolic activity of blood cells, and although the relationship between cellular ATP and viability is less certain with other somatic cell types (Farber, 1973), there is some evidence to suggest that the concentration of ATP in blood plasma may reflect the metabolic status of peripheral tissues.

Table 1 Bacterial Cellular ATP Levels Reported in the Literature

Bacteria	Range of cellular ATP (pmol/10^6 cells)	Reference
7 Species of marine bacteria (exponential phase)	1.0-11.5	Hamilton and Holm-Hansen (1967)
19 Bacterial species (? stationary phase)	0.05-1.8	Chappelle and Levin (1968)
13 Bacterial species (? stationary phase)	0.4-2.1	D'Eustachio and Johnson (1968)
6 Urinary pathogens (stationary phase)	0.25-1.1	Picciolo et al. (1975)
4 Bacterial species (? stationary phase)	0.9-2.3	Thore et al. (1975)
3 Bacterial species:		
a) stationary phase	0.3-2.8	Harber and Asscher (1979)
b) exponential phase	1.2-7.0	

It is the object of this chapter to describe the microbiological and hematological applications of luminescence methodology which have been investigated to date, and to discuss the potential clinical usefulness of these procedures in relation to the alternative techniques currently available.

B. Absolute Cellular ATP Levels

A wide range of bacterial ATP levels has been reported in the literature (Table 1). Care must be exercised whenever absolute cellular ATP levels are quoted because the level will vary not only between different species of bacteria, but also within any particular species depending on the constituents of the growth medium (Franzen and Binkley, 1961; Cole et al., 1967; Bush et al., 1975), the oxygen tension (Strange et al., 1963; Cole et al., 1967), and the growth phase of the organisms at the time of sampling (Bush et al., 1975; Harber and Asscher, 1979).

For the purposes of most clinical microbiological tests absolute ATP values are not required, results being calculated with reference to appropriate bacteriological standards. However, absolute ATP levels do need to be determined whenever ATP per se is the substance of interest, as is the case with studies involving blood cells and plasma. Purified reagents are necessary for the assay of ATP for these applications, and careful standardization is required to negate the effects of inhibitory/quenching substances which may be present in clinical specimens.

II. MEASUREMENT OF BACTERIAL ADHERENCE

A. Polystyrene Tube Test

The adherence of bacteria to mucosal surfaces has been implicated as a virulence factor in a variety of infectious diseases including urinary tract infections, respiratory infections, bacterial endocarditis, gastrointestinal infections, and gonococcal venereal disease. Attachment of gram-negative bacteria is generally mediated by protein fibrils termed fimbriae, although synthesis of these structures is dependent on cultural conditions and may not always occur in vivo (Harber et al., 1982a). In vitro studies on bacterial adhesion mostly utilize isolated epithelial cells as the attachment surface, but there is considerable uncertainty whether sloughed-off cells are representative of intact epithelia, and the presence of mucoid on the surface of uroepithelial cells can present a practical problem by trapping some organisms and not others (Chick et al., 1981). This latter difficulty inevitably introduces inaccuracies in the quantitation of cell-adherent bacteria using radiolabeled cultures, while the alternative detection method involving microscopic examination of epithelial cell preparations is tedious, time-consuming, and subject to many sources of error.

An in vitro method for quantifying adherence has recently been developed in which polystyrene tubes are used as the attachment surface, and firefly bioluminescence ATP analysis is used as a measure of the number of adherent bacteria (Harber et al., 1982b; Harber et al., submitted). Broth cultures are diluted in phosphate-buffered saline pH 6.8, and 300 μl aliquots are introduced into small polystyrene cuvettes designated for use with the LKB luminometer 1250. After 10 min incubation at 37°C the bacterial suspensions are discarded and the tubes washed to remove nonadherent bacteria. ATP is then extracted from the adherent organisms remaining in the tubes using a commercial nucleotide releasing reagent for bacteria (Lumac NRB reagent), measured using firefly bioluminescence, and results are related to the amount of extractable ATP present in the original bacterial suspension to obtain an adherence ratio.

A 200-fold variation was observed in the adherence ratios obtained with 34 urinary isolates of *Escherichia coli* (Fig. 1), and the adherence was virtually abolished on addition of 2.5% D-mannose to the incubation buffer. This indicated that the attachment was mediated by mannose-sensitive (MS) type 1 fimbriae, although flagella were also found to be an essential requirement. A significant correlation was observed between adherence of the uropathogens to polystyrene and their ability to adhere to uromucoid.

Bacterial adherence mediated by MS fimbriae is of potential clinical significance in two respects. First, MS fimbriate bacteria are known to be trapped by uromucoid (Chick et al., 1981), and second, MS fimbriae facilitate attachment of bacteria to professional phagocytes in the absence of antibody and complement (Silverblatt et al., 1979). The polystyrene tube test may therefore give an indication of the susceptibility of pathogens to these host defense mechanisms. Furthermore, the basic method could possibly be applied to the study of bacterial adhesion using certain biological as well as nonbiological surfaces, or with polystyrene coated with appropriate "receptor" substances.

B. Viable Mass in Dental Plaque

Not all bacteria rely on fimbriae for attachment to surfaces. Oral streptococci adhere to tooth enamel by synthesizing insoluble glucan polymers from sucrose (Wenham et al., 1981), and the accumulation of these organisms produces dental plaque. However, the accurate determination of bacterial numbers in samples of plaque is a formidable problem. One major difficulty is that of dispersion of the plaque sample. Sonication is frequently used for this purpose, but this poses the dilemma of inadequate separation of bacterial cells on the one hand and loss of viability induced by cell disruption on the other.

These problems may be overcome by using extractable ATP as an index of the number of bacteria in plaque. Robrish et al. (1978) used boiling Tris buffer to extract ATP from samples of dental plaque obtained from monkeys, and were able to estimate the viable cell mass

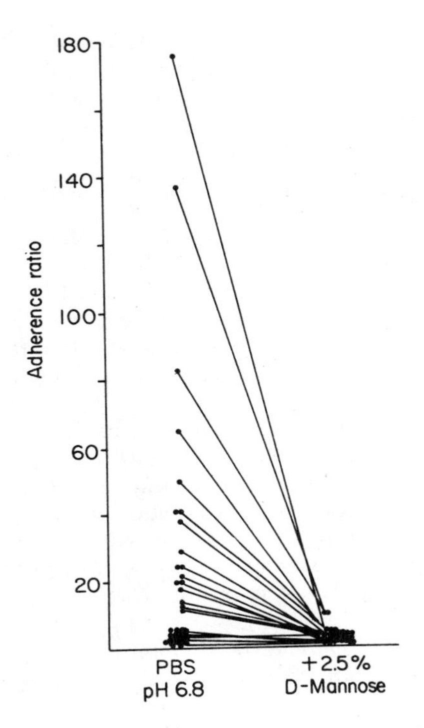

Figure 1 Inhibitory effect of D-mannose on the adherence of 34 strains of *E. coli* to polystyrene.

present by relating their results to the specific ATP and protein contents of *Streptococcus sanguis*. This procedure yielded estimates of viable counts comparable to or greater than those obtained using conventional bacteriological methods. Firefly bioluminescence was also used by Distler et al. (1980) to measure the ATP content in microgram quantities of human dental plaque after sonication, and a high correlation was found with plaque wet weight. These workers demonstrated that the ATP content of plaque was dependent on its age, reaching a maximum at 24 hr and then declining to a stable level after 72 hr.

Because bacterial cellular ATP levels vary with growth phase and environmental conditions, Kemp (1979) used the adenylate energy charge as a measure of bacterial numbers in plaque. ADP and AMP were quantified by firefly bioluminescence after prior enzymatic conversion to ATP using pyruvate kinase and adenylate kinase, and a computer program was devised to process the data. The energy charge undoubtedly gives a more constant parameter on which to base estimates of viable cell mass, but the analysis of all three adenine nucleotides greatly increases the complexity of the method. For most purposes determination of ATP alone is likely to yield satisfactory information concerning the nature of dental plaque.

III. SCREENING TEST FOR BACTERIURIA

A. Principle

Methods for detecting bacteriuria based on firefly bioluminescence have been developed independently in the United States (Picciolo et al., 1975; Alexander et al., 1976), Sweden (Thore et al., 1975), and Great Britain (Johnston et al., 1976). The basic procedure is common to all methods. Nonbacterial cells in the urine are first lysed with a detergent such as Triton X-100 and the released ATP in hydrolysed with apyrase. Bacterial ATP is then extracted with Lumac NRB reagent or an inorganic acid containing ethylenediamine tetraacetic acid (EDTA), and measured using firefly bioluminescence.

The ATP content of clinical urine specimens treated with Triton/apyrase has been shown to correlate with the viable bacterial count. The lower limit of detection is approximately 10^4 bacterial cells per ml urine using a commercial kit, this being sufficient to detect the level of bacteriuria which is generally accepted as representing significant infection in freshly voided specimens (Kass, 1957).

B. Problems

A number of problems have been encountered with the bioluminescence test for bacteriuria. Conn et al. (1975) found that incomplete destruction of ATP from leukocytes or other nonbacterial sources could give rise to false positive results being recorded. Conversely, false negative results were also encountered due to inhibition of the bioluminescence reaction by unidentified urinary substances. A third problem was variability in the cellular ATP level of different species of bacteria. In the light of these difficulties Conn and coworkers concluded that the method had limited applicability to the detection of bacteria in clinical specimens.

Each problem has received some attention in the interests of reducing the potential sources of error. Johnston and Curtis (1979) found that a major cause of the inability of Triton/apyrase treatment to completely destroy nonbacterial ATP, particularly in urine samples with a high count of mammalian cells, was binding of ATP to mitochondria or other membrane systems. The membrane-bound ATP escaped destruction by apyrase but was released on subsequent lysis of the bacterial cells. This difficulty was greatly reduced by substituting a commercial nucleotide-releasing agent for somatic cells (Lumac NRS reagent) for the Triton and increasing the temperature for apyrase treatment from 37 to 55°C.

The problem of false negative results described by Conn et al. (1975) does not appear to have been encountered to such a great extent by other workers. Nonspecific interference in the bioluminescence reaction is considerably less of a problem since the advent of purified and highly sensitive commercial luciferin/luciferase preparations in the late 1970s.

The majority of urinary tract infections are caused by *Escherichia coli* (McAllister et al., 1971), and although in the hands of the author a strain of this organism exhibited relatively minor variations in cellular ATP level throughout the growth cycle, other workers (Bush et al., 1975) have described a 10-fold variation during the exponential phase of growth. This is a factor which could affect the interpretation of results in the bioluminescence test. Clinical bacteriuria can also be caused by a variety of other bacterial genera such as *Proteus, Klebsiella, Streptococcus, Pseudomonas*, and *Providentia*, which will have differing levels of cellular ATP even under identical conditions of culture. Because the bioluminescence test cannot distinguish between different organisms, follow-up with standard bacteriological tests is always necessary with bioluminescence-positive urine specimens. This is not necessarily a disadvantage, however, as negative specimens may be identified rapidly and discarded, allowing laboratory resources to be concentrated on the samples for which they will be of definite value.

C. Comparative Assessment

It is difficult to evaluate the accuracy of the bioluminescence technique in terms of ability to distinguish between positive and negative clinical urinary tract infections since the reference cultural methodology is subject to many sources of error. False negative results may arise with bacterial cultures due to the presence of antimicrobial agents in the urine or due to unusual nutrient or gas tension requirements of a particular organism (Thore et al., 1975). For instance, anaerobic bacteria such as *Lactobacillus* spp. would go undetected in conventional cultural screening programs. Furthermore, unless urine specimens are collected by suprapubic aspiration it is necessary with any test for bacteriuria to choose an arbitrary cutoff point which will designate any single urine specimen as being clinically infected or noninfected. On the basis of published results the bioluminescence test would appear to be at least as reliable as the traditional plate count method or its adaptations in this respect. In terms of accuracy and convenience the bioluminescence method must represent an advance over the various chemical tests for bacteriuria which have been described (Mackinnon et al., 1973). The technique would also appear to suffer less from the problem of false positive results than an equivalent system involving luminol chemiluminescence (Ewetz and Strangert,

1974), which is subject to interference from nonspecific inorganic or organic matter. The bioluminescence test certainly appears to fulfill the main requirement of a bacteriuria screening program, which is the rapid and efficient recognition of negative specimens.

The relative cost efficiency of the bioluminescence test is also difficult to determine accurately because it will depend largely on the number of samples to be processed and the amount of technician time required. The method is likely to prove unjustifiably expensive and elaborate for a laboratory handling small numbers of urine specimens. The use of bacteriological dip-slides is probably adequate for most screening programs, while if a rapid diagnosis is required microscopic examination of the urine deposit by a skilled operator will generally provide as much information as any more elaborate procedure. However, this method is impracticable when a large number of clinical urines are to be screened for bacteriuria, and it is in this situation that the bioluminescence test is likely to be of greatest value. Furthermore, in countries where patients may have to travel considerable distances to attend a clinic, the speed with which a result may be obtained using the bioluminescence assay could be a critical factor in favor of its adoption. In both these circumstances an automated procedure (Curtis and Johnston, 1979; Ansehn et al., 1979) could be of particular benefit. Although comparatively expensive and elaborate to establish in the first instance, an automated bioluminescence test should be capable of providing a rapid and efficient service with the minimum amount of supervision. A semiautomated system is currently available commercially from Lumac B.V.

IV. ANTIMICROBIAL SUSCEPTIBILITY TESTING

A. Basic Procedure

The potential application of the firefly bioluminescence assay for evaluating the susceptibility of microorganisms to selected antibiotics was first described by Vellend et al. (1975) in the United States. These workers studied the sensitivity of 87 strains of 11 bacterial species to 12 antibiotics in common clinical usage. Broth cultures of 10^6 log-phase cells/ml were incubated for 2.5 hr using appropriate concentrations of antibiotic, and the logarithmic ratio of ATP in these cultures to ATP in control antibiotic-free cultures was calculated. An arbitrary endpoint of 0.25 was selected as the ratio which best distinguished between whether an organism should be designated as sensitive or resistant. The method was reproducible and showed a 90% agreement with a reference agar diffusion technique. However, in approximately 7% of instances there was a major disagreement between the two methods which was apparently related to the mode of action of the antibiotic.

Similar methodology applicable to routine antimicrobial susceptibility testing has been reported subsequently by Hojer et al. (1976)

in Sweden. This team found that inhibitors of cell wall synthesis, such as the penicillin group, caused considerable leakage of intracellular ATP into the growth medium even at concentrations below the minimum inhibitory concentration (MIC) (Thore et al., 1977). To overcome this problem apyrase was used to destroy extracellular ATP prior to extraction of endogenous ATP from the bacterial cells. In the most recent report from the Swedish group, heavy inoculates (10^8 cells/ml) of clinically isolated bacteria were exposed to each of ampicillin, doxycycline, gentamicin, and nitrofurantoin for incubation periods of 2 to 3 hr (Hojer et al., 1979). An ATP growth index was calculated which was found to correlate with the diameters of the zones of inhibition observed with a conventional 24-hr agar diffusion technique. The results indicate that the bioluminescence method can successfully differentiate between resistant, intermediate, and susceptible organisms, while at the same time offering the major advantage of rapidity. Current work is aimed at evaluating the clinical relevance and usefulness of this technique and at determining the extent of interference, if any, caused by β-lactamase producing strains.

B. Applicability

Although the bioluminescence approach to antimicrobial susceptibility testing appears to work well in the hands of its originators, it is uncertain as to how acceptable this type of methodology might be to clinicians and hospital laboratories generally. There has been growing concern over the last 18 years regarding the reproducibility and standardization of procedures used to test antimicrobial susceptibility (Blowers and Brown, 1976). A wide variety of factors have been shown to influence results including the nutrient or electrolyte composition of the growth medium, pH, size of inoculum, length of incubation, and the concentration of antibiotic.

These problems become more acute if MIC values need to be quoted and particularly so if the effectiveness of a new drug is being compared with that of other antibiotics (Waterworth, 1972). The introduction of a novel technology into an area already suffering from the effects of poor standardization is therefore a difficult task.

A number of automated systems for antimicrobial susceptibility testing have become available commercially in recent years. These are elaborate and expensive compared with standard microbiological techniques, some instruments possessing the facility for computer analysis of the rate of microbial growth. Turbidimetry is the most common method for growth detection but instruments based on radiometric and impedance measurement are also available. Bioluminescence analysis could possibly provide a more refined growth detection system which might be advantageous in terms of accuracy or rapidity due to its greater sensitivity. However, the bioluminescence test suffers the disadvantage of necessitating samples to be withdrawn from cultures for ATP extraction and analysis, whereas turbidimetric measurements can be made every few minutes, if required, without disturbing the cultures.

The bioluminescence technique is worth further evaluation, but it remains to be seen how well accepted this approach to antimicrobial susceptibility testing will become.

V. ANTIBIOTIC ASSAY

A. Principle

Because the cellular ATP level of bacteria can be used as an index of growth, it is logical to suppose that the concentration of growth-promoting or growth-inhibitory substances added to a bacterial culture will be reflected in the ATP content of the culture. This principle forms the basis of a novel assay technique for antibiotics in biological fluids developed initially by Harber and Asscher (1976,1977) in Great Britain and independently by workers in Sweden (Nilsson et al., 1977; Hojer and Nilsson, 1978). Both groups chose to investigate the applicability of bioluminescence for measuring the aminoglycoside antibiotic gentamicin in the first instance using a fast-growing gentamicin-sensitive organism. The range of the assay technique has been extended subsequently to include antibiotics with different modes of action. The Swedish method has been semiautomated and is currently being evaluated in routine use for the assay of gentamicin in clinical serum specimens (Nilsson et al., 1979).

B. Basic Assay Procedure

Bioluminescence detection is linked to a standard microbiological tube assay design using small (1-ml) culture volumes. Appropriate dilutions of test sera and a range of antibiotic standards prepared in normal human serum are added to broth cultures of a bacterium which is sensitive to the antibiotic under investigation. All sera need to be heat-inactivated to destroy complement activity if the test organism is gram-negative, although this stage can be eliminated if the volume of serum is reduced to 1% or less of the total culture volume (Nilsson et al., 1979). As with any bacteriological procedure the use of replicates is advisable to ensure accuracy.

On incubation, increasing concentrations of antibiotic produce a corresponding suppression of growth which is reflected in a decreased concentration of extractable ATP in the culture (Figure 2). Incubation is continued until a difference in ATP level of 10-fold or

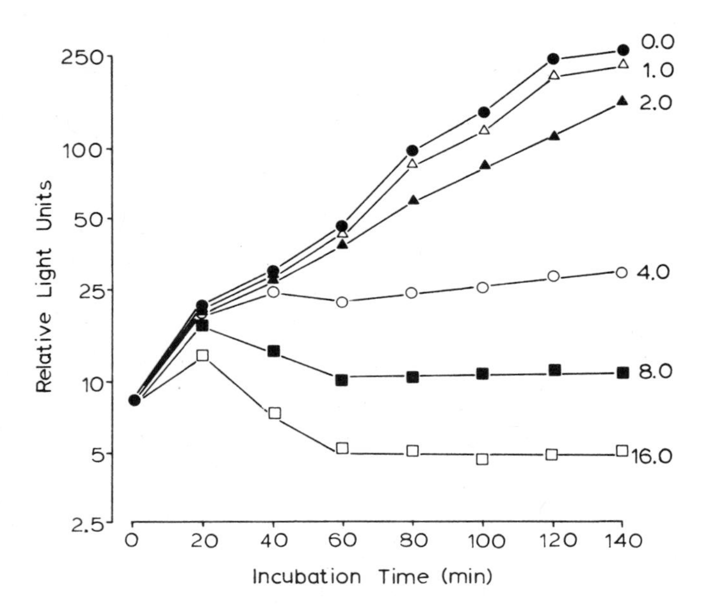

Figure 2 ATP production by *K. edwardsii* var. *atlantae* grown in the presence of gentamicin. Gentamicin levels are recorded as μg/ml serum which was incorporated into nutrient broth at a 5% concentration v/v. (From Harber and Asscher, 1977.)

more is produced between cultures containing the highest and lowest antibiotic standards, the exact period varying according to the generation time of the test organism. Only 90 min incubation is required for the assay of serum gentamicin using *Klebsiella edwardsii* var. *atlantae* (Harber and Asscher, 1977) or a strain of *E. coli* (Nilsson et al., 1977), but periods of 4 to 5 hr may be necessary with other organisms.

Poor differentiation between the concentration of ATP in cultures containing high and low concentrations of antibiotic might arise with certain bacteria if ATP released by cell lysis accumulates in the growth medium. In their early work Nilsson and colleagues added apyrase to all cultures for the final 10 min of incubation to destroy extracellular ATP (Nilsson et al., 1977), but subsequently this stage was found to be unnecessary with the bacteria used in some of the assays of clinical interest (Nilsson et al., 1979). Nilsson (1978) has even described the reverse procedure for the assay of serum gentamicin in which extracellular ATP alone was used as the index of bacterial growth. In contrast, the original technique reported by Harber and Asscher (1976,1977) involved measurement of both intra- and extracellular ATP.

Extraction of ATP is a crucial stage in the assay procedure. Sulfuric acid containing EDTA to inhibit ATP-converting enzymes has been used most commonly as extractant since it is rapid, simple, and efficient, and after neutralization does not interfere with the bioluminescence reaction. The commercial Lumac NRB reagent provides a more simple and efficient method for extraction but this also adds to the cost of the assay. Whichever reagent is used, rapid handling of the tubes is advisable at this stage to overcome the potential problem of changes in cellular ATP level which may occur in some organisms on withdrawal of exponential-phase cultures from the incubation waterbath (Harber and Asscher, 1976). The simplest solution to this problem is to cool all the tubes in an iced waterbath for 5 to 10 min immediately after incubation and before extraction of ATP.

C. Practical Application

The assay system has been validated for antibiotics with different modes of action (Table 2) including inhibitors of cell wall synthesis (cephaloridine; ampicillin), protein synthesis (doxycycline; gentamicin

Table 2 Range of Antibiotics Which Have Been Measured Using the Bioluminescence Assay

Antibiotic	Test organism	Incubation time (min)	Reference
Gentamicin	*Klebsiella edwardsii* var. *atlantae*	90	Harber and Asscher (1976, 1977)
Gentamicin	*Escherichia coli*	90	Nilsson et al. (1977)
Tobramycin	*Klebsiella edwardsii* var. *atlantae*	120	Harber and Asscher (1976, 1977)
Cephaloridine	*Staphylococcus aureus*	240	Harber and Asscher (1977)
Doxycycline	*Enterobacter cloacae*	90	Hojer and Nilsson (1978)
Ampicillin	*Staphylococcus epidermidis*	120	Hojer et al. (1978)
Rifampicin	*Sarcina lutea*	300	Harber (unpublished)

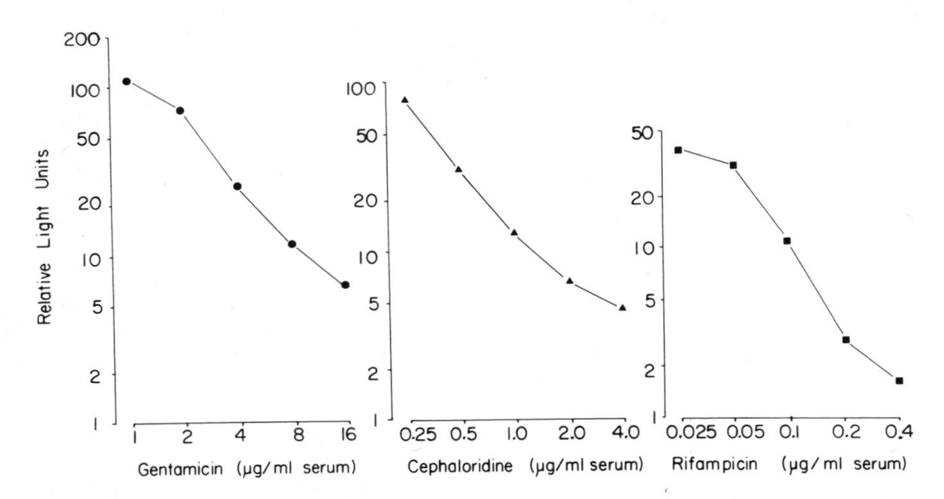

Figure 3 Standard graphs for the assay of gentamicin, cephaloridine, and rifampicin in serum. Culture media contained 5% serum v/v. (Gentamicin and cephaloridine data from Harber and Asscher, 1977.)

and tobramycin), and DNA synthesis (rifampicin). The relationship between observed ATP level and antibiotic concentration is generally slightly sigmoidal rather than linear. Typical standard graphs suitable for the assay of serum levels of gentamicin, cephaloridine, and rifampicin are illustrated in Fig. 3.

Due to the sensitivity of the bioluminescence methods only microliter quantities of serum are required for assay, so that blood samples may be obtained by finger lance should venepuncture be either undesirable or impossible. The sensitivity of the rifampicin assay is particularly noteworthy, the limit of detection being 0.025 µg/ml serum which is equivalent to 1.25 ng/ml culture. Dilutions of test sera of up to 1 in 100 may be necessary in order to bring the rifampicin level within the range of concentration of the assay standards.

Neither high concentrations of urea nor serum from uremic patients produce any interference with the assay of serum gentamicin (Harber, unpublished data), while the use of a multiple antibiotic-resistant test organism makes this assay reasonably specific for gentamicin (Harber and Asscher, 1977; Nilsson et al., 1977). Addition of excess glucose to the growth medium has been proposed to be sure of eliminating any error which might otherwise arise due to differences in glucose level between different serum specimens (Nilsson et al., 1979).

D. Comparative Assessment

The bioluminescence antibiotic assay has been shown to correlate closely with reference agar diffusion techniques when applied to clinical serum specimens (Harber and Asscher, 1977; Nilsson et al., 1977; Hojer and Nilsson, 1978). Furthermore, an appraisal of reproducibility conducted by Harber and Asscher (1977) indicated an overall coefficient of variation of <10% (n = 16), which would fall within the accuracy requirements for clinical antibiotic assays outlined by Reeves and Bywater (1975).

Although the bioluminescence assay is more expensive than conventional microbiological procedures, the cost of reagents is lower than that involved with radioenzymatic or radioimmunoassay systems. Instrumentation costs are also lower than for techniques involving radioisotope or fluorescence detection, and a suitable luminometer can even be purchased for less than the cost of an automated inhibition zone reading system which has been described for use with agar diffusion antibiotic assays (Hallynck and Pijck, 1979). In view of these advantages, along with the versatility of bioluminescence methodology and its adaptability to automation, the ATP method may prove to be a useful technique for the routine assay of antimicrobial agents.

When first reported the bioluminescence assay for antibiotics was probably no more sensitive than using turbidimetric detection of bacterial growth, but purified modern reagents and commercial instruments have increased the sensitivity of ATP detection by 1000-fold or more, and have also improved the accuracy of measurement. The added sensitivity permits the use of more dilute inocula and shorter incubation times, factors which are likely to enhance the applicability of the method.

VI. VITAMIN ASSAY

A logical extension of the bioluminescence assay for antibiotics is to apply the same basic principle to the measurement of substances which promote bacterial growth. Bioluminescence detection has been applied successfully by Harber and Asscher (1979) to the conventional microbiological assay system for folic acid using *Lactobacillus casei* (NCIB 11295) as test organism. Because of a long lag phase associated with the growth of this organism in batch culture, to date it has not been possible to provide a rapid assay procedure. However, Tennant (1977) reported successful use of an automated *L. casei* assay using only a 4-hr incubation period in a continuous-culture system coupled to a dye reduction detection device, and the ATP detection could easily be incorporated into this type of assay design if required to give a more sensitive indicator of bacterial growth.

Using an overnight (18-hr) incubation period, a direct relationship is demonstrable between folic acid level and the concentration of extractable ATP in the culture (Fig. 4). The assay procedure has a remarkable degree of sensitivity, the limit of detection being 25 pg folic acid/ml serum which is equivalent to 1.25 pg (i.e., 3 fmol) per ml culture. The method is therefore even more sensitive than the radioassay techniques for serum folate. Because of this sensitivity, sera need to be diluted up to 20-fold prior to assay, whereas neat serum is generally used in standard microbiological methods. This dilution factor reduces the likelihood of interference from physiological serum factors or antimicrobial agents. In addition, sufficient serum for assay may be collected by finger lance rather than by venepuncture, which is important with regard to pediatric use.

An appraisal of precision has indicated a mean coefficient of variation of 12% (n = 8), which is sufficiently precise for the purposes of a clinical test. Radioimmunoassay techniques may give slight advantages in accuracy or rapidity but are more costly in terms of reagents and instrumentation. In any case, many laboratories prefer microbiological assays for folic acid and vitamin B_{12} since these allow measurement of total biologically active vitamin whereas radioassay protein-binding techniques only detect selected fractions.

At the present time, folic acid is the only vitamin which has been measured using the bioluminescence method, but the technique

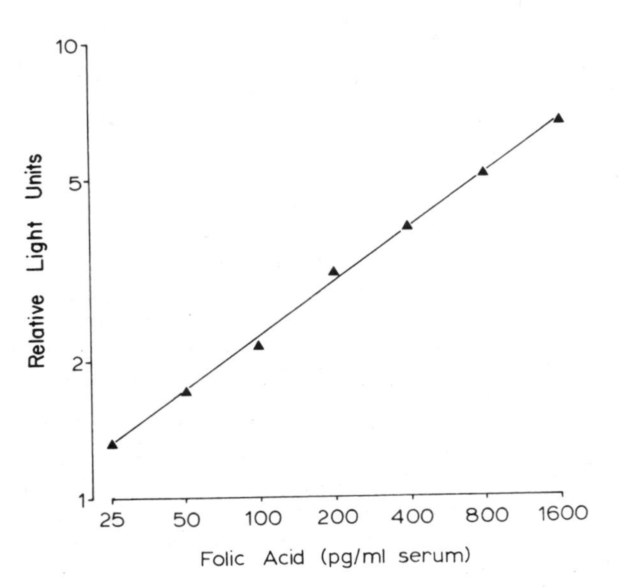

Figure 4 Standard graph for the assay of serum folate. Culture medium contained 5% serum v/v.

is potentially applicable to the assay of B_{12} or any other substance which can specifically stimulate bacterial growth rate.

VII. LUMINESCENCE ANALYSES OF BLOOD CELLS

A. Erythrocytes

Viability Test

Although erythrocytes derive their energy from glycolysis and not from aerobic respiration, there is considerable evidence of a causal relationship between the ATP content and cell viability (Nakao et al., 1962; Akerblom et al., 1967; Haradin et al., 1969). Storage of erythrocytes in media containing adenine, adenosine, or inosine, or incubation of stored erythrocytes with these purine compounds prior to reinfusion, is known to increase the cellular ATP level and simultaneously improve posttransfusion viability (Mollison, 1979). Prolonged red cell survival has been shown to be associated with the ATP-dependent parameters of maintenance of a discoidal shape coupled with an ability to deform (Haradin et al., 1969), although on prolonged storage irreversible loss of lipid from the cell membrane becomes increasingly important as a determinant of cell viability (Mollison, 1979).

Most blood transfusion centers rely on experience to determine the time period for which whole blood may be stored prior to reinfusion. The simplicity of the firefly bioluminescence system allows rapid quantitation of cellular ATP, which might provide a less empirical approach to blood storage. Routine screening of the ATP level in stored erythrocytes may yield little more information than can be gained from experience alone, but the technique is likely to be of value in the quest for improved methods for storing blood intended for transfusion. Since measurement of erythrocyte ATP appears to be a good in vitro correlate of in vivo survival as assessed using ^{51}Cr-labeled cells, this simple test might eliminate the need for the use of radioisotopes in studies relating to erythrocyte viability.

Erythrocyte ATP and Disease

A number of diseases of widely differing origin may produce metabolic defects which result in abnormal levels of adenine nucleotides in erythrocytes. For example, erythrocyte ATP is depressed in fasting diabetics (Windisch et al., 1970) but is elevated in chronic uraemia. The latter observation may be largely dependent on the accumulation of a circulating inhibitor of sodium/potassium ATPase (Cohen, 1974) rather than to an increase in plasma phosphate as is commonly supposed (Barcenas et al., 1977).

The ATP:ADP ratio is frequently of more value than the absolute level of either nucleotide alone for distinguishing between different metabolic disorders. Certain hereditary enzyme deficiency diseases, such as hexokinase and pyruvate kinase deficiency, are characterized by a low erythrocyte ATP level but a normal ADP level, giving rise to a low ATP:ADP ratio. Conversely, autoimmune hemolytic anemia and some other chronic hemolytic diseases can give rise to an abnormally high erythrocyte ATP:ADP ratio (Summerfield et al., 1981).

Both ATP and ADP are easily extracted from erythrocytes using dilute inorganic acids, detergents, organic solvents or the commercial Lumac NRS reagent. ATP is quantified directly after addition of purified firefly luciferin/luciferase reagent, while ADP is measured after conversion to ATP using pyruvate kinase in the presence of phosphoenol pyruvate (Holmsen et al., 1972b; Lundin et al., 1976). The bioluminescent assay of adenine nucleotides in erythrocytes is therefore a straightforward procedure, and although it is not always of diagnostic value, it can provide useful information to supplement other laboratory investigations.

B. Platelet Viability and Function

Platelets contain two separate reservoirs of adenine nucleotides designated the metabolic pool and the storage pool (Holmsen et al., 1972a). Approximately 60% of total adenine nucleotides are contained in the δ storage pool granules from which ATP and ADP are secreted during the platelet release reaction (Holmsen et al., 1974). The ability of platelets to perform their functions appears to be closely linked to the ATP level in the metabolic nucleotide pool. At a given level of stimulation a decrease in the metabolic ATP pool brings about inhibition, in order, of the platelet release reaction (secondary aggregation), primary platelet aggregation, and the shape-change response. ATP analysis should therefore be of value for monitoring the viability of stored platelets intended for transfusion, and may provide a suitable endpoint for comparing the effectiveness of different methods for storing platelets. These conclusions are supported by the observation that the ATP level in platelets has been found to decline rapidly after 48 hr storage (Tarkkanen et al., 1979), this being the maximum period for which platelets may be usefully stored at the present time.

Primary release defects in platelets are characterized by normal nucleotide levels, but with defective secretion and lack of aggregation. The most common acquired release defect is produced by aspirin which inactivates the key enzyme cyclo-oxygenase. Conversely, storage pool disease is associated with a deficiency in adenine nucleotides, serotonin, and calcium which are normally present in the δ storage granules, while the overall platelet ATP:ADP ratio is increased (Holmsen and Weiss, 1972). Storage pool disease may be congenital in origin, or may be acquired in association with circulating antiplatelet antibodies, acute myeloblastic leukaemia, postoperative bleeding following cardiopulmonary bypass, chemotherapy with mithramycin, and myeloproliferative disorders (Shattil and Bennet, 1980). Such acquired storage pool disease may arise from defective granule formation in the megakaryocyte or from reversible activation of circulating platelets.

The most fundamental test for platelet function is the measurement of platelet aggregating ability. Aggregation may be induced by stimulants such as thrombin, collagen, adrenalin, or ADP, and although most aggregometers simply measure the increase in light transmission which accompanies aggregation, the reaction may also be monitored by measuring release of ATP, or ATP plus ADP. Once a defect in platelet function has been established, a distinction may be made between primary release defects and storage pool diseases by measuring the cellular levels of both ATP and ADP. The two nucleotides are commonly extracted from platelets using Lumac NRS reagent or a Triton/EDTA mixture in ethanol (Summerfield et al., 1981), and assayed by the firefly bioluminescence system. This simple test is likely to become an important routine investigation for patients with abnormalities of platelet function.

C. Leukocyte Function

Lymphocyte Transformation

Resting lymphocytes may be stimulated in vitro by a variety of substances of microbial, plant, or animal origin to transform into blast cells that proliferate by mitosis (Cunningham-Rundles et al., 1976). The proliferative phase, which may last for several days, is usually monitored by measuring the incorporation of radioactive thymidine into the DNA of dividing cells. Changes in the energy status of the cells (reported as the adenylate energy charge or simply in terms of the ATP level) may be easily measured at different stages of transformation using the firefly bioluminescence system. A decrease in ATP level may be demonstrated over the first few hours following exposure of human lymphocytes to the plant mitogen phytohemagglutinin (PHA) (Table 3), presumably due to cell metabolism becoming more anaerobic (Cunningham-Rundles et al., 1976). Thereafter large increases in the cellular ATP level occur which coincide with maximal DNA synthesis (Fig. 5; Harber, unpublished data). Conversely, unstimulated control cultures maintain a relatively constant ATP level which would appear to give a better index of cell viability than the notoriously unreliable trypan blue exclusion test. This methodology could provide a useful research tool to complement the use of radioactive tracers in lymphocyte transformation studies.

A particularly interesting approach to the study of lymphocyte activation arises from an investigation by Wrogemann et al. (1978), who observed a luminol-enhanced chemiluminescence associated with rat thymocytes within minutes of the cells being stimulated by the mitogen concanavalin A or the calcium ionophore A23187. The chemiluminescent response was greatly enhanced by preincubation of the

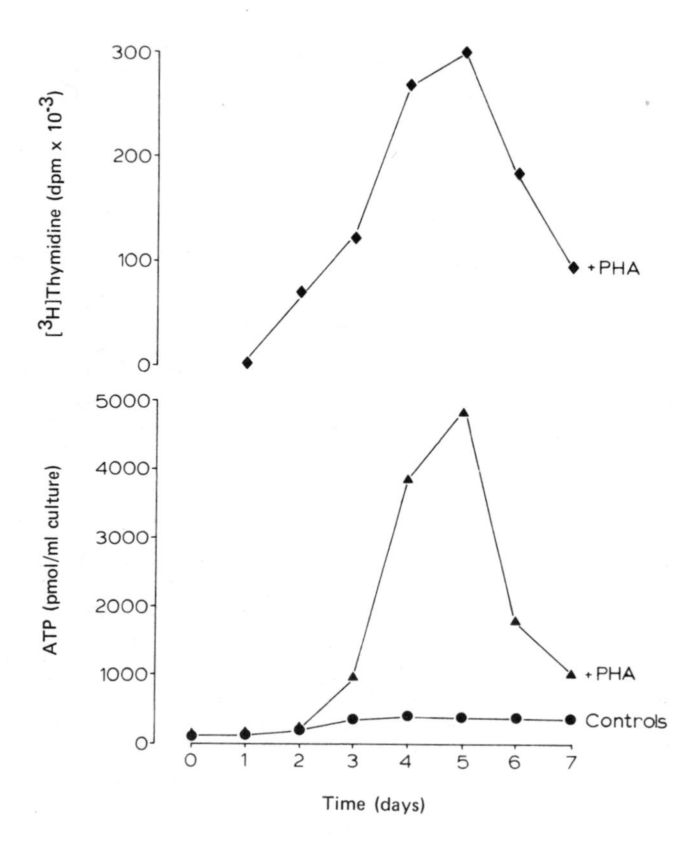

Figure 5 Changes in ATP concentration and [^{3}H]thymidine uptake during PHA-induced lymphocyte transformation. Results are the mean data from triplicate cultures.

Table 3 Mean Lymphocyte ATP Levels (±SD) 4 and 6 hr Following Exposure to PHA (n = 12)

Time interval following PHA stimulation (hr)	pmol ATP/ml culture		
	controls	+ PHA	P
0	460 ± 40	500 ± 40	>.01
4	440 ± 55	193 ± 45	<.001
6	205 ± 65	47 ± 27	<.001

Source: Data from Harber (unpublished).

thymocytes with macrophages (Wrogemann et al., 1980). These observations could have important implications with regard to tissue typing and to the study of cell activation mechanisms, immunosuppressive drug action, and immune cell interactions. Unfortunately, attempts by the author to apply this methodology to the study of mitogen stimulation of peripheral human lymphocytes have so far been unsuccessful since any luminescence which might be associated with the lymphocytes was found to be masked by the luminescence response of granulocyte contaminants.

Phagocytosis

Fluctuations in cellular ATP coinciding with changes in energy demand can be demonstrated in activated or chemically elicited macrophages (Hard, 1970; Michl et al., 1976; Harber, unpublished data) and in phagocytosing granulocytes (Venge et al., 1979). Furthermore, the study of phagocytosis per se has been largely revolutionized in recent years by the advent of chemiluminescence techniques to monitor respiratory burst activity. Electronically excited products of the respiratory burst in human granulocytes (Allen et al., 1972; Webb et al., 1974; Cheson et al., 1976) and monocytes (Sagone et al., 1976; Johnston, Lehmeyer, and Guthrie, 1976) give rise to a native chemiluminescence which is proportional to the magnitude of the phagocytic response. Luminol has been used to enhance the chemiluminescent response of stimulated human granulocytes (Allen and Loose, 1976) and to detect chemiluminescence in phagocytosing macrophages of animal origin (Allen and Loose, 1976; Diaz et al., 1979). A detailed discussion of phagocytosis is beyond the scope of this chapter but it is worth noting that the chemiluminescence technique provides a powerful tool which has been used for investigating clinical disorders of phagocytic function, such as chronic granulomatous disease (Babior, 1978), for detecting immune complexes in clinical sera (Doll et al., 1980; Starkebaum et al., 1981), for studying the role of calcium fluxes in cell activation (Hallett and Campbell, 1982), and in an in vitro model of immunological glomerular disease (Davies, Harber, and Coles, manuscript in preparation).

VIII. ATP IN BLOOD PLASMA

ATP and other nucleotides have been considered historically to be intracellular components that cannot traverse cell membranes, but this is now known not to be the case. Spectrophotometric analysis is insufficiently sensitive to detect ATP in blood plasma, but the presence of this nucleotide in human plasma was demonstrated by Holmsen et al. (1966) using firefly bioluminescence. Other workers have subsequently reported that normal plasma contains a fairly constant ATP level of approximately 1 μmol/L (Forrester and Lind, 1969; Jabs et al., 1977).

Part of the plasma ATP content appears to originate from damage to the formed elements of the blood, particularly platelets, but a significant proportion must arise through leakage from peripheral tissues since the plasma ATP level of the venous effluent of occluded human forearm muscles increases during exercise (Forrester and Lind, 1969; Forrester, 1972).

Measurement of cell levels of adenine nucleotides can provide an index of cellular energy status as both the glycolytic pathway and the tricarboxylic acid cycle are regulated by the ATP to ADP to AMP ratios. The concentration of ATP in whole blood is measured in some hospitals in an attempt to test cellular metabolic activity, but this analysis is only likely to reflect erythrocyte metabolism which is not typical of peripheral tissues generally. It is considered that plasma ATP levels, perhaps expressed in terms of phosphorylation potential or adenylate energy charge, might reflect more accurately the energy status or efficiency of perfusion of peripheral tissues. Drastic reductions in plasma levels of ATP and creatine phosphate have been reported in animals with tracheal clamps or suffering bacterial endotoxin shock (Jabs et al., 1977), while plasma ATP concentrations one-tenth of normal have been observed in patients with impaired peripheral tissue perfusion (Harber, unpublished data). More work is required to assess the clinical relevance of these observations, but preliminary results are sufficiently encouraging to make a more detailed investigation worthwhile. The plasma ATP test is inexpensive, rapid, and simple to perform, and it could become a useful method for clinical investigation.

IX. DISCUSSION

It is apparent that under defined environmental conditions the ATP content of cell types as diverse as bacteria and human leukocytes accurately reflects the degree of intracellular metabolic activity. This principle forms the basis of the tests for cell viability and the assay procedures involving measurement of bacterial numbers described in this chapter. Standardized conditions are required for all tests since the absolute cellular ATP level at any given moment of time is dependent on a number of different parameters of which oxygen tension is probably the most important single factor. The ATP content of Ehrlich ascites tumor cells, for instance, has been shown to parallel changes in oxygen uptake throughout different phases of the growth cycle (Skog and Tribukait, 1979), while large fluctuations in ATP level occur in bacteria (Cole et al., 1967), spermatozoa (Brooks, 1970), and leukocytes (data in this chapter) depending on the balance between aerobiosis and anaerobiosis. It is important, therefore, to consider such possible sources of variation in any procedure in which cellular ATP is correlated with biomass.

To date, luminescence analysis in the fields of medical microbiology and hematology has been confined almost exclusively to measurement of ATP using firefly bioluminescence. Quantitation of bacterial numbers by detecting bacterial porphyrins using luminol chemiluminescence has been described (Oleniacz et al., 1968; Ewetz and Strangert, 1974; Searle, 1975), but despite a sensitivity of approximately 10^4 cells/ml this technique has not gained the same popularity as firefly ATP analysis because it is subject to nonspecific interference. Luminescence immunoassay systems involving pyrogallol/peroxidase chemiluminescence have been reported for *Serratia marcescens* (Halmann et al., 1977) and for Sindbis virus particles (Velan et al., 1979). The former method represents the most sensitive system yet devised for quantitating bacterial numbers, the limit of detection being in the order of 30 to 300 cells. This and other luminescence techniques could become important analytical tools in the future as the demand for more sensitive and precise microbiological methods increases.

As yet, few of the clinical applications described in this chapter have reached the stage of becoming routine procedures, and for this reason the discussion regarding their potential usefulness has been speculative. An attempt has been made to discuss the merits and pitfalls of each system in relation to the alternative methods in current use. All of the techniques described are promising innovations, but their value in clinical practice remains unproven at the present time.

REFERENCES

Akerblom, O., C-H. de Verdier, M. Finnson, L. Garby, C. F. Hogman, and S. G. O. Johansson (1967). Further studies on the effect of adenine in blood preservation. *Transfusion* 7:1-9.

Alexander, D. N., G. M. Ederer, and J. M. Matsen (1976). Evaluation of an adenosine-5'-triphosphate assay as a screening method to detect significant bacteriuria. *J. Clin. Microbiol.* 3:42-46.

Allen, R. C., and L. D. Loose (1976). Phagocytic activation of a luminol-dependent chemiluminescence in rabbit alveolar and peritoneal macrophages. *Biochem. Biophys. Res. Commun.* 69:245-252.

Allen, R. C., R. L. Stjernholm, and R. H. Steele (1972). Evidence for the generation of an electronic excitation state(s) in human polymorphonuclear leukocytes and its participation in bactericidal activity. *Biochem. Biophys. Res. Commun.* 47:679-684.

Ansehn, S., A. Lundin, L. Nilsson, and A. Thore (1979). Detection of bacteriuria by a simplified luciferase assay of ATP. In *Proceedings of the International Symposium on Analytical Applications of Bioluminescence and Chemiluminescence*, E. Schram and P. Stanley (Eds.). State Printing and Publishing, Westlake Village, Cal., pp. 438-445.

Babior, B. M. (1978). Oxygen-dependent microbial killing by phagocytes. *N. Engl. J. Med.* *298*:721-725.

Barcenas, C. G., P. Paez, K. Uyeda, P. C. Peters, and A. R. Hull (1977). Erythrocyte adenosine triphosphate and 2,3-diphosphoglycerate after human renal transplantation: dissociation from hypophosphataemia. *Clin. Sci. Mol. Med.* *52*:413-422.

Blowers, R., and D. F. J. Brown (1976). Control of antibiotic sensitivity testing in Britain. In *Proceedings of the 9th International Congress of Chemotherapy*, J. D. Williams and A. M. Geddes (Eds.). Plenum, New York and London, pp. 7-17.

Brooks, D. E. (1970). Observations on the content of ATP and ADP in bull spermatozoa using the firefly luciferase system. *J. Reprod. Fertil.* *23*:525-528.

Bush, V. N., G. L. Picciolo, and E. W. Chappelle (1975). The effect of growth phase and medium on the use of the firefly adenosine triphosphate (ATP) assay for the quantitation of bacteria. In *Analytical Applications of Bioluminescence and Chemiluminescence*, E. W. Chappelle and G. L. Picciolo (Eds.). National Aeronautics and Space Administration, Washington, D.C., pp. 35-41.

Chappelle, E. W., and G. V. Levin (1968). Use of the firefly bioluminescent reaction for rapid detection and counting of bacteria. *Biochem. Med.* *2*:41-52.

Cheson, B. D., R. L. Christensen, R. Sperling, B. E. Kohler, and B. M. Babior (1976). The origin of the chemiluminescence of phagocytosing granulocytes. *J. Clin. Invest.* *58*:789-796.

Chick, S., M. J. Harber, R. Mackenzie, and A. W. Asscher (1981). Modified method for studying bacterial adhesion to isolated uroepithelial cells and uro-mucoid. *Infect. Immun.* *34*:256-261.

Cohen, B. D. (1974). Uremia and blood cell dysfunction. *Adv. Intern. Med.* *19*:27-39.

Cole, H. A., J. W. T. Wimpenny, and D. E. Hughes (1967). The ATP pool in *Escherichia coli*. 1. Measurement of the pool using a modified luciferase assay. *Biochim. Biophys. Acta.* *143*:445-453.

Conn, R. B., P. Charache, and E. W. Chappelle (1975). Limits of applicability of the firefly luminescence ATP assay for the detection of bacteria in clinical specimens. *Amer. J. Clin. Pathol.* *63*:493-501.

Cunningham-Rundles, S., J. A. Hansen, and B. Dupont (1976). Lymphocyte transformation in vitro in response to mitogens and antigens. In *Clinical Immunobiology*, Vol. 3, F. H. Bach and R. A. Good (Eds.). Academic, London and New York, pp. 151-194.

Curtis, G. D. W., and H. H. Johnston (1979). A rapid screening test for bacteriuria. In *Proceedings of the International Symposium on Analytical Applications of Bioluminescence and Chemiluminescence*, E. Schram and P. Stanley (Eds.). State Printing and Publishing, Westlake Village, Cal., pp. 448-457.

D'Eustachio, A. J., and D. R. Johnson (1968). Adenosine triphosphate content of bacteria. *Fed. Proc.* *27*:761.

Diaz, P., D. G. Jones, and A. B. Kay (1979). Histamine-coated particles generate superoxide (O_2^-) and chemiluminescence in alveolar macrophages. *Nature 278*:454-456.

Distler, W., A. Kroncke, and G. Maurer (1980). Adenosine triphosphate content of human dental plaque as a measure of viable cell mass. *Caries Res. 14*:265-268.

Doll, N. J., M. R. Wilson, and J. E. Slavaggio (1980). Inhibition of polymorphonuclear leukocyte chemiluminescence for detection of immune complexes in human sera. *J. Clin. Invest. 66*:457-464.

Ewetz, L., and K. Strangert (1974). A simple method for detection of bacteriuria with an automated chemiluminescence technique. *Acta. Path. Microbiol. Scand. sec B. 82*:375-381.

Farber, E. (1973). ATP and cell integrity. *Fed. Proc. 32*:1534-1539.

Forrester, T. (1972). An estimate of adenosine triphosphate release into the venous effluent from exercising human forearm muscle. *J. Physiol. 224*:611-628.

Forrester, T., and A. R. Lind (1969). Identification of adenosine triphosphate in human plasma and the concentration in the venous effluent of forearm muscles before, during and after sustained contractions. *J. Physiol. 204*:347-364.

Franzen, J. S., and S. B. Binkley (1961). Comparison of the acid-soluble nucleotides in *Escherichia coli* at different growth rates. *J. Biol. Chem. 236*:515-519.

Hallett, M. B., and A. K. Campbell (1982). Measurement of changes in cytoplasmic free Ca^{2+} in fused cell hybrids. *Nature 295*:155-158.

Hallynck, Th., and J. Pijck (1979). New zone-reading equipment in microbiological diffusion assays. *J. Antimicrob. Chemother. 5*:179-182.

Halmann, M., B. Velan, and T. Sery (1977). Rapid identification and quantitation of small numbers of microorganisms by a chemiluminescent immunoreaction. *Appl. Environ. Microbiol. 34*:473-477.

Hamilton, R. D., and O. Holm-Hansen (1967). Adenosine triphosphate content of marine bacteria. *Limnol. Oceanogr. 12*:319-324.

Haradin, A. R., R. I. Weed, and C. F. Reed (1969). Changes in physical properties of stored erythrocytes. Relationship to survival in vivo. *Transfusion 9*:229-237.

Harber, M. J., and A. W. Asscher (1976). A new assay technique for antibiotics. In *Proceedings of the 9th International Congress of Chemotherapy*, Vol. 11, *Chemotherapy*, J. D. Williams and A. M. Geddes (Eds.). Plenum, New York and London, pp. 125-131.

Harber, M. J., and A. W. Asscher (1977). A new method for antibiotic assay based on measurement of bacterial adenosine triphosphate using the firefly bioluminescence system. *J. Antimicrob. Chemother. 3*:35-41.

Harber, M. J., and A. W. Asscher (1979). Bioluminescence assay for antibiotics and vitamins. In *Proceedings of the International Symposium on Analytical Applications of Bioluminescence and Chemi-*

luminescence, E. Schram and P. Stanley (Eds.). State Printing and Publishing, Westlake Village, Cal., pp. 531-542.

Harber, M. J., S. Chick, R. Mackenzie, and A. W. Asscher (1982a). Lack of adherence to epithelial cells by freshly isolated urinary pathogens. *Lancet* (i):586-588.

Harber, M. J., R. Mackenzie, and A. W. Asscher (1982b). A rapid bioluminescence method for quantifying bacterial adherence. *Clin. Sci. 62*:8P.

Hard, G. C. (1970). Some biochemical aspects of the immune macrophage. *Br. Exp. Pathol. 51*:97-105.

Hojer, H., L. Nilsson, S. Ansehn, and A. Thore (1976). In-vitro effect of doxycycline on levels of adenosine triphosphate in bacterial cultures. *Scand. J. Infect. Dis. Suppl. 9*:58-61.

Hojer, H., and L. Nilsson (1978). Rapid determination of doxycycline based on luciferase assay of bacterial adenosine triphosphate. *J. Antimicrob. Chemother. 4*:503-508.

Hojer, H., L. Nilsson, S. Ansehn, and A. Thore (1978). Evaluation of a rapid semiautomated bioassay of antibiotics. In *Current Chemotherapy*, Vol. 1, W. Siegenthaler. and R. Luthy (Eds.). American Society for Microbiology, Washington, D. C., pp. 523-525.

Hojer, H., L. Nilsson, S. Ansehn, and A. Thore (1979). Possible application of luciferase assay of ATP to antibiotic susceptibility testing. In *Proceedings of the International Symposium on Analytical Applications of Bioluminescence and Chemiluminescence*, E. Schram and P. Stanley (Eds.). State Printing and Publishing, Westlake Village, Cal., pp. 523-530.

Holmsen, H., and H. J. Weiss (1972). Further evidence for a deficient storage pool of adenine nucleotides in platelets from some patients with thrombocytopathia: "storage pool disease," *Blood 39*:197-209.

Holmsen, H., I. Holmsen, and A. Bernhardsen (1966). Microdetermination of adenosine diphosphate and adenosine triphosphate in plasma with the firefly luciferase system. *Anal. Biochem. 17*:456-473.

Holmsen, H., H. J. Day, and C. A. Setkowsky (1972a). Behaviour of adenine nucleotides during the platelet release reaction induced by adenosine diphosphate and adrenaline. *Biochem. J. 129*:67-82.

Holmsen, H., E. Storm, and H. J. Day (1972b). Determination of ATP and ADP in blood platelets: a modification of the firefly luciferase assay for plasma. *Anal. Biochem. 46*:489-501.

Holmsen, H., C. A. Setkowsky, and H. J. Day (1974). Effects of antimycin and 2-deoxyglucose on adenine nucleotides in human platelets. Role of metabolic adenosine triphosphate in primary aggregation, secondary aggregation and shape change of platelets. *Biochem. J. 144*:385-396.

Jabs, C. M., W. J. Ferrell, and H. J. Robb (1977). Microdetermination of plasma ATP and creatine phosphate concentrations with a luminescence biometer. *Clin. Chem. 23*:2254-2257.

Johnston, H. H., and G. D. W. Curtis (1979). Detection of bacteriuria: problems in removal of non-bacterial ATP. In *Proceedings of the International Symposium on Analytical Applications of Bioluminescence and Chemiluminescence*. E. Schram and P. Stanley (Eds.). State Printing and Publishing, Westlake Village, Cal., pp. 446-447.

Johnston, H. H., C. J. Mitchell, and G. D. W. Curtis (1976). An automated test for the detection of significant bacteriuria. *Lancet 2*: 400-402.

Johnston, R. B. Jr., J. E. Lehmeyer, and L. A. Guthrie (1976). Generation of superoxide anion and chemiluminescence by human monocytes during phagocytosis and on contact with surface-bound immunoglobulin G. *J. Exp. Med. 143*:1551-1556.

Kass, E. H. (1957). Bacteriuria and the diagnosis of infections of the urinary tract. *Arch. Intern. Med. 100*:709-714.

Kemp, C. W. (1979). Adenylate energy charge: a method for the determination of viable cell mass in dental plaque samples. *J. Dent. Res. 68*(D):2192-2197.

Lundin, A., A. Rickardson, and A. Thore (1976). Continuous monitoring of ATP-converting reactions by purified firefly luciferase. *Anal. Biochem. 75*:611-620.

Mackinnon, A. E., C. J. L. Stachan, J. D. Sleigh, and M. M. Burns (1973). Detection of bacteriuria with a test for urinary glucose. In *Urinary Tract Infection*, W. Brumfitt and A. W. Asscher (Eds.). Oxford University Press, London, pp. 11-15.

McAllister, T. A., J. G. Alexander, C. Dulake, A. Percival, J. M. H. Boyce, and P. J. Wormald (1971). Part 1: The sensitivities of urinary pathogens—a survey. Multicentric study of sensitivities of urinary tract pathogens. *Postgrad. Med. J. (September suppl.) 47*:7-18.

Michl, J., D. J. Ohlbaum, and S. C. Silverstein (1976). 2-Deoxyglucose selectively inhibits Fc and complement receptor-mediated phagocytosis in mouse peritoneal macrophages. II. Dissociation of the inhibitory effects of 2-deoxyglucose on phagocytosis and ATP generation. *J. Exp. Med. 144*:1484-1493.

Mollison, P. L. (1979). *Blood Transfusion in Clinical Medicine*. Blackwell Scientific Publications, Oxford.

Nakao, K., T. Wada, T. Kamiyama, M. Nakao, and K. Nagano (1962). A direct relationship between adenosine triphosphate level and in vivo viability of erythrocytes. *Nature 194*:877-878.

Nilsson, L. (1978). New rapid bioassay of gentamicin based on luciferase assay of extracellular ATP in bacterial cultures. *Antimicrob. Agents Chemother. 14*:812-816.

Nilsson, L., H. Hojer, S. Ansehn, and A. Thore (1977). A rapid semiautomated bioassay of gentamicin based on luciferase assay of bacterial adenosine triphosphate. *Scand. J. Infect. Dis. 9*:232-236.

Nilsson, L., H. Hojer, S. Ansehn, and A. Thore (1979). A simplified luciferase assay of antibiotics in clinical serum specimens. In *Proceedings of the International Symposium on Analytical Applications of Bioluminescence and Chemiluminescence*, E. Schram and P. Stanley (Eds.). State Printing and Publishing, Westlake Village, Cal., pp. 515-522.

Oleniacz, W. S., M. A. Pisano, M. H. Rosenfeld, and R. L. Elgart (1968). Chemiluminescent method for detecting microorganisms in water. *Environ. Sci. Technol.* *2*:1030-1033.

Picciolo, G. L., E. W. Chappelle, E. A. Knust, S. A. Tuttle, and C. A. Curtis (1975). Problem areas in the use of the firefly luciferase assay for bacterial detection. In *Analytical Applications of Bioluminescence and Chemiluminescence*, E. W. Chappelle and G. L. Picciolo (Eds.). National Aeronautics and Space Administration, Washington, D.C., pp. 1-26.

Reeves, D. S., and M. J. Bywater (1975). Quality control of serum gentamicin assays: experience of national surveys. *J. Antimicrob. Chemother.* *1*:103-116.

Robrish, S. A., C. W. Kemp, and W. H. Bowen (1978). Use of extractable adenosine triphosphate to estimate the viable cell mass in dental plaque samples obtained from monkeys. *Appl. Environ. Microbiol.* *35*:743-749.

Sagone, A. L. Jr., G. W. King, and E. N. Metz (1976). A comparison of the metabolic response to phagocytosis in human granulocytes and monocytes. *J. Clin. Invest.* *57*:1352-1358.

Searle, N. D. (1975). Applications of chemiluminescence to bacterial analysis. In *Analytical Applications of Bioluminescence and Chemiluminescence*, E. W. Chappelle and G. L. Picciolo (Eds.). National Aeronautics and Space Administration, Washington, D. C., pp. 95-103.

Shattil, S. J., and J. S. Bennett (1981). Platelets and their membranes in hemostasis: physiology and pathophysiology. *Annals Int. Med.* *94*:108-118.

Silverblatt, F. J., J. S. Dreyer, and S. Schauer (1979). Effect of pili on susceptibility of *Escherichia coli* to phagocytosis. *Infect. Immun.* *24*:218-223.

Skog, S., and B. Tribukait (1979). ATP-content and oxygen uptake during the cell cycle of diploid and tetraploid Ehrlich ascites tumour cells. In *Proceedings of the International Symposium on Analytical Applications of Bioluminescence and Chemiluminescence*, E. Schram and P. Stanley (Eds.). State Printing and Publishing, Westlake Village, Cal., pp. 392-400.

Starkebaum, G., D. L. Stevens, C. Henry, and S. E. Gavin (1981). Stimulation of human neutrophil chemiluminescence by soluble immune complexes and antibodies to neutrophils. *J. Lab. Clin. Med.* *98*:280-291.

Strange, R. E., H. E. Wade, and F. A. Dark (1963). Effect of starvation of adenosine triphosphate concentration in *Aerobacter aerogenes*. *Nature* *199*:55-57.

Summerfield, G. P., J. P. Keenan, N. J. Brodie, and A. J. Bellingham (1981). Bioluminescent assay of adenine nucleotides: rapid analysis of ATP and ADP in red cells and platelets using the LKB luminometer. *Clin. Lab. Haemat.* 3:257-271.

Tarkkanen, P., K. H. Sturner, R. Driesch, and H. Greiling (1979). A simplified bioluminescent method for the determination of intracellular and extracellular ATP in the platelets. In *Proceedings of the International Symposium on Analytical Applications of Bioluminescence and Chemiluminescence*, E. Schram and P. Stanley (Eds.). State Printing and Publishing, Westlake Village, Cal., pp. 411-420.

Tennant, G. B. (1977). Continuous-flow automation of the *Lactobacillus casei* serum folate assay. *J. Clin. Pathol.* *30*:1168-1174.

Thore, A., S. Ansehn, A. Lundin, and S. Bergman (1975). Detection of bacteriuria by luciferase assay of adenosine triphosphate. *J. Clin. Microbiol.* *1*:1-8.

Thore, A., L. Nilsson, H. Hojer, S. Ansehn, and L. Brote (1977). Effects of ampicillin on intracellular levels of adenosine triphosphate in bacterial cultures related to antibiotic susceptibility. *Acta. Pathol. Microbiol. Scand. Sect. B.* *85*:161-166.

Velan, B., H. Schupper, T. Sery, and M. Halmann (1979). Solid-phase chemiluminescent immunoassay. In *Proceedings of the International Symposium on Analytical Applications of Bioluminescence and Chemiluminescence*, E. Schram and P. Stanley (Eds.). State Printing and Publishing, Westlake Village, Cal., pp. 431-437.

Vellend, H., S. A. Tuttle, M. Barza, L. Weinstein, G. L. Picciolo, and E. W. Chappelle (1975). A rapid method for the determination of microbial susceptibility using the firefly luciferase assay for adenosine triphosphate (ATP). In *Analytical Applications of Bioluminescence and Chemiluminescence*, E. W. Chappelle and G. L. Picciolo (Eds.). National Aeronautics and Space Administration, Washington, D.C., pp. 43-44.

Venge, P., L. Hakansson, R. Hallgren, and C. Pettersson (1979). Continuous measurement of ATP and chemiluminescence in migrating and phagocytizing neutrophil granulocytes. In *Proceedings of the International Symposium on Analytical Applications of Bioluminescence and Chemiluminescence*, E. Schram and P. Stanley (Eds.). State Printing and Publishing, Westlake Village, Cal., pp. 560-581.

Waterworth, P. M. (1972). The in-vitro activity of tobramycin compared with that of other aminoglycosides. *J. Clin. Pathol.* *25*:979-983.

Webb, L. S., B. B. Keele Jr., and R. B. Johnston Jr. (1974). Inhibition of phagocytosis-associated chemiluminescence by superoxide dismutase. *Infect. Immun.* *9*:1051-1056.

Wenham, D. G., R. M. Davies, and J. A. Cole (1981). Insoluble glucan synthesis by mutansucrase as a determinant of cariogenicity of *Streptococcus mutans*. *J. Gen. Microbiol.* *127*:407-415.

Windisch, R. M., P. R. Pax, and M. M. Bracken (1970). Variations in blood ATP after oral administration of glucose, in individuals diagnosed as normal, equivocal, or diabetic according to the glucose tolerance sum principle. *Clin. Chem. 16*: 941-944.

Wrogemann, K., M. J. Weidemann, B. A. Peskar, H. Staudinger, E. T. Rietschel, and H. Fischer (1978). Chemiluminescence and immune cell activation. 1. Early activation of rat thymocytes can be monitored by chemiluminescence measurements. *Eur. J. Immunol. 8*: 749-752.

Wrogemann, K., M. J. Weidemann, U-P. Ketelsen, H. Wekerle, and H. Fischer (1980). Chemiluminescence and immune cell activation. II. Enhancement of concanavalin A-induced chemiluminescence following in vitro preincubation of rat thymocytes; dependency on macrophage-lymphocyte interaction. *Eur. J. Immunol. 10*: 36-40.

11

INSTRUMENTATION

PHILIP E. STANLEY* The Queen Elizabeth Hospital, Woodville, South Australia

I. Introduction 220

II. Reaction Vessel 221

III. Detector 229

IV. Detector Chamber 230

V. Temperature Control 232

VI. Injection 233

VII. Signal 234
- A. Delay 235
- B. Peak 235
- C. Integration 235
- D. Zero Adjustment 236
- E. Overflow or Overload 236
- F. Signal Autostart 237
- G. Output Device 237

VIII. Sensitivity 238

IX. Flow Monitoring 239

X. Microprocessor Control, Automation, and Data Processing 239

Note Added in Proof 241

References 259

*Dr. Stanley is currently a consultant scientist based in Cambridge, England.

I. INTRODUCTION

Although analytical luminescence has obvious potential advantages, in particular, sensitivity and linearity, this methodology had not played a role in the clinical laboratory until very recently.

The last 20 years has seen substantial advances in our understanding of the processes underlying chemiluminescence and bioluminescence, and a number of workers in specialized fields have developed analytical procedures based on this new knowledge. However, the lack of availability of quality-controlled reagents as well as specialized yet flexible instrumentation has thwarted the growth of the technique in the clinical as well as other laboratories.

Recently, a distinct change in this situation occurred when a number of quite sophisticated instruments (Figs. 1-21) appeared on the market and a variety of good-quality reagents became available.

In this chapter I wish to discuss the topic of instrumentation and have taken the view that the availability in the clinical laboratory of a commercially available luminometer is likely to be more attractive than a home-built device. Therefore, in the discussion that follows concerning the various facets of the instrumentation, I have placed some emphasis on those features available on current commercial instrumentation.

I make no claim that this overview is complete or comprehensive but hope that it will enable the reader to become familiar with this rapidly growing area of instrumentation.

Having had but little experience with build-it-yourself instrumentation, I do not intend to comment further other than to indicate two recent publications (Wampler, 1975; Anderson et al., 1978) for the reader interested in this aspect.

The use of the liquid scintillation counter (LSC) for this technique will likewise receive little comment, save to say that it has been discussed by the present author (Stanley, 1974) and now with the availability of flexible luminometers at moderate cost, it seems likely that LSC use for this purpose will wane.

There have been but few recent (since 1975) papers concerned with the performance of commercial luminometers and their capabilities, the group at NASA being the main contributors (Chappelle et al., 1975; Picciolo et al., 1977,1978).

I believe it is true to say that some of the instruments available today were not designed with the clinical laboratory specifically in mind. Instead they were probably intended primarily for measuring stat ATP levels, e.g., in the determination of biomass in the laboratory and in the field. Indeed, for this requirement three instruments have been designed to be readily portable and battery-operated (Lumac Lumatec (Model 1010), Lumac Luminometer (Model 1070), and SAI ATP-Photometer (Model 100)].

The instrument for use in the clinical laboratory is likely to require a number of rather advanced facilities including good sensitivity and long-term stability with flexibility to measure stat levels and both reaction rate and endpoint determinations. In addition, it should be able to cope in both capacity and capability with a clinical laboratory load, perhaps have automation, good temperature control for both reagents and the detector, and probably data-processing capability.

I have collated in Table 1 the models known to me in October 1980 and have included some brief comments on each instrument. Instruments introduced since that date are described in the Note Added in Proof. Company names and addresses, updated in July 1982, appear in Table 1.

II. REACTION VESSEL

Reaction vessels, sometimes called cuvettes, range in size from around 50 µl to 20 ml, although most are between 1 and 2 ml. The majority are cylindrical and constructed from glass or clear polystyrene and are usually disposable. Volumes of reactants and analytes used for work in clinical luminescence generally range from 10 to 200 µl and so a reaction vessel in which from 200 µl to 1 ml can be contained comfortably is likely to cover many applications.

Because some assays involve reactions which take place very quickly (1 to 5 sec), the shape and size of the vessel together with the mode of addition of reactants or analytes are important design parameters since adequate and rapid mixing is a prerequisite. Poor mixing can often be the cause of variable or spurious results and one should not assume that a mixing scheme which is suitable for a particular assay will also be suitable for any other assay. Attention must be paid to the volumes being mixed as well as their densities and viscosities. Seitz and Neary (1976) discussed this problem in some detail.

The SAI ATP-Photometers (Models 2000 and 3000) and New Brunswick Lumitran (Models L-2000 and L-3000) can utilize liquid scintillation vials of 20-ml capacity, although reaction volumes are usually only a few milliliters. These instruments can also be configured with other reaction vessels so that as little as 10 µl reaction volume can be used. The Packard Pico-Lite (model 6100) unit requires 50 × 6 mm tubes accommodating up to 400 µl and tubes of a similar size may be used in the ALL Monolight (Models 301 and 401), the Turner ATP-Photometer (Model 20), and Aminco Chem-Glow instruments, although the latter can be configured to accept 35 × 12 mm tubes. A standard size cuvette holding 100 to 200 µl is used in the Jobin-Yvon Photometre PICO-ATP model while the Skan Bioluminescence Analyzer (Model XP-2000-2) has a 3-ml cuvette holding 200-µl

Table 1 Manufacturers (names and addresses updated July 1982)

Company	Address	Model	Comments
1. Luminometers			
Alpkem Corporation	Box 1260 Clackamas OR 97015 U.S.A.	Automated Luminescence Analyzer	For measuring cells in fluids. Automatic (40) sampler, peristaltic movement of sample and reagents (luciferase and extracting reagent) through coils. Flow cell in photometer. Recorder output.
American Instrument Co.	8030 Georgia Avenue Silver Spring MD 20910 U.S.A.	Chem-Glow Photometer	Meter, manual, or semiautomatic injection, two models for different cuvette sizes, integrator timer available, analog output.
Analytical Luminescence Laboratory Inc. (ALL)	1180 Roselle St. Units D and E San Diego CA 92121	Autolight, Model 101	Microprocessor, automatic preparation unit for 40 samples, printer, flow cell with 3 ports, use with other luminometers, data processing.
		Monolight, Model 201	Microprocessor, printer, auto-dual injection, photon or current measuring modes, integrator peak measurement, data processing.
		Monolight Model 301	Meter, manual, or automatic injection integrator timer available, analog output.

		Monolight Model 401	Microprocessor, current measuring, manual injection with autostart available, automatic safety shutter for photomultiplier, push button sensitivity control, optional printer. Output suitable for computer processing.
Berthold Laboratorium Prof. Dr. Berthold	Postfach 160 D-7547 Wildbad 1 West Germany	Biolumat Model LB 9500	Photon counter, temperature-controlled cuvette, automatic injection, peak, rate, or integral measurements, digital display. When interfaced with HP 97S microcomputer, unit programmable and wide range of data processing, printer.
		Six-Channel Luminescence Analyzer Model LB 9505	Photon counter. Six samples measured simultaneously. Temperature control of samples, cooling of photomultipliers. Interfacing to computer available.
Du Pont Company	Wilmington DE 19898 U.S.A.	Biometer Model 760	Digital readout, autoranging, semiautomatic.
Jobin Yvon S.A.	19, Rue de la Gare 94230 Cachan France	Photometer Pico-ATP	Pneumatic injection, digital display.

Table 1 (Continued)

Company	Address	Model	Comments
LKB Wallac	P.O. Box 10 20101 Turku 10 Finland	Luminometer Model 1250	Modular, printer, built-in standard photon source, manual or semi-automatic injection, temperature-controlled cuvette available (water circulation), mixer for multiple reagent applications available. Integration 1 or 10 sec.
Lumac/3M B.V. and Medical Products Division/3M	P.O. Box 31101 6370 AC Schaesberg The Netherlands	Lumacounter Model 2080	Photon counter, temperature-controlled cuvette, automatic injection, peak, rate, or integral measurements, digital display. When interfaced with HP97S microcomputer, unit programmable and wide range of data processing, printer.
	3M Center St. Paul MN 55144 U.S.A.	Biocounter Model 2010	Photon counter, temperature-controlled cuvette, automatic or manual injection, rate or integral measurements, digital display. When interfaced with microcomputer, unit programmable with wide range of data-processing capability, video display. Three autoinjectors for reagents and analytes.

Biocounter Model 2000	Photon counter, temperature-controlled cuvette, automatic or manual injection, rate or integral measurements, digital display. When interfaced with microcomputer, range of data processing, printer.
Luminometer Model 1070	Solid state detector, battery or ac operation, portable, autoranging, semiautomatic injection, integral measurement, automatic zeroing, digital display.
Celltester Model 1060	Solid state detector, automatic or manual injection, peak or integral measurement, automatic zeroing, signal displayed every 2.4 sec. When interfaced with HP 97S, data processing.
Celltester Model 1030	As model 1060 but without interfacing for HP 97S.
Lumatec Model 1010	Solid state detector, portable, battery-operated, manual zeroing, meter display.

Table 1 (Continued)

Company	Address	Model	Comments
Marwell International AB	Kyrkbacken 27 S-17150 Solna Sweden	Model 302 Kinetic Luminescence Analyzer	Semiautomatic, digital display, photon counter, pneumatic injection, temperature-controlled cuvette to 0°C (electrical).
		Model 340/50 Luminescence Analyzer	Automatic, microprocessor-controlled, photon counter, data processing, up to 500 samples, temperature-controlled cuvette to 0°C (electrical), printer.
New Brunswick Scientific Co. Inc. (See Note Added in Proof)	P.O. Box 986 44 Talmadge Road Edison NJ 08817 U.S.A.	Lumitran Model 2000	Digital display, three sizes of detector assembly, semiautomatic injection available, peak and integral measurement (timed delay and count time).
		Lumitran Model 3000	As above but with variable time delay and count time.
Packard Instrument Co. Inc.	2200 Warrenville Rd. Downers Grove, IL 60515 U.S.A.	Pico-Lite Model 6100	Separate detector (6 samples) and microprocessor-controlled analyzer, photon counter, temperature-controlled cuvette (adjustable), 7 measuring programs and data processing, printer, manual injection, peak and integral measurements.

SAI Technology Inc. (See Note Added in Proof)	4060 Sorrento Valley Blvd. San Diego CA 92121 U.S.A.	ATP-Photometer Model 100	Portable, semiautomatic injector, disposable tips (no further information available at present).
		ATP-Photometer Model 2000	Digital display, three sizes of detector assembly, semiautomatic injection available, peak and integral measurement (timed delay and count time).
		ATP-Photometer Model 3000	As above but with variable time delay and count time.
Skan AG	Baslerstrasse 354 Allschwil CH-4009 Basel Switzerland	Bioluminescence Analyzer Model XP-2000-2	Semi automatic or manual injection, digital display, printer. Peak and integral measurement.
Turner Designs	2247 Old Middlefield Way Mountain View CA 94043 U.S.A.	ATP-Photometer Model 20	Microprocessor, current measuring. Manual injection with autostart available, automatic safety shutter for photomultiplier, push button sensitivity control, optional printer. Output suitable for computer processing.
Vitatect Corporation	8510 Conover Place Alexandria VA 22308 U.S.A.	Vitatect II	For measuring cells in fluids. Cells filtered and ATP extracted on movable filter tape, photomultiplier, digital and meter display.

Table 1 (Continued)

Company	Address	Model	Comments
		Vitatect IIS	As above, but with different electronic circuitry.
		Vitatect III	Automatic version of above.
2. Flow Monitors			
Berthold Laboratorium	Postfach 160 D-7547 Wildbad 1 West Germany	Model LB 503	Meter, autoranging rate meter, analog integration.
Radiomatic Instruments and Chemical Co.	5102 S. Westshore Blvd. Tampa FL 33611 U.S.A.	FLO-ONE	Microprocessor, built-in reagent pump, digital integration, peak collection, printer.

to 1.5-ml reactants. LKB recommends a cuvette 55 × 10-12.5 mm to accommodate 200 μl to 2 ml in their Luminometer (Model 1250). Lumac supplies disposable cuvettes (plastic or glass) for their instrumentation which may be used for final volumes between 10 μl and 1 ml. Du Pont supplies special cuvettes for their Biometer unit. The flow cell on the ALL Autolight (Model 101) accommodates 1-ml reactants and has three ports for reactants and waste. The Alpkem Automated Luminescence Analyzer also has a flow cell.

It is important that reaction vessels or cuvettes be stored under clean conditions since dust and sometimes microorganisms can interfere with luminescence assays. In addition, they should be kept away from any light sources (fluorescent lighting, sun) which might cause them to phosphoresce and thus produce spurious background signals.

III. DETECTOR

The detector is of course the heart of the instrument and its characteristics determine the ultimate sensitivity that can be achieved. All luminometers currently marketed sense the entire spectrum of light emitted from the reaction vessel; however, their quantum efficiency varies in the spectral range (400 to 600 nm) where luminescence occurs. Solid-state detectors (silicon photodiodes) are employed by Lumac in some of their instrumentation and such detectors provide moderate sensitivity over the range 420 to 600 nm as well as ruggedness, portability, and rapid warmup. Stimson (1974) discussed the characteristics of these detectors. The other type of detector, the photomultiplier, is used on all other luminometers. It has excellent sensitivity but requires a stable, high-voltage power supply. Warmup time is only a few seconds for photon counters but may be several minutes for current measuring devices.

Photomultipliers have a wide dynamic range of linear operation and a number of luminometers will operate over 5 decades of sensitivity. The light-sensitive surface (the photocathode) of commonly used photomultipliers exhibits an excellent quantum efficiency of around 25% at lower wavelengths (350 to 450 nm). Unfortunately, most spectra of interest in clinical luminometry occur in the wavelength range 400 to 600 nm [firefly (ATP), λ_{max} = 560 nm; *Photobacterium* [NAD(P)H], λ_{max} = 490 nm; luminol λ_{max} in water (H_2O_2) = 425 nm] where the efficiency is noticeably less. Photocathodes produce a definite background (dark noise) which is temperature-dependent, and photomultipliers for luminometers are selected for low noise at ambient temperature and high efficiency since they must be capable of measuring very small amounts of light.

Present photomultipliers which have good red sensitivity (more suited to the ATP-firefly assay) have somewhat high backgrounds unless they are cooled to rather low temperatures. Berthold has employed Peltier elements to cool such photomultipliers of which they employ six in their Six-Channel Luminescence Analyzer (Model LB 9505), which is used for studies of phagocytosis (Berthold et al., 1981).

The properties and sensitivities of some photomultipliers are given by Stimson (1974). The types of photocathodes used include the following: S4 [Aminco, ALL Monolight (Model 301)], S11 (Du Pont), S20 (Marwell) and bialkali [SAI, Lumac Lumacounter (Model 2080) and Biocounter (Models 2000 and 2010), Berthold Biolumat (Model LB 9500), Vitatect, Packard]. In the Marwell (Model 340/50) the type of photomultiplier may be chosen by the purchaser. In choosing the type of photocathode, care must be taken to assess not only the peak quantum efficiency but also the quantum efficiency at the wavelengths to be measured as well as its noise and the temperature coefficient of the noise signal.

Some units, e.g., Packard Pico-Lite (Model 6100), Lumac Lumacounter (Model 2080) and Biocounter (Models 2000 and 2010), and Berthold Biolumat (Model LB 9500), are fitted with a 1-in. diameter photomultiplier rather than the more commonly used 2-in. device. The photosensitive area is around 4 times smaller but is still more than adequate to monitor the sample. With 4 times less photocathode the background noise is substantially reduced thereby producing an improved signal-to-noise ratio.

The sensitivity of photomultipliers can be changed by adjusting the high-voltage supply, and among those with this feature are luminometers made by Du Pont, LKB, SAI, Vitatect, and Skan. In others, the highest sensitivity is available at all times [Lumac Lumacounter (Model 2080) and Biocounter (Models 2000 and 2010), Packard Pico-Lite (Model 6100), Berthold Biolumat (Model LB 9500)]. In the Turner ATP-Photometer (Model 20) and ALL Monolight (Model 401), however, variations in photometer sensitivity are corrected for 10 times every second after exposing the photomultiplier to a stabilized reference light source and by making an appropriate small change in the high-voltage supply.

In most instruments in which photomultipliers are utilized, the signal is derived as current which may be displayed on an analog meter or in digital form perhaps as a number directly associated with the analyte concentration. The Lumac Lumacounter (Model 2080) and Biocounter (Models 2000 and 2010), the Berthold Biolumat (Model 9500), and Marwell instruments (Models 302 and 340/50) as well as the Packard Pico-Lite (Model 6100), however, are photon (pulse or quantum) counters in that the signals at the anode are presented as a small pulse of electricity, each of which has been derived from a single photon detected by the photocathode.

IV. DETECTOR CHAMBER

The detector chamber should be designed to optimize light collection by the detector from the reaction vessel. As the detector is placed to one side of the reaction vessel, light emitted on the other side will be lost to the detector if it is not reflected back toward it. A reflective surface on the chamber wall will facilitate this, although a certain number of photons will be directed back into the reaction vessel where they may be either absorbed, refracted, or passed through to the detector. This is a complicated situation because reflection will vary with wavelength and physical size of the chamber, its surface character, and, of course, the optical characteristics of the reagent solution (especially if colored) and shape and size of reagent vessel.

I know little about the efficiency of light collection in any of the available instruments and to obtain information is technically difficult. Carter et al. (1979) and Bullock et al. (1979) used computer modeling to look at the effect of detector size and shape, mirrors, lenses, and shape of reaction vessel. The present author has used computer modeling in an effort to understand the liquid scintillation counting process and the geometry of the scintillation vessel, and the form of the detector and its chamber were found to be important design features when measuring very low light levels (quenched tritium) (Malcolm and Stanley, 1976a,b).

Some instruments incorporate a design in which the reaction vessel is placed into the detector chamber which is an integral part of a rotating drum device. Rotation of the drum through 180° brings the sample adjacent to the detector. Such a system is used in the Aminco, Du Pont, LKB, Skan, and ALL Monolight (Model 301) instruments. The LKB instrument also incorporates in the drum assembly a radioluminescent photon standard which may be used to calibrate the luminometer. Most other instruments have the detector chamber fixed in close proximity to the detector with a light shutter that must be operated when the sample vessel is introduced or removed. The shutter and drum assembly protect the detector from ambient light which may damage it irreparably. The Packard Pico-Lite (Model 6100) has a separate detector unit (Model 601) with six sample positions which may be rotated manually to bring each reaction vessel in turn adjacent to the photomultiplier.

It is important that the detector chamber be completely light-tight so that extraneous light cannot reach the detector. Tubing from the exterior to the reaction vessel should not act as a light pipe. The same applies to syringes and syringe needles used for injection of reagents. Septa through which such needles pass must be thick and tough enough to withstand a reasonable number of injections before light passes through the hole caused by repeated needle penetrations. Septa have been dispensed with in the Lumac instrumentation and replaced with an automatic shutter assembly and precision reagent dispenser or, more recently, up to three dispensers (Biocounter, Model 2010). It is important to keep such dispensers clean.

V. TEMPERATURE CONTROL

Control of temperature of the detector chamber and hence the reaction vessel and its contents had not received much attention from commercial instrument manufacturers until recently. Operation at ambient temperatures for relatively short periods of time allowed workers using the firefly system for measuring ATP to obtain acceptable results. However, more recently the use of the bacterial luciferase system to measure NADH and dehydrogenases, phagocytosis, and enzymes utilizing or degrading ATP have made temperature control essential because they are temperature-sensitive systems and sometimes require long-term incubation. When such temperature control is required I would advise checking this parameter in the instrument. The heat generated over a period of time by the electronics may also be a problem in that it may elevate the detector temperature over the preset value. Of course, this is important when operating near ambient temperatures or in control units where excess heat cannot be removed.

The instruments marketed presently by SAI, Skan, New Brunswick, Aminco, Alpkem, and Du Pont have no temperature control facility and neither do the Models 1010, 1030, and 1070 of Lumac or the ALL Monolight (Model 301). However, it must be pointed out that such units are mainly used for stat ATP assays or are portable field instruments (Lumac, Models 1010 and 1070). The LKB Luminometer (Model 1250) can be fitted with a special cuvette holder through which temperature-regulated fluid can be passed. The detector chamber in the Turner ATP-Photometer (Model 20) and ALL Monolight (Model 401) can be temperature-controlled from an external circulating waterbath via appropriate hose. The Packard Pico-Lite detector chamber (Model 601) is provided, under microprocessor control, with a heater to give control from just above ambient to 44°C. Temperature-regulated fluid may be circulated through the chamber for subambient operation. The Lumac Lumacounter (Model 2080) and Berthold Biolumat (Model LB 9500) are supplied with chambers set at 25, 30, and 37°C while on the Lumac Biocounter (Models 2000 and 2010) the user can select temperatures 25, 30, and 37°C by a front panel switch. Temperature control is an optional extra on the Vitatect instrumentation and standard on the flow cell of the ALL Monolight (Model 201) (heating only, ambient to 45°C) and of the ALL Autolight (Model 101) (heating only, ambient to 45°C and preselected at 30, 32, and 37°C). The Marwell instruments have detector chambers fitted with Peltier elements which allow temperature control down to 0°C while the Berthold (Model LB 9505) has detector chambers that may be set for operation between 25.6 and 51.2°C in 0.1°C steps.

In using temperature control facilities it is of course important to allow the reagents in the reaction vessel to come to thermal equilibrium before the reaction is initiated. If the added reagent is at a considerably different temperature or is of a volume comparable with that of the solution in the vessel, then the actual temperature at the time of mixing may be substantially removed from that set on the instrument. The use of small injection volumes therefore facilitates good temperature control of the assay reaction.

The actual temperature of the detector chamber can be displayed or printed in some instruments, e.g., Packard Pico-Lite, ALL Autolight (Model 101) and Monolight (Model 201), Marwell (Model 340/50). In others, e.g., Lumac Lumacounter (Model 2080) and Biocounter (Models 2000 and 2010) and Berthold Biolumat (Model LB 9500), an indicator is lit when the heater is operational.

Temperature control of reagents prior to use is available on some fully automatic instruments. Thus the sample preparation table on the ALL Autolight (Model 101) can be adjusted from ambient to 45°C and in the Marwell (Model 340/50) injected reagents are kept in a cooled reservoir prior to use. For automatic instrumentation, cooling of temperature-labile reagents prior to their use seems mandatory if a substantial time elapses over the cycle time.

When changing the operating temperature of the detector, chamber care must be taken to note any change in the background signal of the detector.

At present two instruments have the facility to maintain at temperature a number of reaction vessels with their contents. The detector (Model 601) of Packard Pico-Lite (Model 6100) can do so for six vessels concurrently and the Berthold (Model LB 9505) can also handle six and in addition measure light output from each simultaneously. However, for long-term incubations, e.g., some enzyme determinations or phagocytosis, a typical experimental run may require multiple sequential readings of, say, 20 samples over a period of an hour.

VI. INJECTION

The assay of ATP by the flash height procedure requires that ATP or firefly luciferase be added to the reaction vessel in the darkened detector chamber because the signal to be measured occurs within a second or so. The addition may be accomplished by reproducible injection, often via a septum using a syringe or device operated externally to the chamber. Almost all units marketed currently can be fitted for this maneuver.

Injection of internal standard, inhibitors, or other reagents may also need to be performed without removal of the reaction vessel from the chamber.

The purpose of injection is not only to add reagents but also to mix them quickly and reproducibly. When changing to a new assay or procedure, ensure that good mixing occurs by viewing outside the detector. Difficulties may be encountered with reagents of different density and/or viscosity (Seitz and Neary, 1976).

The units marketed by Jobin-Yvon and Skan utilize a special injector assembly whereas SAI and New Brunswick instrumentation can be fitted with an optional injection pipette which plugs in electrically to the unit so that delivery of reagent and assay sequence may be synchronized. The LKB Luminometer (Model 1250) has as an option a manually operated injection device fitted with a syringe at the head of the luminometer. A repeating Hamilton syringe dispenser is used on the Aminco Chem-Glow and Du Pont Biometer units. A similar syringe that is electrically coupled to permit autostart is available for the Turner unit and the ALL Monolight (Model 401). These latter two instruments may also be fitted with an electrically coupled pipette dispenser which can be used to inject 50, 100, or 200 μl per operation via a disposable tip. The fitting of the dispenser does not involve the use of a septum. The SAI ATP-Photometer (Model 100) uses a spring-loaded pipetting device fitted with a disposable tip. Lumac (Models 1030, 1060, 1070, 2000, 2010, 2080) and Berthold (Models LB 9500 and LB 9505), on the other hand, use a high-precision reagent injector dispensing 100 μl directly into the reaction vessel and have abandoned the septum because the injector incorporates a light-tight fitting around the cuvette. Lumac recently introduced a feature for their Biocounter (Model 2010) unit which allows it to be fitted with three independent injector/mixers, one for ATP standard (internal standard injection), one for reagent (e.g., firefly luciferase), and one for releasing reagent (to extract ATP from cells). The injection sequence may be controlled by a computer.

Automatic pneumatic injection is used on the Marwell (Model 302) and mixing occurs in less than 20 msec while a manually operated syringe is used for addition of reagents in the Packard detector head (Model 601).

Although working on quite a different concept, the Vitatect instrumentation makes use of a repeating Hamilton syringe or Vitatect dispensing pump to dispense reagents (10 to 50 μl) onto the tape containing the extracted ATP.

Flow mixing is utilized on the ALL Autolight (Model 101) and Alpkem Automated Luminescence Analyzer, and dual injection (sample and standard) is available on the ALL Monolight (Model 201).

Injection volume is generally only changeable by manual adjustment; however, the Marwell (Model 340/50) provides for this injected volume to be programmed.

VII. SIGNAL

The two kinds of detector, the silicon photodiode and the photomultiplier, produce a signal which is linearly related to the amount of light detected, and this relationship holds for four or five decades.

Basically the signal which is measured is either current or pulses. I will consider the latter first.

When operated at high gain, the photomultiplier will produce a fast but small pulse of electricity when the photocathode responds to a single photon. By scaling the pulses or measuring their rate of production, a measure of light incident on the photocathode is obtained. Among those instruments utilizing this method are the Lumac Lumacounter (Model 2080) and Biocounter (Models 2000 and 2010), the Marwell (Models 302 and 340/50), the Berthold Biolumat (Model LB 9500), and the Packard Pico-Lite (Model 6100).

Photomultipliers may also be employed so that the signal for analysis is in the form of anode current. This may be displayed on an analog meter (with or without time constant) as in the Aminco Chem-Glow and ALL Monolight (Model 301) or as in most other instruments in a digital fashion. On all nonportable instruments, a signal is available for a chart recorder.

A. Delay

On a number of instruments a facility is available to delay the recording of signal for a specified time following injection. This may be necessary if integration mode is used after the light flash has occurred or if light production is delayed. Thus the SAI ATP-Photometer (model 2000) gives a fixed 15-sec delay while the model 3000 provides set delays of 0.5, 5, 10, 15, 20, and 30 sec as required. The Turner ATP-Photometer (Model 20) and ALL Monolight (Model 401) have 12 settings between 0 and 30 sec. Vitatect IIs provide for delay periods within a set range of values. The Packard Pico-Lite and Lumac Lumacounter (Model 2080 with HP97S) and Biocounter (Model 2000 with HP97S and Model 2010 fitted with Pet or Apple computer) and Berthold Biolumat (Model LB 9500) can be programmed for a delay selectable as required.

B. Peak

Most instruments can be set to detect and measure the peak of light output (flash height) and record this value on a digital display or a printer. An accessory is required for the Aminco Chem-Glow and ALL Monolight (Model 301).

C. Integration

All units can perform integration of light output in the form of either total pulses or current. This can be done simply over a fixed time period, e.g., 60 sec for SAI ATP-Photometer (Model 2000) and

New Brunswick Lumitran (Model L-2000), or over a selection of fixed times, e.g., 1 and 10 sec (repeated) on LKB Luminometer (Model 1250), 5, 10, and 40 sec on Lumac Celltester (Models 1030 and 1060), and 10, 30, and 60 sec on the Lumac Lumacounter (Model 2080) and Biocounter (Models 2000 and 2010) and Berthold Biolumat (Model LB 9500). The Turner ATP Photometer (Model 20) and ALL Monolight (Model 401) have 15 preset integration times between 0 and 120 sec and these instruments can also display the integral for half the elapsed integration time. A constant ratio between the half and full integral indicates that the reaction is proceeding as expected.

The Lumac Lumacounter (Model 2080), Biocounter (Model 2000), and Berthold Biolumat (Model LB 9500) can be fitted with a microcomputer, HP97S, and the Biocounter (Model 2010) with a computer such as the Pet or Apple unit, and in this configuration any integration time interval may be selected up to 60 sec. Similarly, the microprocessor on the Packard Pico-Lite (Model 6100) allows the user to set any integration time.

D. Zero Adjustment

Both the detector and the reagents will produce a signal in the absence of sample and this may be nulled electrically by a manual adjustment on a number of instruments including SAI ATP-Photometer (Models 2000 and 3000), New Brunswick Lumitran (Models L-2000 and L-3000), ALL Autolight 101, ALL Monolight (Model 301), LKB Luminometer (Model 1250), Vitatect II and IIs, Skan Bioluminescence Analyzer (Model XP-2000-2), Du Pont Biometer (Model 760), Lumac Lumatec (Model 1010). Nulling of background signal is performed automatically on the Lumac Celltester (Models 1030 and 1060) and Lumac Luminometer (Model 1070). The Turner ATP-Photometer (Model 20) and the ALL Monolight (Model 401) have automatic background subtraction of the photomultiplier noise during the 10 times a second sensitivity adjustment. Subsequent automatic digital subtraction of a background value is possible on the ALL Monolight (Model 201), the Packard Pico-Lite (Model 6100), and the Lumac Lumacounter (Model 2080), Lumac Biocounter (Model 2000), and Berthold Biolumat (Model LB 9500) when interfaced with HP97S microcomputer and when the Lumac Biocounter (Model 2010) is interfaced with HP97S, Apple, or Pet computers.

E. Overflow or Overload

Should the light level be too high for the detector or the display to indicate, several instruments have a signal light to show this condition (most Lumac instruments, SAI). The high voltage to the photomultiplier is automatically switched off on SAI, New Brunswick, and LKB instruments and must be manually reset. In the Turner ATP-Photometer (Model 20) and ALL Monolight (Model 401) the presence of excessive light is indicated by all four digits of the display flashing on and off. An automatic shutter, electrically operated, is fitted to these instruments and this operates to protect the photomultiplier from excessive light from the sample or surroundings.

F. Signal Autostart

It may be convenient at times for the instrument to start monitoring the signal automatically when the signal reaches a preset value. The Packard Pico-Lite (Model 6100) only has provision for manual injection which is not electrically coupled to the detector. However, it may be programmed to commence counting when the signal has reached a value 5 times a selected background value.

The Lumac Lumacounter and Biocounter have automatic injection which is coupled to the detector and scaler. However, autostart may also be used with a suitable program for the HP97S (Model 2080) or other computer (Model 2010) and the Berthold Biolumat (Model LB 9500) when fitted with the HP97S.

The Turner ATP-Photometer (Model 20) and ALL Monolight (Model 401) can be fitted with semiautomatic injection units which are electrically coupled to the unit so as to enable autostart when the injection occurs.

G. Output Device

The signal for analysis may be output on a range of devices, the simplest being an analog meter. This is used on the Aminco Chem-Glow, ALL Monolight (Model 301), Lumac Lumatec (Model 1010), and Vitatect II, although the last mentioned also has digital display.

Other instruments have digital displays. A few have printers. These include the LKB Luminometer (Model 1250), Packard Pico-Lite (Model 6100), and the Lumac Lumacounter (Model 2080), Biocounter (Model 2010), and Celltester (Model 1060) when interfaced with the HP97S microcomputer, the Berthold Biolumat (Model LB 9500) when interfaced with the HP97S microcomputer, and ALL (Models 101 and 201), Marwell (Model 340/50), and Turner ATP-Photometer (Model 20) and ALL Monolight (Model 401) when interfaced with a suitable printer. The printer may also provide hard copy of experimental conditions, e.g., temperature, correction factors, average values, etc.

Recently, the Lumac Biocounter (Model 2010) was interfaced to Pet and Apple computers with video screen, printers, and floppy disc as output devices.

VIII. SENSITIVITY

Useful sensitivity is the difference between the blank (zero control, or instrument noise + reagent) and the activity (signal). An often used measure of sensitivity* is that amount of analyte required to give twice the background value.

It is difficult to compare and contrast commercially available instrumentation at the present time with regard to their sensitivity because the author has not had an opportunity to test each under identical conditions. Manufacturers usually specify their instruments in terms of ATP sensitivity with firefly luciferase. Care should be exercised in interpreting and comparing stated sensitivities. Thus, be careful to distinguish between moles of ATP and ATP molarity. The latter is sometimes used together with volume of solution employed. In addition, volumes of analyte may be different. Certainly firefly luciferase may vary from source to source not only in specific activity but also in background light level (under assay conditions) and contaminating enzymes such as adenylate kinase. Be aware that the enzyme blank and light output from the sample may decay at different rates.

Generally speaking, photomultipliers of today are not the limiting factor on sensitivity; rather, it is the specific activity and background of the firefly luciferase. Fortunately, several manufacturers now market good preparations of firefly luciferase and subpicogram amounts of ATP can be measured with most instruments.

ATP is an unstable molecule (both in solution and as a solid) and care should be taken in using it as reference material. Standardization by an enzymatic procedure is advisable prior to sensitivity testing.

When purchasing an instrument in which sensitivity is an important operational feature, an effort should be made to measure sensitivities prior to purchase. This would be especially important where assays other than that for ATP are to be performed because the spectrum of photons and thus the characteristics of the sample chamber may affect the sensitivity.

Seliger (1980) recently discussed methods for absolute calibration of single-photon detectors.

Sensitivity can be changed on the current measuring photometers by adjusting the high voltage to the photomultiplier. However, on the Turner ATP-Photometer (Model 20) and ALL Monolight (Model 401), sensitivity is adjusted downward from the maximum by means of a light shutter. Having set the latter, three decades of sensitivity can then be selected (1, 10, 10^2, 10^3 steps). These are automatically set and adjusted by reference to light-emitting diodes.

---

*Sensitivity: that amount of analyte required to give a signal equal to twice the standard deviation of a series of blank determinations.

IX. FLOW MONITORING

In the clinical chemistry laboratory there is considerable potential for flow monitoring of bioluminescence and chemiluminescence-based assays. Several workers have published reports and the most recent ones include Seitz (1979) and Schram et al. (1979). The former reference includes some of the theory and rationale of flow cell design.

Presently two commercially available instruments are known to the author (see Table 1) which are designed for flow monitoring of luminescent solutions, and these may be of interest to the clinical chemist. They are in fact designed to monitor soft β-emitting radionuclides using the liquid scintillation technique. However, when operated out-of-coincidence and without the scintillator solution they are suited to flow monitoring of bio- and chemiluminescence. The scintillator solution pump may be used to add reagent, e.g., luminol, luciferase, immediately prior to monitoring. Several different size flow cells are available.

X. MICROPROCESSOR CONTROL, AUTOMATION, AND DATA PROCESSING

During the last 3 years a new generation of luminometers has appeared. Such features as microprocessor control and data processing are available on a number of different instruments, including the ALL Monolight (Model 101), ALL Autolight (Model 201), and ALL Monolight (Model 401), the Lumac Celltester (Model 1060 + HP97S), Lumac Lumacounter (Model 2080 + HP97S), Lumac Biocounter (Model 2000 + HP97S), Lumac Biocounter (Model 2010 + Apple or Pet computer), Berthold Biolumat (Model LB 9500 + HP97S) and Marwell (Model 340/50), and Turner ATP-Photometer (Model 20). The ALL Autolight (Model 201) and Marwell (Model 340/50) are also capable of handling automatically up to 40 and 500 samples, respectively.

I believe it is worth discussing each of these instruments in some detail.

The Lumac Biocounter (Model 2010) and Berthold Biolumat (Model LB 9500) are available with the option of a Hewlett-Packard HP97S microcomputer which can be programmed by the user through the keyboard or with prewritten manufacturer-supplied programs on magnetic strips. This allows a vast range of activities, such as automatic and multiple injections, multiple readings of the sample, and variable delay and integration times. In data processing the user may automatically fit a standard curve, calculate results for samples in a wide range of units, perform internal standard calculations and simple statistical calculations, e.g., standard deviation. Printout or display of all or selected parts of the information is possible as designated by the programmer or worker. The Lumac Biocounter (Model 2000) can be interfaced with a HP97S, the latter giving such commands as start to count, but it is mainly used as a data-processing device. The newer version of this unit, the Lumac Biocounter (Model 2010), may be interfaced with a range of microcomputers, including the HP97S and the Apple and Pet units. The microcomputer may be programmed to have full control of the Biocounter, including its configuration of three separate injectors (standard, reagent, releasing reagent), and is also employed for a wide range of data processing, including the determination of adenylate charge.

The Packard Pico-Lite (Model 6100) has an analyzer module (Model 605) with fixed software, allowing the user default-counting parameters or a selection of seven counting programs. In the latter the user supplies the required parameter values via a touch-sensitive keyboard. These include delay time, count time, temperature for assay, background subtraction, normalization of results, averaging of replicate counts, calculation based on internal standard, and automatic start when signal is 5 times a designated value. Each printout contains the parameter information, the sample position number (1 to 6), and the detector temperature as well as the assay result.

The author has only limited information on the ALL Autolight (Model 201) and Monolight (Model 101). Although the former instrument is microprocessor-controlled, standard values such as delay and count times and quenching factor are set on thumb wheels. Capacity to read and adjust for a reagent blank is incorporated as is that for normalization to a standard value. The printer not only gives the set parameters but also provides a curve of the light output of the reaction. The Monolight (Model 101) has a microprocessor which controls the automatic sample-handling unit and luminometer and receives digital data from the latter. These data are analyzed by the microprocessor, and the result is printed for each sample. A standard deviation program is included in the software. Volumes of reagents are set using manual syringe stops. Up to 40 samples may be processed sequentially.

At the time of writing the author has only limited information concerning the Marwell (Model 340/50). The unit is a fully automated system for up to 500 samples, with background subtraction, internal generation of a standard curve based on peak height or integrated time interval. Printout of results may be in any concentration unit together with date, time, and sample number. Injection volume is fully programmable.

The Vitatect III is an automated instrument designed to measure cell numbers in fluids, e.g., urine, by virtue of their ATP content. Cells are filtered on a moving tape of filter paper, the ATP is extracted in situ, and the tape is placed in a position where firefly luciferase is added automatically. An extensive evaluation of the nonautomated version of this instrument has been published by Picciolo et al. (1977).

The Alpkem Automated Luminescence Analyzer is also designed to measure cell numbers in fluids and consists of an automatic sampler (up to 40 samples) and a peristaltic pump to move the measured sample and reagents (firefly luciferase, extractant, etc.) Suitable delay coils allow sufficient time for ATP extraction from the cells and sample carryover is minimized by an intervening wash sequence. A recorder is used as the output device.

LKB has announced a software package to process data concerning creatine kinase assays which may be carried out with the LKB Luminometer (Model 1250).

NOTE ADDED IN PROOF

The instrumentation used for luminometry has advanced at a fast pace. It has become desirable to bring this chapter up to date just prior to publication in order that the reader has access to recent information.

Changes in Companies and Instrumentation

The New Brunswick Scientific Company Inc. has, I understand, discontinued the Lumitran 2000 and 3000. However, New Brunswick U.K. Ltd. (London, U.K.) has just announced the Lumitran II. Lumac B.V. of The Netherlands has been acquired recently by the 3M Company and, I understand, will be known as Lumac/3M B.V. in The Netherlands. In the United States the products will be marketed through Medical Products Division/3M (see Table 1). The luminescent products division SAI Technology Company, Inc., of San Diego, has been acquired by a new company, Diagnostic Sciences, Inc., also of San Diego. I understand that they will continue to market the Model 3000 and have also introduced a new model, the Model 101 Photometer.

New Instruments

I am indebted to the following companies who have provided me with very up-to-date information concerning their instrumentation which in some cases has been introduced within the last few months.

There has been a distinct move toward automation in the new instruments both in cuvette loading and sample processing. This kind of instrumentation should have marked advantages for the clinical laboratory.

Berthold

The Model LB 950 was shown at the Analytica-82 exhibition at Munich in April 1982. It is a photon counting instrument capable of processing up to 400 samples which are held in a free-standing chain, the elements of which can be separated. It is microprocessor controlled and has an external processor in the form of an Apple II microcomputer. Up to three reagents can be injected automatically (volumes 10 to 500 μl). Temperature control (20 to 43°C) can be set independently for the incubation area and the measuring position. A wide range of data reduction programs are available including one for phagocytosis studies for up to 12 samples held in a closed loop.

LKB/Wallac

The Model 1251 Luminometer was introduced in early 1982 and is a microprocessor-based instrument. It has capacity for 25 samples housed in a temperature controlled circular turntable. Temperature control (20 to 45°C) is stated to be within 0.1°C, and four cuvettes are accessible at any one time with others accessible after indexing the turntable. Cuvettes are elevated into the detector chamber which is very easily accessed. The signal from the 2 in. diameter photomultiplier is in the form of anode current, and a time constant for signal smoothing can be selected. Integration and delay times can be selected. Up to three optional peristaltic dispensers can be used to add automatically reagents to the cuvette. Volumes are adjustable with a built-in thumb wheel, and actuation of the dispensers may be controlled from the push buttons on the control panel. There are 23 labeled push buttons each with an incorporated LED, which allow a wide range of control parameters to be set. An interface (IEEE) may be used to connect the instrument to accessories such as a printer, video screen, or external computer. Factorization and background subtraction are built into the luminometer's data handling capacity. Mixing of the contents of the cuvette in the detector chamber can be accomplished by cuvette rotation either continuously or by pulses.

Lumac/3M B.V.

At Analytica-82 at Munich in April 1982, Lumac B.V. showed a prototype of the Biocounter M3050. This is a microprocessor controlled instrument which functions in the photon counting mode. All functions are said to be automated. It has capacity for 48 samples held in a removable circular tray. The tray is held in a temperature controlled chamber and can be moved in both directions. Up to five dispensers for automatic addition of reagents may be used. Automatic disposal of processed samples may be selected. Control of the instrument is via a full alphanumeric keyboard (touchpad) and an in-built video screen. Hard copy is produced on the built-in printer. An interface (RS232) is fitted for data transport to an external device.

New Brunswick Scientific

The Lumitran II is a microprocessor-based instrument of modular design with computer compatibility. It comprises a control module and a detector module. The control module can be set for manual or automatic injection via the single Hamilton type unit (50 aliquots). Flash height or integral measurements can be made and variable delay and measure times can be selected. Factorization and background subtraction is available. An interface (RS232C) is available for external devices and control. The other module consists of a temperature controlled circular carrier with a capacity for six cuvettes together with the Hamilton-type dispenser.

Packard

The Autolite Model 6200 was shown at FASEB in New Orleans and Analytica-82 at Munich in April 1982. The unit comprises an automatic rotary sample changer holding up to 48 cuvettes and an analyzer module housing the microprocessor, the timers, scaler, printer, and memory. The turntable and cuvette movements are actuated by compressed air, and three cuvette sizes may be accommodated in the turntable by changing the individual reflectors at each of the 48 positions. Mixing of the contents for all samples is by magnetic stirring. Software control of up to three dispensers for reagents is possible and temperature control (4 to 43°C) is included. Like the Packard Picolite a number of fixed programs are available and accessed via a touchpad keyboard. Several new programs have been added including those for phagocytosis studies. An interface (RS232) is available.

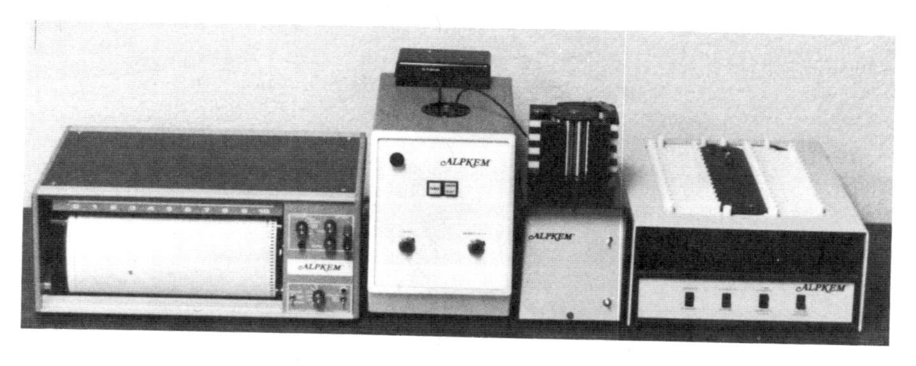

Figure 1 Automated Luminescence Analyzer. (Courtesy of Alpkem Corp., Clackamas, Oregon.)

Figure 2 Monolight Model 201. (Courtesy of Analytical Luminescence Laboratory, Inc., San Diego, California.)

Figure 3 Berthold Biolumat Model LB 9500. (Courtesy of Berthold Laboratorium, Wildbad, West Germany.)

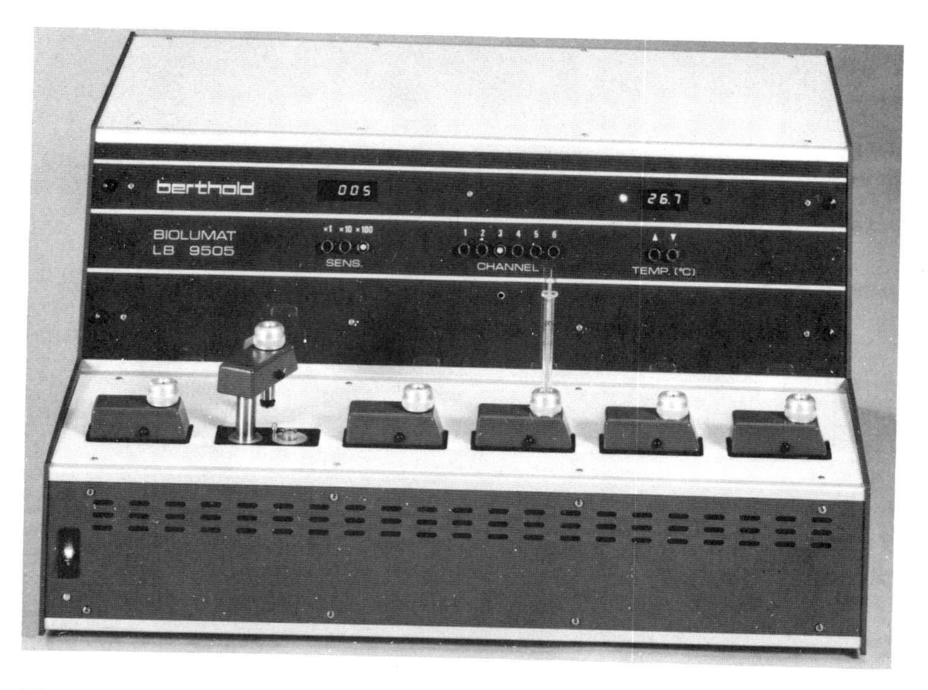

Figure 4 Berthold Six Channel Luminescence Analyzer Model LB 9505. (Courtesy of Berthold Laboratorium, Wildbad, West Germany.)

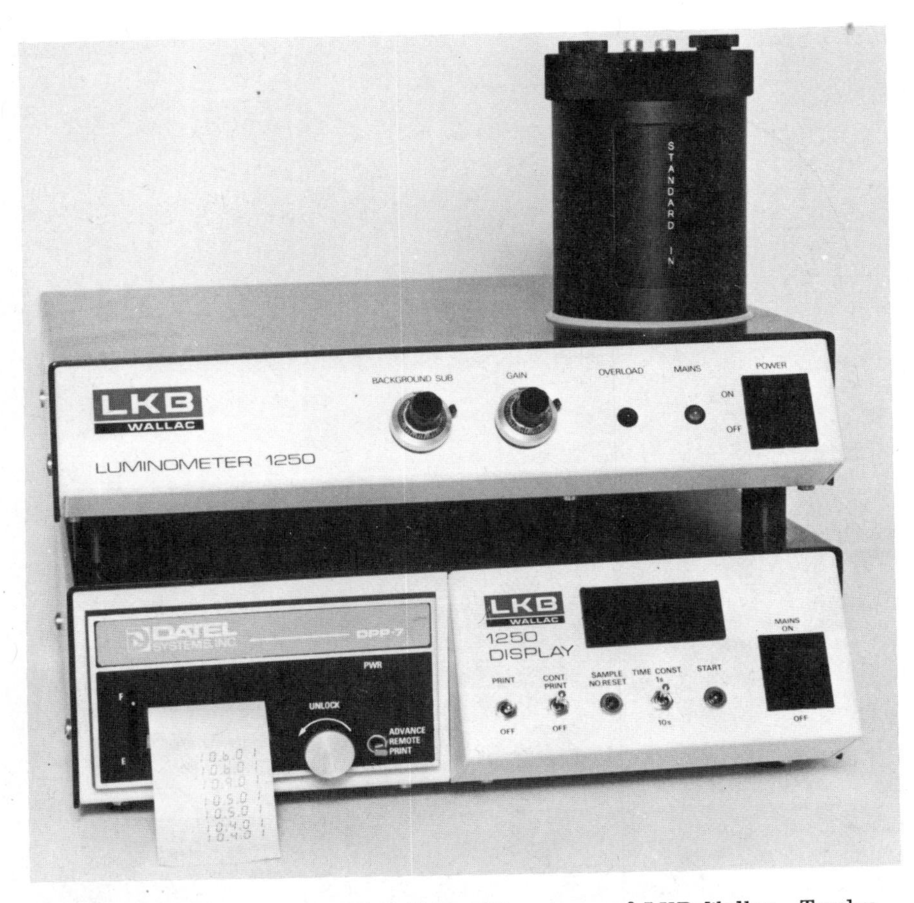

Figure 5 Luminometer Model 1250. (Courtesy of LKB Wallac, Turku, Finland.)

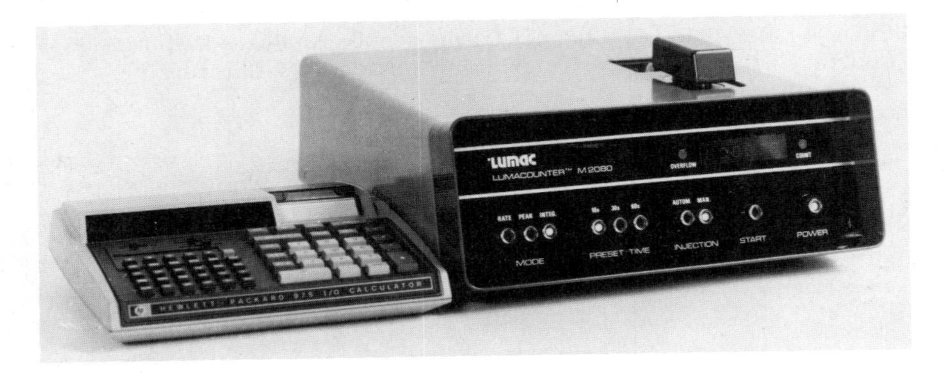

Figure 6 Lumac Lumacounter Model 2080 with HP97S microcomputer. (Courtesy of Lumac B. V., Schaesberg, The Netherlands.)

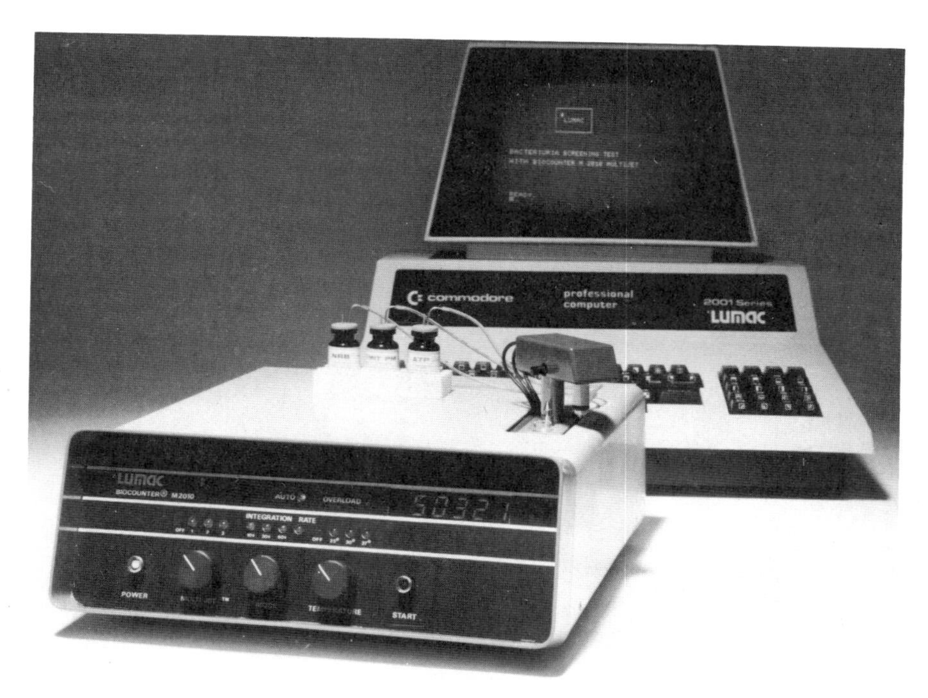

Figure 7 Lumac Biocounter Model 2010 with PET computer and video. (Courtesy of Lumac B.V., Schaesberg, The Netherlands.)

Figure 8 Lumac Biocounter Model 2000. (Courtesy of Lumac B.V., Schaesberg, The Netherlands.)

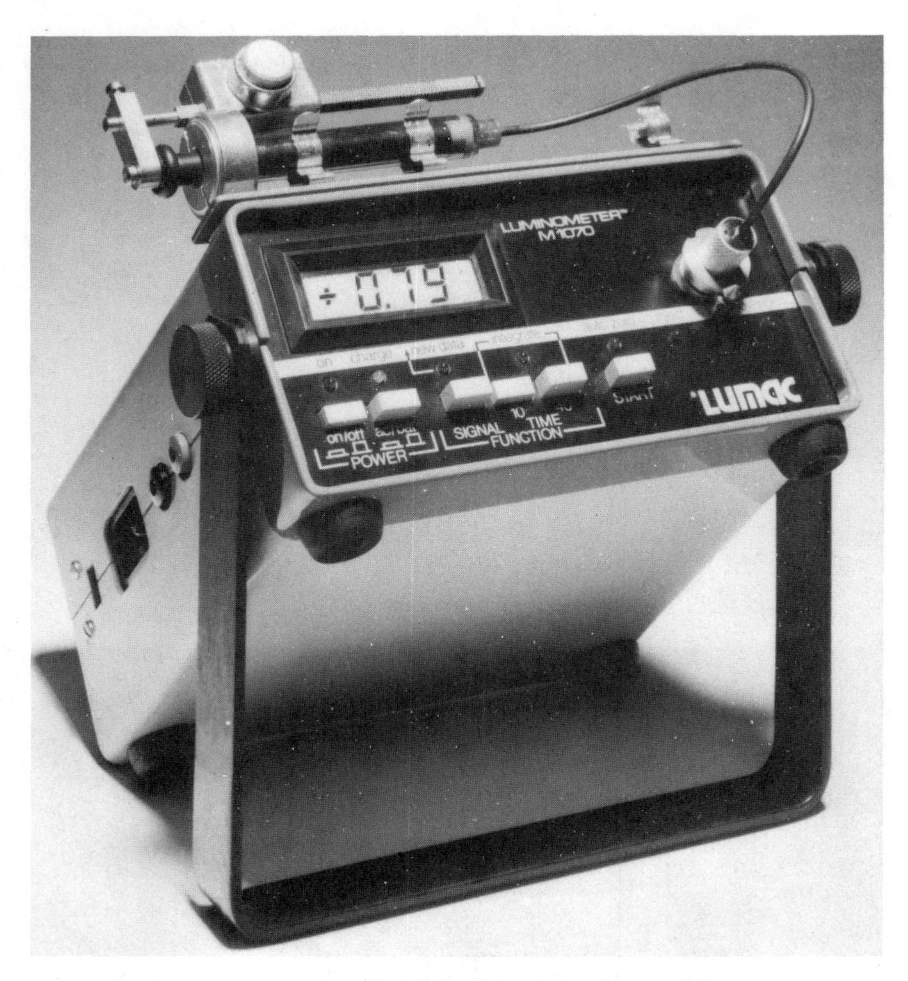

Figure 9 Lumac Luminometer Model 1070 with Hamilton dispenser. (Courtesy of Lumac B.V., Schaesberg, The Netherlands.)

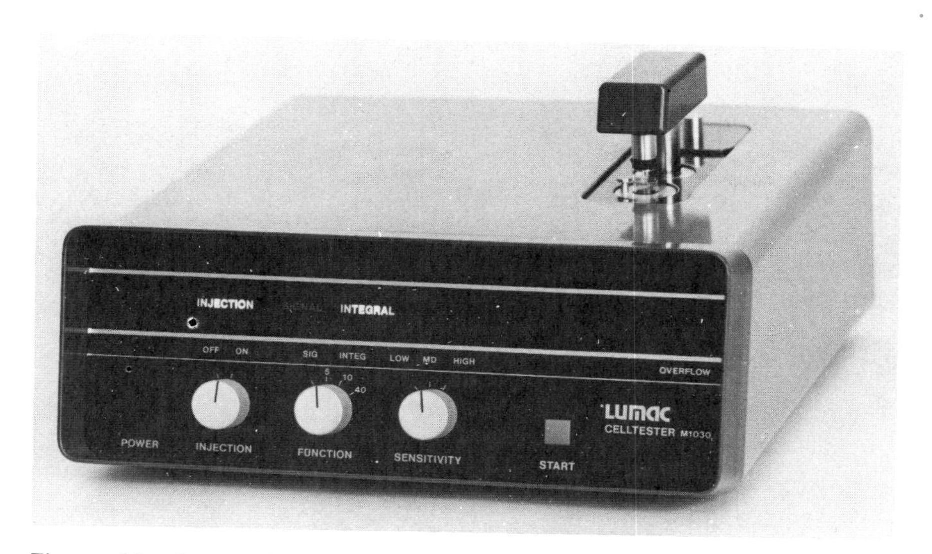

Figure 10 Lumac Celltester Model 1030. (Courtesy of Lumac B.V., Schaesberg, The Netherlands.)

Figure 11 Lumac Lumatec Model 1010. (Courtesy of Lumac B.V., Schaesberg, The Netherlands.)

Figure 12 Marwell Kinetic Luminescence Analyzer Model 302. (Courtesy of Marwell International AB, Solna, Sweden.)

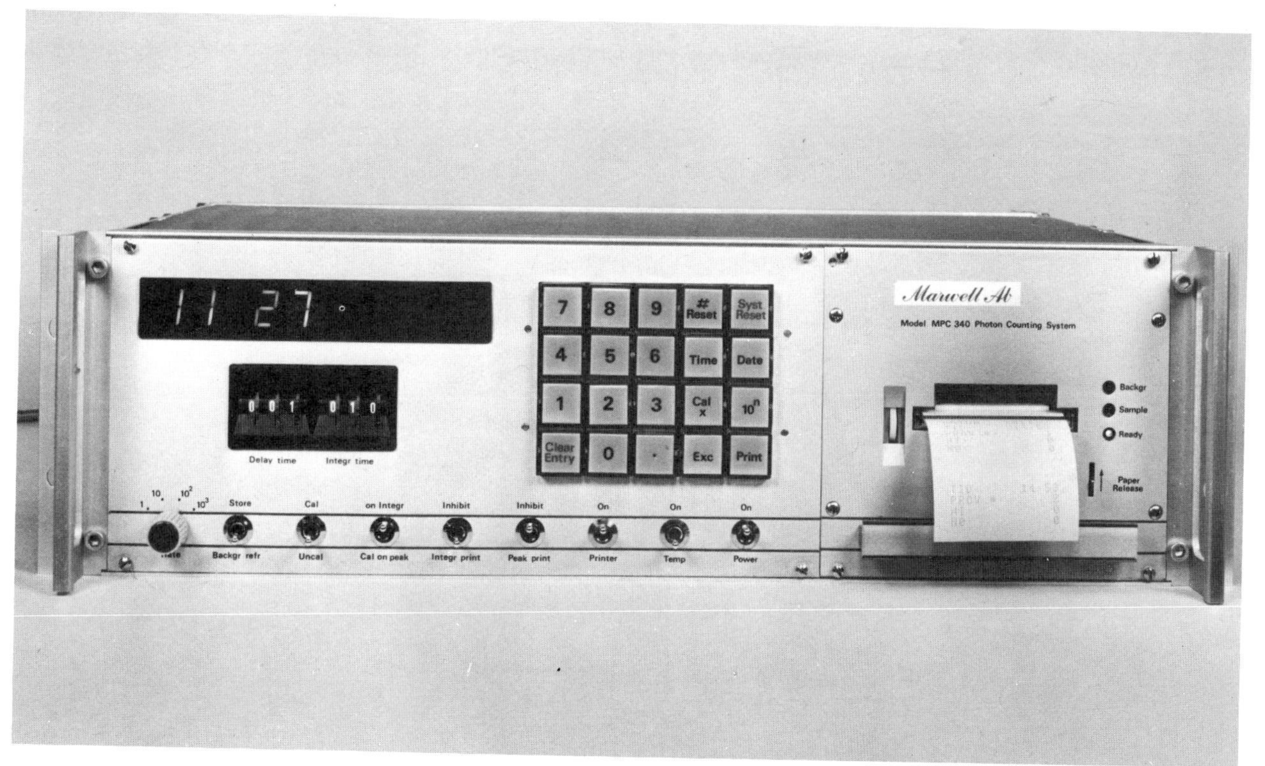

Figure 13 Marwell Luminescence Analyzer System Model 340/50. (Courtesy of Marwell International AB, Solna, Sweden.)

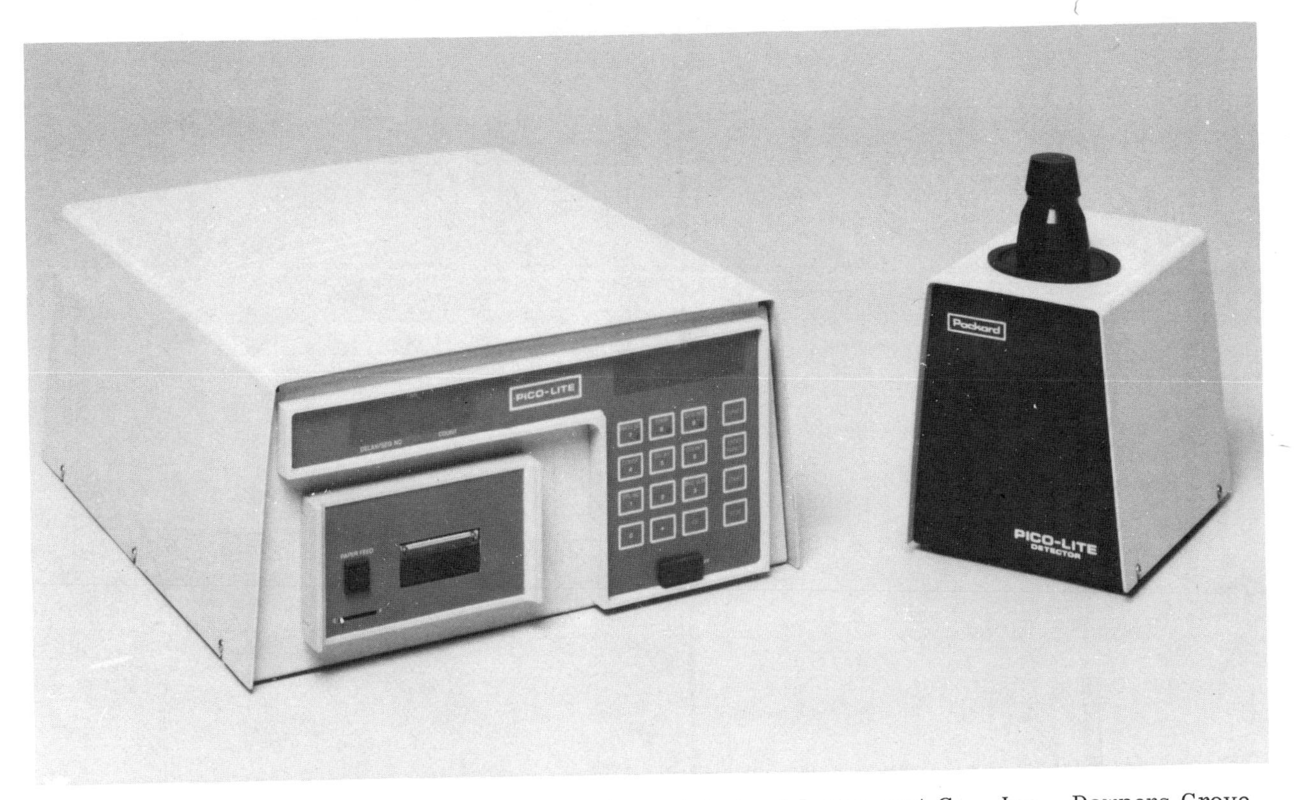

Figure 14 Packard Pico-Lite Model 6100. (Courtesy of Packard Instrument Co., Inc., Downers Grove, Illinois.)

Figure 15 SAI ATP-Photometer Model 100. (Courtesy of SAI Technology, Inc., San Diego, California.)

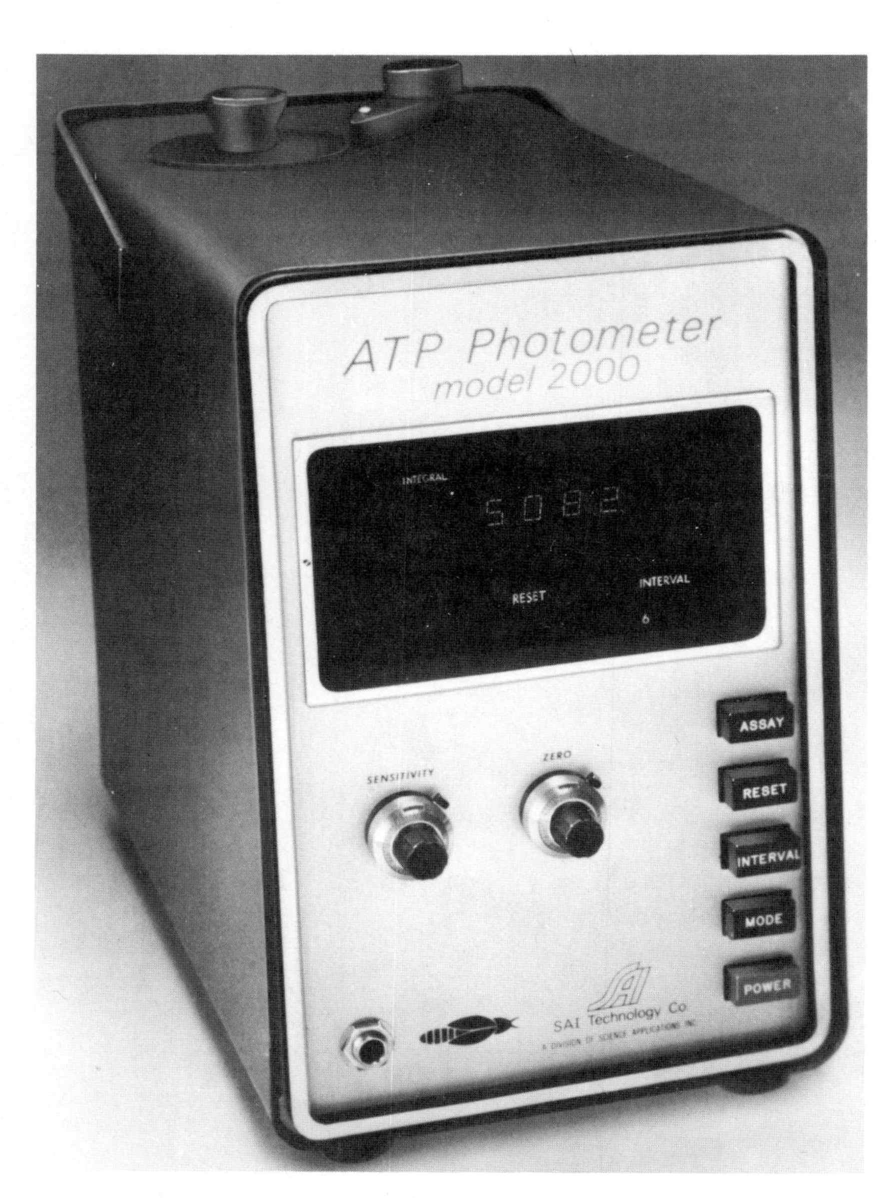

Figure 16 SAI ATP-Photometer Model 2000. (Courtesy of SAI Technology, Inc., San Diego, California.)

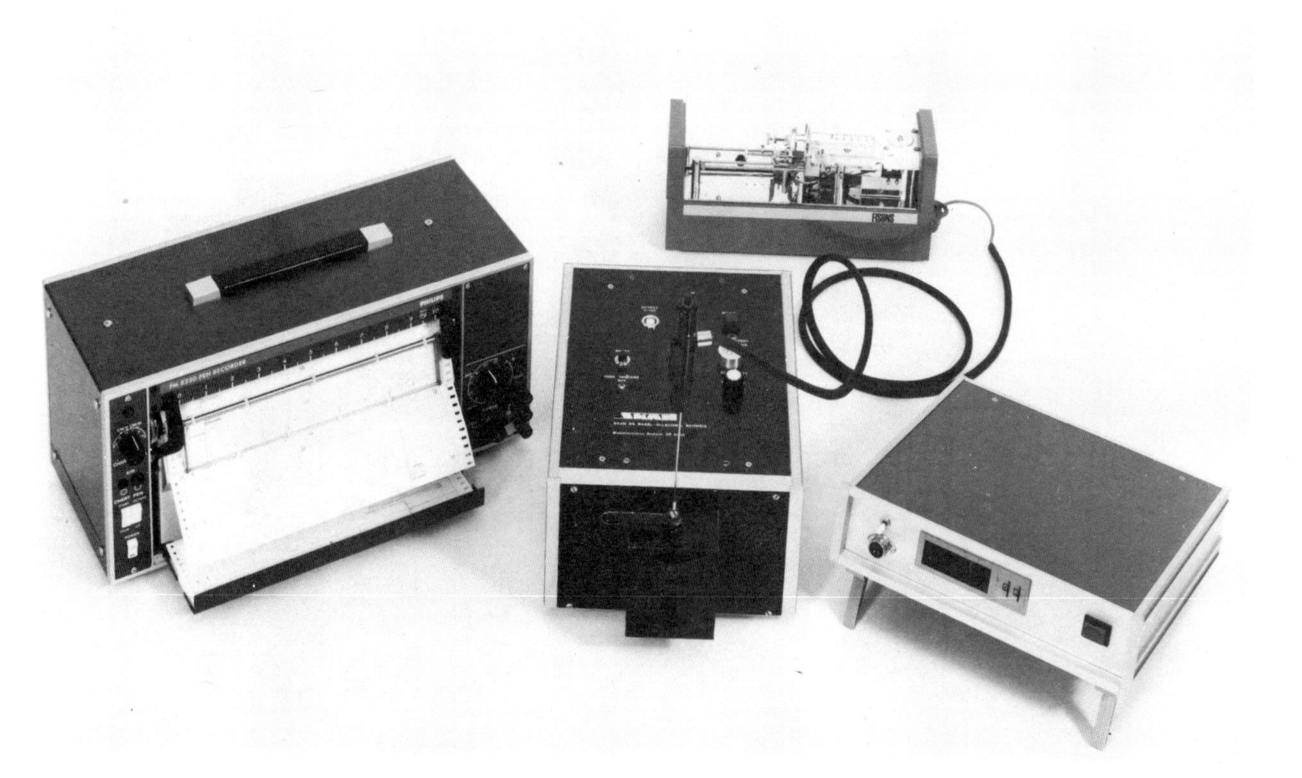

Figure 17 Skan Bioluminescence Analyzer Model XP-2000-2. (Courtesy of Skan AG, Basel, Switzerland.)

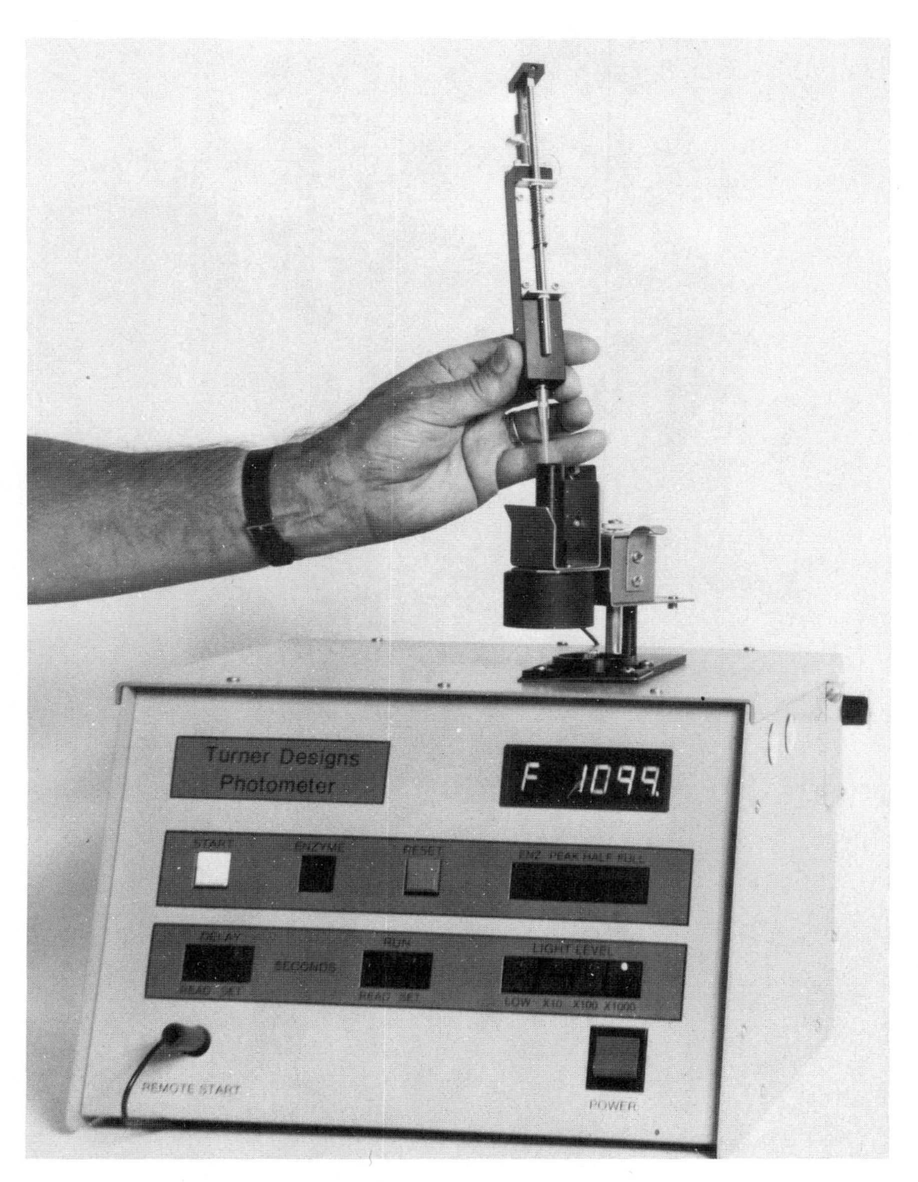

Figure 18 Turner ATP-Photometer Model 20. (Courtesy of Turner Designs, Mountain View, California.)

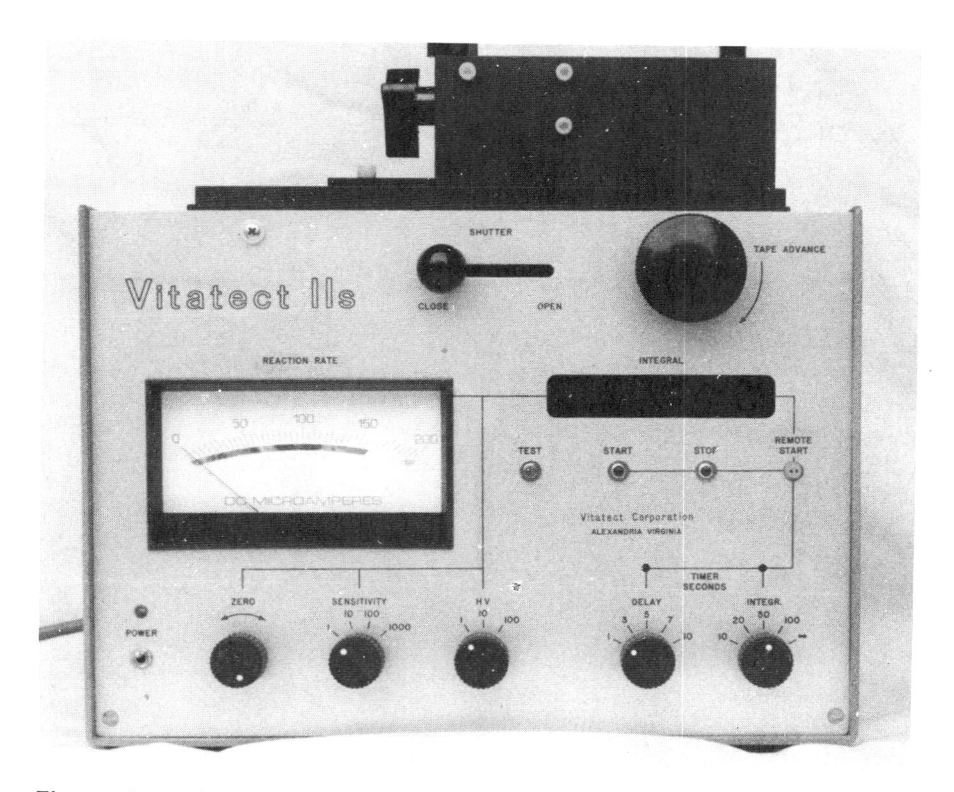

Figure 19 Vitatect Model IIs. (Courtesy of Vitatect Corporation, Alexandria, Virginia.)

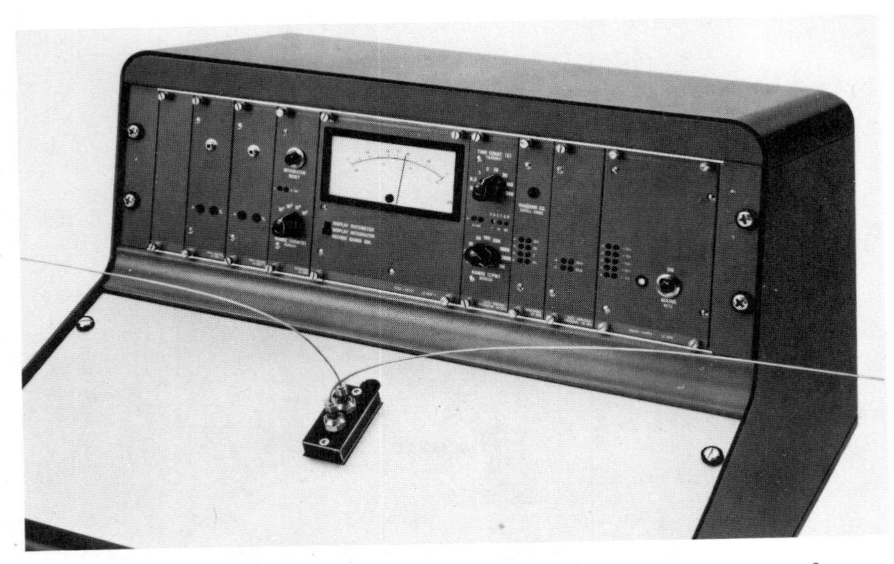

Figure 20 Berthold Model LB 503 (Flow Monitor.) (Courtesy of Berthold Laboratorium, Wildbad, West Germany.)

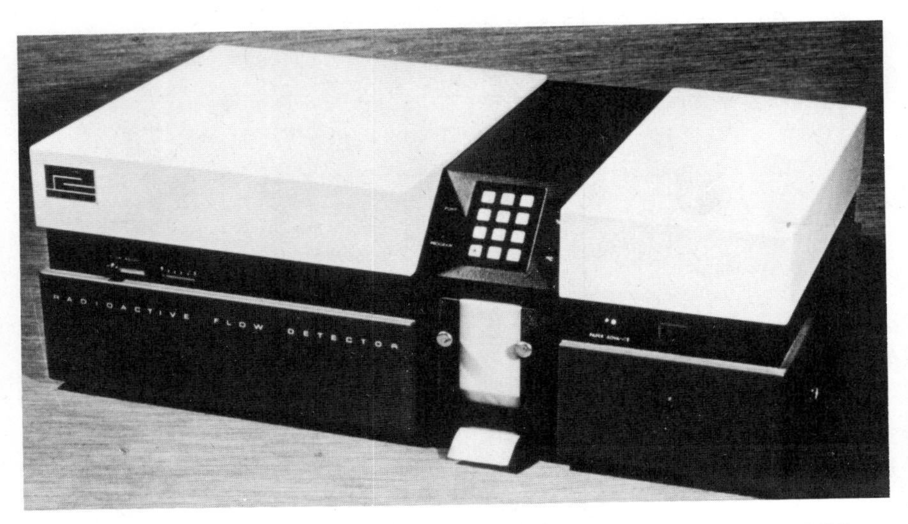

Figure 21 Radiomatic Instruments & Chemical Co. Model FLO-ONE, (Flow Monitor). (Courtesy of Radiometric Instruments Company, Inc., Tampa, Florida.)

ACKNOWLEDGMENTS

I thank the manufacturers who have readily made available to me the various details that I have sought about their instrumentation and those who in addition have provided photographs of their units for publication in this chapter. In particular, I thank Dr. A. Lundin (LKB), Dr. L. Everett (Packard), and Dr. V. Tarkkanen (Lumac) for making instrumentation available for my perusal.

REFERENCES

Anderson, J. M., G. J. Faini, and J. E. Wampler (1978). Construction of instrumentation for bioluminescence and chemiluminescence assays. *Meth. Enzymol. 57*:529-540.

Berthold, F., H. Kubisiak, M. Ernst, and H. Fischer (1981). Six-channel luminescence analyzer for phagocytosis applications. In *Bioluminescence and Chemiluminescence: Basic Chemistry and Analytical Applications*. M. DeLuca and W. D. McElroy (Eds.). Academic, New York, pp. 699-703.

Bullock, D. G., R. A. Bunce, and T. J. N. Carter (1979). The measurement of extremely low light levels in analytical chemi- and bio-luminescence: A mathematical and practical evaluation of some physical factors influencing detector output and instrument design. *Anal. Lett. 12*:841-854.

Carter, T. J. N., L. J. Kricka, D. G. Bullock, R. A. Bunce, and T. P. Whitehead (1979). Optimisation of luminescence instrumentation. In *Proc. International Symposium on Analytical Applications of Bioluminescence and Chemiluminescence*, E. Schram and P. Stanley (Eds.). State Printing and Publishing, Westlake Village, Cal., pp. 637-651.

Chappelle, E. W., G. L. Picciolo, C. A. Curtis, E. A. Knust, D. A. Nibley, and R. B. Vance (1975). Laboratory procedures manual for the firefly luciferase assay for adenosine triphosphate (ATP). Publication No. X-726-75-1. Goddard Space Flight Center, Greenbelt, Ma 20771.

Malcolm, P. J., and P. E. Stanley (1976a). A unified approach to the liquid scintillation counting process. 1. A stochastic computer model. *In. J. Appl. Rad. Isotop. 27*:397-414.

Malcolm, P. J., and P. E. Stanley (1976b). A unified approach to the liquid scintillation counting process. 2. Optimizing optical design. *Int. J. Appl. Rad. Isotop. 27*:415-430.

Picciolo, G. L., E. W. Chappelle, R. P. Thomas, and M. A. McGarry (1977). Performance characteristics of a new photometer with a moving filter tape for luminescence assay. *Appl. Environ. Microbiol. 34*:720-724.

Picciolo, G. L., J. W. Deming, D. A. Nibley, and E. W. Chappelle (1978). Characteristics of commercial instruments and reagents for luminescent assays. *Meth. Enzymol.* 57:550-559.

Schram, E., H. Roosens, and H. Van Esbroeck (1979). Contribution to the automation of luminescence measurements. In *Proc. International Symposium on Analytical Applications of Bioluminescence and Chemiluminescence*, E. Schram and P. Stanley (Eds.). State Printing and Publishing, Westlake Village, Cal., pp. 689-691.

Seitz, W. R. (1979). Flow cell design for chemiluminescence analysis. In *Proc. International Symposium on Analytical Applications of Bioluminescence and Chemiluminescence*, E. Schram and P. Stanley (Eds.). State Printing and Publishing, Westlake Village, Cal., pp. 663-686.

Seitz, W. R., and M. P. Neary (1976). Recent advances in bioluminescence and chemiluminescence assay. *Meth. Biochem. Anal.* 23:161-188.

Seliger, H. H. (1980). Single photon counting and spectroscopy of low intensity chemiluminescent reactions. In *Liquid Scintillation Counting: Recent Applications and Development*, Vol. 2, C.-Z. Peng, D. L. Horrocks, and E. L. Alpen (Eds.). Academic, New York, pp. 281-319.

Stanley, P. E. (1974). Analytical bioluminescence assays using the liquid scintillation spectrometer: A review. In *Liquid Scintillation Counting*, Vol. 3, M. A. Crook and P. Johnson (Eds.). Heyden and Son, London, pp. 253-272.

Stimson, A. (1974). *Photometry and Radiometry for Engineers*. Interscience, New York.

Wampler, J. E. (1975). Simple photometer circuits using modular electronic components. In *Analytical Applications of Bioluminescence and Chemiluminescence*, E. W. Chappelle and G. L. Picciolo (Eds.). NASA Publication No. SP-388, National Aeronautics and Space Administration, Washington, D.C., pp. 105-115.

ACRONYMS

ADP	adenosine 5'-diphosphate
AENAD	nicotinamide 6-(2-aminoethyl)purine dinucleotide
AENADH	reduced form of AENAD
AMP	adenosine 5'-monophosphate
cAMP	adenosine 3',5'-cyclic monophosphate
APS	adenosine phosphosulfate
ATP	adenosine 5'-triphosphate
BFP	blue fluorescent protein
BSA	bovine serum albumin
CFU	colony-forming unit
CTP	cytidine 5'-triphosphate
DTT	dithiothreitol
EDTA	ethylenediaminetetraacetic acid
EGTA	ethylene glycol bis(β-aminoethyl ether)-N,N'-tetraacetic acid
EMIT	enzyme multiplied immunoassay technique
FDNB	fluorodinitrobenzene
FMN	flavin mononucleotide

$FMNH_2$	reduced form of FMN
GDP	guanosine 5'-diphosphate
GMP	guanosine 5'-monophosphate
cGMP	guanosine 3',5'-cyclic monophosphate
GFP	green fluorescent protein
GTP	guanosine 5'-triphosphate
LCIA	luminescent cofactor immunoassay
LEIA	luminescent enzyme immunoassay
LEMIT	luminescent enzyme multiplied immunoassay technique
LIA	luminescent immunoassay
LIETS	luminescence immobilized enzyme test system
LSC	liquid scintillation counter
MIC	minimum inhibitory concentration
NAD	nicotinamide-adenine dinucleotide
NADH	reduced form of NAD
NADP	nicotinamide-adenine dinucleotide phosphate
NADPH	reduced form of NADP
NMN	nicotinamide mononucleotide
PAP	3',5'-phosphoadenosine phosphate
PAPS	3'-phosphoadenosine-5'-phosphosulfate
PHA	phytohemagglutinin
PP_i	pyrophosphate
SDS	sodium dodecyl sulfate
TCPO	bis(2,4,6-trichlorophenyl)oxalate
TES	N-tris(hydroxymethyl)methyl-2-aminoethane sulfonic acid
TMAE	tetrakis(dimethylamino)ethylene
TNT	trinitrotoluene
UTP	uridine 5'-triphosphate

AUTHOR INDEX

Italic numbers give the page on which the complete reference is listed.

Adair, J.A., 59, *72*
Adam, W., 27, *37*
Adler, F. L., 155, *174*
Ågren, A., 64, *67*, 81, 83, 84, 85
Ahkong, Q. F., 109, *124*
Akerblom, O., 201, *207*
Albrecht, H. O., 5, *8*, 136, *148*
Alexander, D. N., 55, *66*, 192, *207*
Alexander, J. G., 193, *211*
Allen, D. G., 91, 95, 102, 105, 107, 108, 113, 115, 116, 118, 121, *124*, *126*
Allen, R. C., 205, *207*
Allen, R. D., 107, 119, 121, *132*
Allman, G. J., 92, *124*
Altas, D., 182, *187*
Ambrose, J., 30, *40*
Anderson, J. M., 91, 92, 93, 101, 110, *124*, *128*, 216, *253*
Anderson, J. W., 98, 99, *126*
Anderson, N. G., 137, *148*
Andersson, A., 59, *66*
Andrade, J. D., 180, *187*
Anfinsen, C. B., 158, *175*, 183, *187*
Ånsehn, S., 55, 56, 58, *66*, *69*, *72*, *73*, 194, 195, 196, 197, 198, 199, *207*, *209*, *210*, *211*, *212*
Arakawa, H., 168, *174*, *177*
Arganov, K., 37, *37*
Aristotle, 3, *8*
Armstrong, D. E., 55, *70*
Ashley, C. C., 90, 91, 93, 94, 95, 97, 98, 101, 103, 104, 106, 107, 111, 115, 117, 122, 123, *125*, *127*, *130*, *131*
Asscher, A. W., 56, 58, *68*, 191, 196, 197, 198, 199, 200, *209*
Auses, J. P., 140, 142, *148*
Avrameas, S., 147, *150*, 158, 161, 166, 168, *174*, *176*, *177*
Axen, R., 180, *188*
Ayers, V. E., 202, *209*
Azari, J., 161, 166, *177*
Azzi, A., 98, *125*

Babior, B. M., 205, *208*
Babko, A. K., 137, *148*
Bachofen, R., 55, *70*
Baker, P. F., 103, 106, 108, 115, 116, 117, 121, 122, 123, *125*

Baldwin, T. O., 78, 79, 82, *85*, *86*
Balharry, G. J. E., 62, 64, *66*
Baltscheffsky, M., 49, 65, *66*, *68*, *71*
Bard, A. J., 34, *37*, *39*
Barker, S. A., 180, *186*
Barrett, J., 105, *125*
Barza, M., 194, *212*
Bastiani, R. J., 169, *174*
Baucom, T. L., 77, 78, 80, 83, *86*
Baulch, D. L., 34, 35, *38*
Becher, M. J., 180, *188*
Becvar, J. E., 79, *85*
Beijerinck, M. W., 4, *8*
Benson, S. W., 35, *38*
Beny, M., 50, *66*
Berezin, I. V., 183, 185, *188*
Bergman, S., 55, 56, *73*
Berlman, I. B., 100, *125*
Berne, C., 81, 82, 83, 84, *84*, *85*
Bernhardsen, A., 206, *210*
Berridge, M. J., 97, *125*
Berry, M. G., 37, *39*
Berson, S. A., 154, *178*
Berthold, F., 226, *253*
Binkley, S. B., 191, *209*
Blasse, G., 36, *38*
Bles, E. J., 94, *126*
Blinks, J. R., 91, 95, 102, 105, 106, 107, 108, 113, 115, 116, 118, 121, 122, *124*, *126*, *129*, *130*, *133*
Bloome, W. J., 155, *176*
Blowers, R., 195, *208*
Blumberg, S., 182, *187*
Blumenthal, R., 109, *126*
Bogan, D. J., 35, *38*
Boguslaski, R. C., 44, *67*, 82, 83, *87*, 155, 159, 160, 161, 163, 167, 171, 172, 173, *175*, *176*
Boitieux, J.-L., 180, *186*
Bollyky, L. J., 28, 30, *38*, *40*
Bolton, P. M., 32, *38*
Booth, C. R., 55, *69*
Borglund, E., 59, 64, *66*, *67*, 81, 83, *85*
Bossuyt, R., 55, *66*
Bostick, D. T., 6, *8*, 140, 142, *149*
Bowers, L. D., 180, *186*
Bowling, J. L., 137, *149*
Boyce, J. M. H., 193, *211*
Bracken, M. M., 202, *213*
Brezonik, P. L., 55, *67*
Broadus, A. E., 62, *69*
Brolin, S. E., 54, 56, 59, 64, *66*, *67*, *74*, 78, 81, 83, *84*, *85*, *87*
Brooks, D. E., 59, *67*, 206, *208*
Brote, L., 58, *73*, 195, *212*
Brovko, L. Y., 183, 185, *188*
Brown, D. F. J., 195, *208*
Brown, J. E., 105, 107, *126*
Brown, K. T., 105, *126*
Bruice, T. C., 25, *39*
Brundrett, R. B., 5, *9*, 22, *38*
Brunius, G., 168, 169, *176*
Buckler, R. T., 44, *67*, 160, 163, 172, *175*, *176*
Buckler, S. A., 32, *41*
Bullock, D. G., 227, *253*
Bunce, R. A., 227, *253*
Burguera, J. L., 32, *38*
Burnell, J. N., 98, 99, *126*
Burns, M. M., 194, *211*
Bursey, M. M., 17, *20*, 30, *41*
Bush, V. N., 191, 193, *208*
Bygrave, F. L., 97, *126*
Bywater, M. J., 199, *211*

Cais, M., 155, *175*
Caldwell, P. C., 104, 115, 117, *125*, *126*, *130*
Campbell, A. K., 30, *41*, 90, 91, 93, 94, 95, 97, 98, 100, 101, 103, 104, 109, 110, 111, 112, 113, 115, 116, 117, 121, 123,

[Campbell, A. K.]
125, 126, 127, 128, 129, 130, 133, 161, 166, 167, *176, 177*
Campbell, I. M., 34, 35, *38*
Cantarow, W., 82, 84, *85*
Carbone, E., 106, *129*
Caricco, R. J., 82, 83, *87*
Carlsson, H. E., 168, 169, *176*
Carr, P. W., 139, *149*, 180, *186*
Carrico, J., 44, *67*
Carrico, R. J., 155, 159, 160, 163, 167, 171, 172, 173, *175, 176*
Carter, T. J. N., 30, *41*, 54, *74*, 144, *149*, 183, *187*, 227, *253*
Cass, M. W., 27, *39*
Catt, K. J., 180, *187*
Cavari, B. Z., 63, *67*
Chan, C. S., 55, *70*
Chan, K. W., 25, *39*
Chance, B., 83, *85*, 98, *125*
Chang, J. J., 105, *127*
Chappelle, E. W., 54, 55, 57, 59, *67, 70, 71*, 79, *86*, 190, 191, 192, 193, 194, *208, 211, 212*, 216, 236, *253, 254*
Charache, P., 55, *67*, 192, 193, *208*
Charlier, J., 79, *85*
Charo, I., 59, *68*
Cheson, B. D., 205, *208*
Chiu, S. Y., 55, *67*
Christensen, R. L., 205, *208*
Christner, J. E., 44, *67*, 82, 83, *87*, 155, 159, 167, 171, 172, *175, 176*
Clark, L. C., 180, *187*
Clarke, R. A., 23, 28, *40*
Cobbold, P. H., 107, 113, *127*
Cocking, E. C., 109, *124*
Cohen, B. D., 202, *208*
Cole, H. A., 59, *67*, 191, 206, *208*
Colley, C. M., 108, *133*
Conklin, C. J., 139, 142, *150*
Conn, R. B., 55, *67*, 192, 193, *208*
Cook, J. L., 140, 142, *148*
Cooke, I. M., 94, *130*
Cooney, C. L., 180, *188*
Cooreman, W. M., 155, *176*
Cormier, M. J., 22, 24, 27, 35, *38*, 91, 92, 93, 94, 97, 98, 99, 100, 101, 110, *124, 127, 128, 129, 133*, 136, 143, 144, *149*
Coulombre, J., 44, *72*
Cowden, M., 109, *130*
Cramp, F. C., 109, *124*
Crawford, A. C., 117, *125*
Crawford, A. M., 54, 55, *71*
Crispen, R. G., 58, *70*
Csaba, P., 34, *37*
Cuatrecasas, P., 158, *175*, 183, *187*
Cueto, O., 27, *37*
Cunningham-Rundles, S., 203, *208*
Curtis, C. A., 191, 192, *211*, 216, *253*
Curtis, G. D. W., 55, *67, 69*, 192, 194, *208, 210*

Dahlgren, U., 95, *128*
Daigneault, R., 58, *67*
Dale, J. M., 137, *149*
Dandliker, W. B., 155, *175*
Dani, S., 155, *175*
Dark, F. A., 191, *212*
Davenport, D., 94, *128*
Davey, M. R., 109, *124*
David, J. L., 59, *67*
Davies, C. J., *127*
Davis, G., 55, *70*
Daw, R. A., 97, 100, 110, 111, 112, 113, 116, 121, *127*
Dawes, E. A., 59, *72*
Day, H. J., 59, *69*, 202, 203, *210*
Dean, J. A., 137, *149*
DeLuca, M., 6, *8*, 44, 45, 48,

[DeLuca, M.]
64, *67*, *68*, *70*, *74*, 77, 79, 80, 81, 82, 83, 84, *86*, 161, 166, *177*, 183, 184, 185, *187*, *188*
Delumyea, R., 137, *149*
Deming, J. W., 79, *86*, 216, *254*
Demuylder, F., 80, *87*
Denburg, J. L., 45, 46, 51, 52, 53, *67*, *70*
Derycker, J., 80, *87*
de Saussure, V. A., 155, *175*
Desmet, G., 180, *186*
Detwiler, T. C., 58, 59, *67*
D'Eustachio, A. J., 190, 191, *208*
de Verdier, C-H., 201, *207*
Diaz, P., 205, *208*
Dixon, B., 22, 33, 34, *40*
Doeg, K. A., 63, *72*
Dolivo, M., 50, *66*
Doree, M., 107, *130*
Dormer, R. L., 109, 110, 111, 112, 113, 116, *126*, *128*
Douglas, W. W., 97, *128*
Driesch, R., 202, 203, *212*
Duane, W., 79, *85*
Dubois, R., 3, 4, *8*, 92, *128*
Dulake, C., 193, *211*
Dupont, B., 203, *208*
Durham, A. C., 107, *131*

Eckert, R., 105, 107, 116, 117, *128*
Eckstrom, D. J., 35, *38*
Edelstein, S. A., 35, *38*
Eden, Y., 155, *175*
Eder, J. M., 4, *8*
Ederer, G. M., 55, *66*, 192, *207*
Egghart, H., 81, 83, *85*
Eisen, A., 119, *129*
Elgart, R. L., 139, *150*, 207, *211*
Engasser, J.-M., 183, *187*
Erickson, E. E., 55, *67*
Eriksson, G., *176*
Erlanger, B. F., 184, *187*
Ernst, M., 226, *253*
Ettienne, E. M., 106, *128*
Eusebi, F., 107, *128*
Ewetz, L., 139, *149*, 169, *175*, 194, 207, *209*

Faini, G. J., 77, 78, 80, 83, *86*, 216, *253*
Fan, L. T., 55, *67*
Farber, E., 190, *209*
Faulkner, L. R., 33, *38*
Fay, F. S., 107, 108, *128*
Fehrenbach, F.-J., 59, *67*
Feinman, R. D., 58, 59, *67*, *68*
Felton, S. P., 97, *129*
Feo, C. J., 59, *68*
Ferrell, W. J., 63, *69*, 206, *210*
Fertel, R., 63, 64, *68*
Finnson, M., 201, *207*
Fischer, H., 205, *213*, 226, *253*
Fisher, D., 109, *124*
Flaming, D. G., 105, *126*
Foerder, C. A., 121, *128*
Fontijn, A., 136, *149*
Foote, C. S., 34, *38*
Forbes, E., 91, *128*
Forrester, T., 206, *209*
Fårskal, P., 90, 91, *128*
Forus, F., 22, *38*, 54, *68*, 79, 80, *85*, 142, 143, *149*
Franzen, J. S., 191, *209*
Freeman, T. W., 145, 148, *149*
Freese, E. J., 57, 59, *70*
Fulton, S. P., 180, *188*

Gallati, H., 145, *149*
Gandolfi, O., 155, *175*
Garby, L., 201, *207*
Gates, B. J., 45, *68*
Gawronski, T. H., 81, 83, *85*
Gelperin, A., 105, *127*
Gerlo, E., 79, *85*
Gesner, C., 3, *8*

Gibson, Q. H., 77, *85*
Gilkey, J. C., 107, 119, *129, 131*
Gilles, R., 51, *68*
Gilmore, J., 24, *39*
Gleu, K., 5, *8*, 136, *149*
Golde, M. F., 36, *41*
Goldstein, A., 155, *175*
Goldstein, G., 137, *149*
Goldstein, L., 182, *187*
Golomb, D., 136, *149*
Goodman, D. B. P., 97, *131*
Goto, T., 45, *68*
Gough, D. A., 180, *187*
Granger, W. C., 107, 108, *128*
Gray, D. N., 180, *187*
Green, A. A., 50, *68*
Greenwalt, T. J., 202, *209*
Gregory, J. D., 99, *129*
Greiling, H., 202, 203, *212*
Gruner, S. M., 121, *131*
Guerrier, P., 107, *130*
Guinchant, J., 4, *8*
Gundermann, K. D., 17, *20*, 30, 31, 32, 34, *38*
Gunsalus-Miguel, A., 79, *85*
Gutekunst, R. R., 55, *68*
Guthrie, L. A., 205, *210*

Hackenbrock, C. R., 48, 61, 65, *70*
Haggerty, C., 6, *8*, 81, 83, 84, *86*, 183, 184, *187*
Hakansson, L., 205, *212*
Hales, C. N., 97, *129*
Hall, J. M. M., 17, *20*
Hallett, M. B., 95, 97, 100, 106, 109, 110, 111, 112, 113, 121, *126, 127, 128, 129*
Hallgren, R., 205, *212*
Hallynck, T., 200, *209*
Halmann, M., 168, *177*, 207, *209, 212*
Hamaguchi, Y., 158, *175*
Hamilton, R. D., 55, *68*, 190, 191, *209*
Hammar, H., 59, 64, *68*, 78, *85*
Hansen, J. A., 203, *208*
Haradin, A. R., 201, 202, *209*
Harber, M. J., 56, 58, *68*, 191, 196, 197, 198, 199, 200, *209*
Hard, G. C., 205, *209*
Hardman, J-G., 62, *69*
Harrer, G. C., 95, 102, 105, 116, *126*
Harris, D. A., 65, *68*
Harris, R. F., 55, *70*
Hart, R. C., 22, 24, 27, *38*, 94, *127, 129*
Hartkopf, A. V., 137, *149*
Hartlaub, C., 92, *129*
Harvey, E. N., 3, 4, 5, *8*, 91, 92, 94, *129*, 136, 143, *149*
Hastings, J. W., 44, *72*, 76, 77, 78, 79, 83, 84, *85, 86, 87*, 91, 93, 100, 101, 122, *129, 130*
Heath, T. P., 108, *133*
Helfrich, W., 37, *38*
Hemmila, I., 155, *177*
Hendley, D. D., 65, *73*
Henkart, P., 109, *126*
Hercules, D. M., 6, *8*, 35, *38*, 63, *69*, 137, 138, 139, 140, 142, *149, 150*
Herion, F., 59, *67*
Herring, P. J., 91, 92, 93, 94, 95, *127, 129*
Hersch, L. S., 30, *39*, 160, 163, *175*
Herschbach, D. R., 35, *38, 41*
Herzenberg, L. A., 155, *175*
Hess, W. C., 55, *70*
Hill, J. H. M., 136, *151*
Hincks, T., 91, *129*
Hjerten, S., 78, 83, *85*
Hobson, P. N., 59, *69*
Hodgeson, J. A., 136, *149*
Hodgkin, A. L., 103, 106, 108, 115, 116, 117, 121, *125*
Hogman, C. F., 201, *207*

Hoijer, B., 65, *71*
Hojer, H., 58, *66*, *69*, *72*, *73*, 195, 196, 197, 198, 199, *209*, *210*, *211*, *212*
Holm-Hansen, O., 55, *68*, *69*, 190, 191, *209*
Holmsen, H., 59, *69*, 202, 203, 206, *210*
Holmsen, I., 206, *210*
Hopkins, T. A., 27, *39*, *41*
Hori, K., 92, 94, *128*, *129*
Horn, M., 155, *175*
Horvath, C., 183, *187*
Howell, J. I., 109, *124*
Hoyt, S. D., 137, 142, *149*
Huestis, D. L., 35, *38*
Huff, G. F., 6, *9*, 136, 140, 141, 142, *151*
Hughes, D. E., 59, *67*, 191, 206, *208*
Huser, H., 59, *67*
Hysert, D. W., 55, *69*

Iannotta, A. V., 28, *40*
Impraim, C. C., 111, *129*
Ingle, J. D., 137, 142, *149*
Irie, M., 168, *177*
Isaacs, E. E., 155, *175*
Isacsson, U., 22, 30, 36, 37, *39*, 137, 139, *149*
Isambert, M. F., 184, *187*
Ishikawa, E., 158, *175*
Izutsu, K. T., 97, *129*

Jablonski, E., 6, *8*, 77, 79, 80, 81, 82, 83, 84, *86*, 183, 184, 185, *187*, *188*
Jablonski, I., 44, *70*
Jabs, C. M., 63, *69*, 206, *210*
Jaeschke, W., 32, *41*
Jaffe, L. F., 107, 119, *129*, *131*
Jaschinski, C., 59, *67*
Jewell, B. R., 95, 102, 105, 116, *126*
Jobsis, F. F., 97, 98, *129*
Johansson, S. G. O., 201, *207*
Johnson, D. R., 190, 191, *208*
Johnson, F. H., 92, 94, 98, 101, 105, 108, 113, *127*, *131*, *132*
Johnson, R. A., 62, *69*
Johnson, R. D., 173, *175*
Johnston, H. H., 55, *67*, *69*, 192, 194, *208*, *210*
Johnston, R. B., Jr., 205, *210*, *212*
Jones, D. G., 205, *208*
Jones, H. P., 97, *129*
Jones, J. G., 55, *69*
Josephy, Y., 155, *175*

Kagi, H. M., 17, *20*, 136, *151*
Kamiyama, T., 201, *211*
Kao, I. C., 55, *67*
Kaplan, M. L., 34, *40*
Karl, D. M., 55, 62, *69*, *70*
Kass, E. H., 192, *210*
Kato, K., 158, *175*
Kato, M., 168, *177*
Katunuma, N., 158, *175*
Kawaoi, A., 163, *176*
Kay, A. B., 205, *208*
Kay, I., 180, *186*
Kearns, D. R., 32, *38*
Keele, B. B., Jr., 205, *212*
Kees, U., 55, *70*
Kelley, B. C., 99, *132*
Kennedy, J. H., 166, *175*, 181, *187*
Kessler, M., 180, *187*
Keszthelyi, C. P., 34, *39*
Keszthelyi, T. H., 34, *37*
Ketelsen, U-P., 205, *213*
Keyes, M. H., 180, *187*
Khan, A. V., 34, *41*
Kiehart, D. P., 119, *129*
Kim, J. B., 161, 167, *175*
Kimelberg, H. K., 109, *130*
King, G. W., 205, *211*
Kishi, Y., 45, *68*

Kitagawa, T., 35, *41*
Kjellin, K. G., 64, *73*
Klebanoff, S. J., 121, *128*
Kleeman, J., 64, *70*
Klofat, M., 57, 59, *70*
Knight, D. E., 115, *125*
Knust, E. A., 191, 192, *211*, 216, *253*
Kobayashi, K., 158, *175*
Kohen, F., 161, 167, *175*
Kohler, B. E., 205, *208*
Koo, J., 22, 33, 34, *40*
Kovecses, F., 55, *69*
Kricka, L. J., 30, *41*, 54, *74*, 166, *175*, 181, 183, *187*, 227, *253*
Kubisiak, H., 226, *253*
Kubota, J., 45, *68*
Kusano, K., 105, 106, *130*

Laekeman, G. M., 155, *176*
Landon, J., 155, *177*
LaRock, P. A., 55, *70*
Larouche, A., 58, *67*
Larsson, R., 59, *70*
Lau, C.-T., 155, *174*
Laugton, D., 55, *73*
Lea, T. J., 90, 91, 93, 94, 95, 101, 103, 104, 111, 115, 117, *125*, *127*
Leblond, P. F., 59, *68*
Lechtken, P., 23, *39*
Lee, C. C., 55, *70*
Lee, J., 17, *20*, 35, *38*, 77, 78, 80, 83, *86*
Lee, K., 25, *39*
Lee, T. R., 45, 46, 52, 53, *70*
Lee, Y. S., 44, 58, *70*, 183, 184, *188*
Leeson, P. D., 22, 30, *39*
Legg, K. D., 25, *39*
Lehmeyer, J. E., 205, *210*
Lehne, R., 64, *74*
Lemasters, J. J., 48, 61, 65, *70*
Leute, R. L., 155, *175*
Levin, G. V., 55, *70*, 190, 191, *208*
Levin, Y., 182, *187*
Levine, L., 161, 166, *177*
Lewenstein, A., 55, *70*
Li, R. T., 137, 142, *149*
Lind, A. R., 206, *209*
Lindberg, K., 51, 64, *71*
Lindner, H. R., 161, 167, *175*
Line, W. F., 180, *188*
Linschitz, H., 37, *39*
Lipmann, F., 99, *129*
Llinas, R., 105, 106, *130*
Loose, L. D., 205, *207*
Lowenstein, W. R., 107, 119, 121, *131*
Lowery, S. N., 139, *149*
Loy, M., 28, *40*
Loyter, A., 111, *133*
Lubbers, D. W., 180, *187*
Lubowsky, J., 59, *68*
Lucy, J. A., 109, *124*
Lukovskaya, N. M., 137, *148*
Lundin, A., 44, 45, 46, 49, 50, 51, 52, 53, 55, 56, 61, 62, 64, 65, *66*, *71*, *72*, *73*, 194, *207*
Lundin, U. K., 65, *71*
Luzio, J. P., 97, 100, 110, 111, 112, 113, 116, 121, *127*, *129*

Macaire, J., 3, *8*
McAllister, T. A., 193, *211*
McCapra, F., 22, 24, 26, 27, 30, 31, *39*, *40*, *127*, 136, 140, 143, *149*, *150*
Macartney, J., 92, *130*
McCoy, G. D., 63, *72*
McDermotte, F. A., 4, *8*
MacDonald, A., 25, *39*
McElroy, W. D., 4, *9*, 44, 45, 46, 49, 50, 51, 52, 53, 63, *67*, *68*, *70*, *72*, *73*
McGarry, M. A., 216, 236, *253*
Mackinnon, A. E., 194, *211*
MacLeod, N. H., 54, 55, *71*

Maeda, M., 168, *174*, *177*
Maier, C. L., 160, 163, 166, *175*
Makela, O., 155, *176*
Malcolm, P. J., 227, *253*
Malnic, G., 77, *86*
Maloy, J. T., 140, 142, *148*
Manandhar, M. S. P., 62, *72*
Maskiewicz, R., 25, *39*
Massa, J., 76, *86*
Matsen, J. M., 55, *66*, 192, *207*
Matsuoka, K., 168, *177*
Matthews, J. C., 94, 97, *129*
Mattingly, P. H., 95, 102, 105, 116, 122, *126*, *129*
Maulding, D. R., 30, *39*
Medina, V. J., 25, *41*, 142, 143, *151*
Meighen, E. A., 77, 79, *85*, *86*
Mendenhall, G. D., 32, *40*
Metz, E. N., 205, *211*
Michaliszyn, G. A., 79, *86*
Michelson, A. M., 79, *87*, 147, *150*, 158, 161, 166, 168, *176*, 184, *187*
Michl, J., 17, 19, *20*, 205, *211*
Micklem, K. J., 111, *129*
Milch, J. R., 121, *131*
Miledi, R., 105, 106, 107, *128*, *130*
Mitchell, C. J., 55, *69*, 192, *210*
Mitchell, G., 122, *129*
Moisescu, D. G., 93, 104, 115, *125*, *130*
Mollison, P. L., 201, 202, *211*
Momsen, G., 63, *72*
Montague, M. D., 59, *72*
Moreau, M., 107, *130*
Morin, J. G., 91, 93, 94, 95, 100, 101, *130*
Morrison, N. M., 55, *69*
Murphy, C. L., 77, 78, 80, 83, *86*
Murray, R. W., 34, *40*
Myhrman, A., 50, 52, *72*

Nagano, K., 201, *211*
Nakane, R. K., 163, *176*
Nakao, K., 201, *211*
Nakao, M., 201, *211*
Naruse, H., 168, *177*
Nealson, K. H., 78, 79, *85*, *86*
Neary, M. P., 6, *9*, 35, *41*, 135, 136, 137, 142, *150*, 217, 229, *254*
Nederbragt, C. W., 136, *150*
Neufeld, H. A., 139, 142, *150*
Newman, G. R., 109, *128*
Nibley, D. A., 79, *86*, 216, *253*, *254*
Nicholas, D. J. D., 62, 64, *66*, 99, *132*
Nicholson, C., 105, 106, *130*
Nicol, J. A. C., 91, 94, *128*, *130*
Nicoli, M. Z., 77, 78, 79, *85*, *86*
Nielsen, R., 50, *72*
Nieman, T. A., 25, *39*, *41*
Nilsson, L., 55, 57, 58, 59, *66*, *69*, *72*, *73*, 194, 195, 196, 197, 198, 199, *207*, *209*, *210*, *211*, *212*
Norlander, R., 51, 64, *71*
Nungester, W. J., 59, *72*
Nyquist, O., 51, 64, *71*

O'Connor, M. J., 97, 98, *129*
Ogryzlo, E. A., 36, *40*
Ohlbaum, D. J., 205, *211*
Okawa, S., 158, *175*
Oleniacz, W. S., 139, *150*, 207, *211*
Olsson, P., 59, *70*
Olsson, T., 168, 169, *176*
Oshino, R., 83, *85*
Oyamburo, G. M., 142, 143, *150*

Pagano, R. E., 108, *130*

Panceri, P., 91, 95, *130*
Papahadjopoulos, D., 109, 110, *130, 132*
Paradise, L. J., 59, *72*
Parsons, S. M., 64, *70*
Pasternak, C. A., 111, *129*
Patel, A., *127*
Patterson, J. W., 55, *67*
Pattni, R. D. P., 115, *125*
Pax, P. R., 202, *213*
Pazzagli, M., 161, 167, *175*
Pearson, A. E., 36, *40*
Pecht, M., 182, *187*
Peng, C. T., 1, *9*
Pequeux, A., 51, *68*
Percival, A., 193, *211*
Peskar, B. A., 205, *213*
Petersson, B., 54, 56, *74*
Petsch, W., 5, *8*, 136, *149*
Pettersson, C., 205, *212*
Phelps, G., 63, *67*
Phillips, R. C., 169, *174*
Picciolo, G. L., 57, 59, *70*, 79, *86*, 191, 192, 193, 194, *208, 211, 212*, 216, 236, *253, 254*
Pijck, J., 200, *209*
Pisano, M. A., 139, *150*, 207, *211*
Pollack, H., 97, *130*
Porath, J., 180, *188*
Portzehl, H., 117, *130*
Poste, G., 109, *130*
Pratt, J. J., 161, *176*
Prego, C. E., 142, 143, *150*
Prendergast, F. G., 91, 105, 115, 116, 121, *124, 126*
Prichard, P. M., 136, 143, 144, *149*
Prodanov, E., 142, 143, *150*
Puget, K., 79, *87*, 147, *150*, 158, 161, 166, 168, *176*

Radziszewski, B., 4, *9*
Rapaport, E., 27, *41*
Rasmussen, A., 97, *131*
Rasmussen, H., 50, *72*, 97, *131*
Rauhut, M. M., 22, 23, 24, 25, 27, 28, 30, 32, 34, 35, 36, *38, 40*
Reed, C. F., 201, 202, *209*
Reeves, D. S., 199, *211*
Reimann, L. V., 37, *37*
Reynolds, G. T., 95, 107, 119, 121, *129, 130, 131, 132*
Richardson, A., *127*
Rickardsson, A., 44, 45, 51, 61, 62, 64, *71*
Ridgway, E. B., 90, 98, 103, 106, 107, 108, 115, 116, 117, 119, 121, *125, 128, 129, 131*
Riemann, B., 49, 55, *72*
Rietschel, E. T., 205, *213*
Riley, W. H., 76, *86*
Robb, H. J., 63, *69*, 206, *210*
Roberts, B. G., 28, 30, *39, 40*
Roosens, H., 80, *87*, 235, *254*
Rose, B., 107, 119, 121, *131*
Rosenfeld, M. H., 139, *150*, 207, *211*
Rosengren, A., 59, *70*
Rosewell, D. F., 17, *20*, 22, 30, 31, *38, 40*
Rubenstein, K. E., 155, 169, *176, 177*
Rudel, R., 105, 106, 122, *126, 131, 133*
Ruegg, J. C., 117, *130*
Russell, F. S., 92, *131*
Ryall, M. E. T., 30, *41, 127*, 161, 166, *176, 177*
Ryman, B. E., 108, *133*

Saar, Y., 155, *175*
Sagone, A. L., Jr., 205, *211*
Saiga, Y., 92, 101, *132*
Saive, J. J., 51, *68*
Sander, U., 32, *41*
Schaap, A. P., 23, *42*

Schapira, A. H. V., 122, *125*
Scharpe, S. L., 155, *176*
Schatten, G., 107, *132*
Schmidt, S. P., 22, 27, 33, 34, *40*
Schneider, R. S., 169, 174, *176*
Schneider, W. G., 37, *38*
Schram, E., 22, *38*, 54, *68*, 78, 79, 80, *85*, *87*, 142, 143, *149*, 235, *254*
Schroeder, H. R., 31, *40*, 44, *67*, 82, 83, *87*, 147, *150*, 158, 159, 160, 163, 167, 172, *175*, *176*
Schrot, J. R., 55, *70*
Schuler, P., 180, *188*
Schupper, H., 168, *177*, 207, *212*
Schuster, G. B., 22, 27, 33, 34, *40*
Schwarzberg, M., 155, *177*
Schweitzer, D., 37, *39*
Scoseria, J. L., 25, *41*, 142, 143, *151*
Scott, G., 30, *40*
Searle, N. D., 207, *211*
Seitz, W. R., 6, *9*, 30, 35, *40*, *41*, 135, 136, 137, 138, 139, 140, 141, 142, 145, 148, *149*, *150*, *151*, 217, 229, 235, *254*
Seliger, H. H., 17, *20*, 27, *39*, *41*, 45, 49, *72*, *73*, 93, *133*, 234, *254*
Semsel, A. M., 23, 28, *40*
Sery, T., 168, *177*, 207, *209*, *212*
Setkowsky, C. A., 203, *210*
Shapiro, B. M., 121, *128*
Sharrow, S. O., 109, *126*
Shaw, E. J., 155, *177*
Sheehan, D., 23, *40*, *41*
Sheehan, T. L., 63, *69*, 139, 142, *150*
Sheinson, R. S., 35, *38*
Shimomura, O., 92, 94, 98, 100, 101, 105, 108, 113, *131*, *132*
Shlevin, H. H., 107, 108, *128*
Siddle, K., 97, *127*, *129*
Silver, I. A., 180, *187*
Silverstein, S. C., 205, *211*
Siman, W., 180, *187*
Simon, B. M., 55, *69*
Simpson, G. A., 27, *37*
Simpson, J. S. A., 30, *41*, 97, 100, *127*, 161, 166, 167, *176*, *177*
Singer, L. A., 25, *39*
Singh, P., 59, *74*
Skog, S., *212*
Slayman, C. L., 59, *73*
Sleigh, J. D., 194, *211*
Slivin, E., 155, *175*
Smith, D. S., 155, *177*
Smith, J. P., 22, 33, 34, *40*
Snarsky, L., 155, *175*
Snead, J. L., 37, *41*
Snoke, E. O., 160, *176*
Sobel, S. E., 48, 64, *74*
Sogah, D., 25, *39*
Soini, E., 155, *177*
Sollot, G. P., 37, *41*
Sonnenfled, V., 44, *72*
Soto, H., 142, 143, *150*
Sperling, R., 205, *208*
Sprecht, W., 139, *151*
Spronck, A. C., 51, *68*
Spudich, J. A., 77, *86*
Stachan, C. J. L., 194, *211*
Stanbridge, B. R., 183, *187*
Stanley, P. E., 6, *9*, 63, *73*, 77, 78, 80, 81, 83, *87*, 99, *132*, 159, 170, *177*, 216, 227, *253*, *254*
Statham, M., 55, *73*
Staudinger, H., 205, *213*
Stauff, J., 32, *41*
Stav, L., 6, *8*, 81, 83, 84, *86*, 183, 184, *187*
Steele, R. H., 205, *207*
Steinhardt, R., 107, *132*
Stendahl, O., 109, *132*
Stimson, A., 225, 226, *254*
Stinnakre, J., 105, 106, 122, *130*, *132*

Stjernholm, R. L., 205, *207*
Stollar, B. D., 82, 84, *85*
Storm, E., 59, *69*, 202, *210*
Strada, S., 64, *74*
Strange, R. E., 191, *212*
Strangert, K., 194, 207, *209*
Strecker, R. A., 37, *41*
Strehler, B. L., 4, *9*, 44, 48, 59, 62, 63, 64, 65, *73*
Strive, W. S., 35, *41*
Stryer, L., 100, *132*
Sturner, K. H., 202, 203, *212*
Stymne, H., 54, 56, *74*
Styrelius, I., 50, 51, 64, *71*
Sufwat, F., 109, *124*
Summers, R., 59, *69*
Sutherland, E. W., 62, *69*
Suydam, W. W., 137, 138, 139, *150*
Suzuki, N., 45, *68*
Syers, J. K., 55, *70*
Szoka, F., 110, *132*

Tagesson, C., 109, *132*
Takashi, T., 107, *128*
Tampsion, W., 109, *124*
Tannenbaum, S. R., 180, *188*
Tarkkanen, P., 202, 203, *212*
Tauc, L., 105, 106, 122, *132*
Taylor, D. L., 107, 119, 121, *132*
Taylor, S. R., 105, 106, 107, 108, 122, *126*, *128*, *131*, *133*
Tegner, L., 81, 83, *85*
Tennant, G. B., 200, *212*
Ternynck, T., 166, *177*
Thibault, G., 58, *67*
Thomas, D., 180, *186*
Thomas, R. P., 216, 236, *253*
Thome-Lentz, G., 51, *68*
Thore, A., 44, 45, 46, 49, 50, 51, 52, 53, 54, 55, 56, 58, 59, 61, 62, 64, 65, *66*, *69*, *71*, *72*, *73*, 139, *149*, 168, 169, *175*, *176*, 194, 195, 196, 197, 198, 199, *207*, *209*, *210*, *211*, *212*
Thorpe, G. H. G., 30, *41*, 54, *74*
Thrush, B. A., 36, *41*
Tillotson, D., 105, 107, 116, 117, *128*
Tokel, N., 34, *37*
Tokel-Takyoryan, N. E., 34, *39*
Topping, R. M., 26, 27, *40*, 136, 140, 143, *149*, *150*
Totter, J. R., 25, *41*, 44, 48, 59, 64, *73*, 142, 143, *151*
Towner, K. D., 139, 142, *150*
Townshend, A., 32, *38*
Tregear, G. W., 180, *187*
Tribukait, B., *212*
Trinder, P., 144, *151*
Trush, M. A., 121, *133*
Tsuji, A., 168, *174*, *177*
Tuovinen, O. H., 99, *132*
Turro, N. J., 23, *39*
Tutt, D. E., 26, 27, *40*, 136, 140, 143, *149*, *150*
Tuttle, S. A., 191, 192, 194, *211*, *212*
Tyrrell, D. A., 108, *133*

Ugarova, N. N., 183, 185, *188*
Ulitzur, S., 83, 84, *87*
Ullman, E. F., 155, 169, *174*, *175*, *176*, *177*

Vance, R. B., 216, *253*
Van der Horst, A., 136, *150*
Van Duijn, J., 136, *150*
Van Dyke, K., 62, *72*, 121, *133*
Van Esbroeck, H., 235, *254*
Van Leeuwen, M., 95, 102, 105, 116. 122, *126*, *129*
Vann, W. P., 30, *39*, 160, 163, *175*
Van Zee, R. J., 34, *41*
Varl, W. J., 109, *130*
Veazey, R. L., 25, *41*
Velan, B., 168, *177*, 207, *209*, *212*
Vellend, H., 194, *212*

Venge, P., 205, *212*
Verrinder, M., 109, *124*
Villerius, L., 161, *176*
Vogelhut, P. O., 160, 172, *176*
Volsky, D. J., 111, *133*

Wada, T., 201, *211*
Wade, H. E., 191, *212*
Wahlgren, N. G., 64, *73*
Walling, C., 32, *41*
Walters, P., 59, *74*
Wampler, J. E., 216, *253, 254*
Wannlund, J., 161, 166, *177*
Ward, H. H., 93, *133*
Ward, W. W., 100, *133*
Warner, A. F., 106, 123, *125*
Waterworth, P. M., 195, *212*
Watson, B., 180, *187*
Watson, R. A. A., 155, *177*
Weaver, J. C., 180, *188*
Webb, L. S., 205, *212*
Weed, R. I., 201, 202, *209*
Weidemann, M. J., 205, *213*
Weinstein, J. N., 108, 109, *126, 130*
Weinstein, L., 194, *212*
Weiss, B., 63, 64, *68, 74*
Weiss, H. J., 203, *210*
Weitlaner, P., 4, *9*
Wekerle, H., 205, *213*
Westwood, S. A., 112, *133*
Wettermark, G., 22, 30, 36, 37, *39,* 54, 56, 64, *68, 74,* 78, 81, 83, *85, 87,* 137, 139, *149*
Wexler, S., 34, *38*
White, E. H., 5, *9,* 17, *20,* 22, 27, 30, 31, *38, 39, 40, 41,* 49, *72,* 136, *151*
Whitehead, T. P., 30, *41,* 54, *74,* 227, *253*
Whitlock, D. G., 105, *125*
Whitman, R. H., 28, *40*
Wide, L., 180, *188*
Wilding, P., 166, *175,* 181, *187*
Wilhelm, S. A., 30, *39,* 160, 163, *175*
Williams, D. C., 6, *9,* 136, 140, 141, 142, *151*
Williams, F. W., 35, 36, *38, 41*
Williams, J. D. H., 55, *70*
Wilson, M. E., 121, *133*
Wilson, T., 23, *41, 42*
Wimpenny, J. W. T., 59, *67,* 191, 206, *208*
Windisch, R. M., 202, *213*
Wing, S. S., 79, *86*
Wisdom, G. B., 155, *177*
Witteveen, S. A., 48, 64, *74*
Wladimiroff, W., 64, *68*
Woldring, M. G., 161, *176*
Wolf, P. L., 59, *74*
Woodhead, J. S., 30, *41,* 100, *127,* 161, 166, 167, *176, 177*
Wormald, P. J., 193, *211*
Wrogemann, K., 205, *213*

Yalow, R. S., 154, 157, *178*
Yany, F., 27, *37*
Yates, D. W., 112, *133*
Yeager, F. M., 31, *40,* 147, *150,* 158, 160, *176*
Yekta, A., 23, *39*
Yeung, K-K., 44, *67,* 155, 171, 172, *175*

Zabinski, M. P., 59, *68*
Zaborsky, O. R., 180, 182, *188*
Zafiriou, O. C., 17, *20,* 136, *151*
Zaklika, K., 24, *39*
Zucker, R., 107, *132*

SUBJECT INDEX

Absorptiometry, 6
Acanthoptilum gracile, 101
Acetone, 24, 142
p-Acetylphenyl-10-methyl-acridinium phenyl carboxylate fluorosulfonate, 26
Acridine derivatives, 25-27
Acridinium phenyl carboxylates, 26, 140
Activated sludge processes, 57
Activation assays, 136-139
Acylhydrazides, 30-31
  noncyclic mono, 31
Adenine, 205
Adenosine, 205
Adenosine 5'-diphosphate, 206, 207
  assay, 61, 62, 194
  ion-exchange chromatography, 51
Adenosine 5'-monophosphate
  assay, 61, 62
Adenosine phosphosulfate, 62, 99
Adenosine phosphosulfate kinase, 99
Adenosine tetraphosphate 62
Adenosine 5'-triphosphatase (ATPase)
[Adenosine 5'-triphosphatase]
  assay, 185
Adenosine 5'-triphosphate, 4
  ADP ratio, 205, 206
  assay, 184, 185
    advantages, 47
    analytical interferences, 50
    blood plasma, 209-210
    detection limit, 6, 52, 54
    end-point, 48
    kinetic, 47
    leukocytes, 125
    linear range, 6, 52-53
  binding to mitochondria, 195
  cellular levels, 57, 191
  degrading enzymes, 46, 50
  detection limit, 6, 52, 54
  extraction, 193, 201, 206, 207
  in blood, 60
  in cells, 191-192
  label, 170, 171-173
  levels
    absolute, 192
    bacterial species, 192
    magnesium complex, 44
Adenylate energy charge, 194, 207, 210
Adenylate kinase, assay, 64

Adipocyte, 109
  membrane, 112
Aequorea, 94, 100
Aequorea aequorea
  collection, 100
  isolated cells, 95
Aequorea forskalea, 90, 101
  calcium and, 92
Aequorin, 92, 93, 98
  extraction 102-103
  purification, 102-103
Aerobiosis, 210
Aesculin, 4
  structure, 5
Agar diffusion assays, 190, 203
Air pollution, monitoring, 34, 35
Alazarin sulfonate, 97
Albumin, 143
Alcohol (*see* Ethanol)
Alcohol dehydrogenase, 171, 172
  assay 81, 83, 184
  immobilization, 84
Aldehyde dehydrogenase, 78, 79
Aldehydes, bacterial luciferase and, 78
Alkali metal vapors, 35
ALL
  Autolight 101, 222, 229
  Monolight 201, 222, 242
  Monolight 301, 221, 222
  Monolight 401, 221, 223
Alpkem Automated Luminescence Analyzer, 222, 229, 241
Aminco Chem-Glow Photometer, 80, 221, 222
Amino acids, 139
2-Amino-4,6-dichloro-triazine, 166
5-Amino-2,3-dihydrophthalazine-1,4-dione (*see*Luminol)
6-Amino-2,3,dihydrophthalazine-1,4-dione (*see* Isoluminol)
Aminoglycosides determination, 57
3-Aminophthalate, 17
  doubly excited state, 17
  fluorescence yield, 17
  singlet excited state, 17
Ampicillin, 198, 201
Anaerobiosis, 210
Androsterone, assay, 81, 83
Anthozoa, 91, 92, 101
Anthracene crystals, 37
Antibiotics
  assay, 55, 58, 199-203
  quantitation, 56
  susceptibility assay, 56, 58
Antigen-antibody
  binding, 100
  energy transfer, 100
Antimicrobial susceptibility testing, 197-199
Aplysia, 105, 106
Apple computer, 235, 236, 237, 240
Apyrase, 195
  assay, 64
Argon, 35
Aryl sulfatase, 99
Ascorbic Acid, 143
Asparagine, 4
ATPase, 185
ATP kinase, 64
ATP:NMN adenyl transferase, assay, 82, 84
ATP phosphoribosyl transferase, 64
ATP sulfurylase, 64, 99
Atomic absorption, 155
Attola, 92, 94, 101
Automation, 239-241
Autooxidation reactions, 32
Avidin, 167
  binding assay, 82
2,2'-Azino-di(3-ethylbenzthiazoline-6-sulfonate), 145

Bacillus subtilis, 57, 59
Bacteria, 4 (*see also* specific species)
Bacterial
  adherence, 192-195
  contamination, 139
  endotoxin shock, 210
  porphyrins, 211
Bacteriological dip-slides, 197
Bacteriophage, 155

Bacteriuria screening test, 55, 195-197
Balanoglossus, 144
Balanus nubilus, 103
Beneckea harveyi:
  growth, 78
  oxidoreductases, 76
Benzene, isomeric derivatives, 139
p-Benzoquinone, 166
Benzoyl peroxide, 32
Benzyl luciferin, 100
Benzyl luciferyl disulfate, 99
Benzyl luciferyl sulfate, 99
Beroe spp., 101
Berovin, 93
Berthold Laboratorium:
  Model LB 503, 228, 256
  Model LB 9500, 223, 242
  Model LB 9505, 223, 229, 230, 243
Bioluminescence, definition, 2
Bioluminescent ATP monitoring, 45
  requirements, 46
Biomass, 54-55, 56, 210
  activated sludge, 55
  beer, 55
  milk, 55
  sediment, 55
  water, 55
Biotin
  binding assay, 160, 167
  conjugates, 82, 164, 167, 170, 171, 172
Bis[2,4,6-trichloro-6-carbopentoxylphenyl)oxalate:
  oxidation, 30
  peroxide assay, 30
Bis(2,4,6-trichlorophenyl)oxalate, 27, 28
  quantum yield, 28
  rubrene and, 27
Blast cells, 207
Blood cells, assay, 205-209
Blood stains, 139
Blue-fluorescent protein, 92
Boyle, 3
Cadmium ion, 49
Calcium ion (free)
  intracellular, 98
  release, 98
  tissue, 98
Calcium ionophore, 207
Calcium silicide, 36
Calmodulin, 97
Candida albicans, 56
Carbazole, free radicals, 37
Carbenes, 32
Carbodiimide, 166, 181
Carbonate radical anion, 32
Catalase, 147
  assay, 139, 142
Cell
  -cell communication, 97
  fertilization, 97
  microinjection technique, 105-111
  viability, 98, 207
Cellophane, 183, 185
Cellular energy status, 210
Cellulose acetate tubes, 115
Cephaloridine, 201
  serum, 202
Ceric sulfate, 36
Chelex 100 resin, 115
Chemically initiated electron exchange reaction, 27
Chemiluminescence
  defintion, 2, 22
  electron transfer, 33-34
  gas phase, 34-36
  liquid phase, 23-34
  luminol-enhanced, 207
  peroxidase-mediated, 143-148
  peroxide decomposition
  quantum mechanics, 11-20
  quantum yield, 17, 23
  singlet oxygen, 34
  solid phase, 36-37
Chemiluminescent reaction
  activation energy, 12
  classification, 23
  enthalpy, 12

[Chemiluminescent reaction]
entropy, 12
mechanisms, 17, 22
quantum yield, 17, 23
Chironomus, 107, 119
Chloroplasts, 65
Cholesterol, 109
assay, 7
Cholesterol oxidase, 7
Chromic acid, 36
Chromium (III)
assay, 137, 138, 142
EDTA complex, 137
lophine, 25
Clam, 3
Clytia spp., 101
Clytin, 93
Cnidaria, 90, 91, 93
Cobalt (II), catalyst, 139
Cobalt (III), catalyst, 139
Coelenteramide, 92
Coelenterates, 91
luminescence, 90-97
regulation, 94
stimulation, 94
potassium and, 94
Coelenterazine, 92, 98
Coenzyme A, assay, 63
Cofactor, 155
Colony forming unit, 56
Concanavalin A, 207
Conjugates
ATP, 161, 166, 170, 171-173
bacterial luciferase, 161, 166
cyclic AMP, 167
cytochromes, 169
firefly luciferase, 161, 166
glucose oxidase, 167, 168
glucose-6-phosphate dehydrogenase, 169
isoluminol, 160, 161, 162, 167
lucigenin, 163
luminol, 160, 161, 162, 163
NAD, 82, 170, 171, 172
peroxidase, 167, 168
pyruvate kinase, 168
Controlled pore ceramics, 183
alumina, 183
silica, 183
titania, 183
Coordination complex, 155
Copper (II) catalyst, 30, 139, 161
Cortisol, immunoassay, 168
Coumarin derivatives, 32
Coupling reagents, bifunctional, 166, 181
Creatine kinase
assay, 50, 64
B subunit, 61
immunoinhibition, 61
in myocardial infarction, 61
pH optimum, 51
Creatine phosphate, 51
assay, 63
Ctenophora, 90, 91, 93, 101
Cyclic aminophthaloyl peroxide, 17
potential energy surface, 18
Cyclic AMP, 99
assay, 61, 62, 167
label, 167
Cyclic AMP phosphodiesterase, assay, 64
Cyclic GMP, assay, 63
Cyclic GMP phosphodiesterase, assay, 64
Cyclic hydrazides, 30-31
Cytocholasin B, 110
Cytochrome c, 139, 142
Cytochromes, labels, 169

D600, calcium channel blockers, 95
Dark noise, 229
DEAE-cellulose, oxidoreductase purification, 79
DeAnima, 3
Decanal, 77
Dental plaque, 193-195
Degenerate states, 16
Deoxycyline, 198-201
Detector, 229-230
chamber, 230-231

[Detector]
reflective surface, 231
quantum efficiency, 229
spectral range, 230
Deuterohemin, 147
Diaphorase, 155
Diatomic molecule, 13
dissociation energy, 13
energy, 13
ground state, 13
potential energy curve, 14
Diazaquinones, 30
Di-π-cyclopentadienyl ion, 155
Digoxin, 156, 160, 163
Dimethylbisacridinium nitrate (*see* Lucigenin)
Dimethyl-3-nitro-6-vinylphthalate, 31
Dimethyl sulfoxide, 30
N(2,4-Dinitrophenyl)-β alanine
assay, 171, 173
detection limit, 172, 173
2,4-Dinitrophenyl residue, 171, 172
antibody, 82
Dinoflagellates, 56
1,2-Dioxetane
decomposition, 24
intermediates, 24, 30
1,2-Dioxetanes, 23-27, 30, 35
singlet excited state, 23
triplet excited state, 23
1,2-Dioxtanediones, 27-30, 140
fluorescers and, 27
1,2-Dioxetanones, 27
9,10-Diphenylanthracene, 33, 34
1,3-Diphosphoglycerate, assay, 63
Discharge
electrical, 35
microwave, 35
DNA, 207
synthesis, 202
Dubois, 3,4,92
Dupont Biometer Model 760, 223, 229
E. coli, 56, 57, 59, 196, 200
EDTA, 53
Ehrlich ascites cells, 59, 210
Electrochemiluminescence, 2
Electrogenerated chemiluminescence, 34
Electroluminescence, 34, 36
Electron transfer, 23, 30, 33-34
Electrophoretic injection, 105
EMIT, luminescent, 156, 169-170
Endocytosis, 110
Energy transfer, 100
Enthalphy, 12
Entrapment, 180
Entropy, 12
Enzyme electrode 180
Enzyme immunoassay, 158
luminescent, 156, 167-169
Enzyme multiplied immunoassay technique (*see* EMIT)
Enzyme-protein conjugates (*see* Conjugates)
Epidermal cells, 59
Erythrocyte, 58, 59, 155, 205
chromium labeled, 205
ghosts, 109, 110, 112
immunolysis, 58
label, 155
post-transplant viability, 205
Ethanol, assay, 6, 83, 84, 185
Ethylene, 35
Ethylene imine, 172
Euphysa, 92
Excited states, 2, 12, 13, 22
life times, 15
singlet, 15
vibrational, 15

Fermentation, 57
Ferricyanide catalyst, 30
Ferritin, 139, 142
Ferrocene, 155
Fiber-optic probe, 145, 148
Fibroblasts, embryonic lung, 56
Fimbriae, 192, 193
Firefly, 3 (*see* Luciferase, firefly)

Flavin mononucleotide
  assay, 81, 83
  detection limit, 81
Flavin mononucleotide, reduced
  assay, 184
  half-life, 77
Flow cell, 137, 225
  design, 138, 239
Flow monitoring, 228
Flow monitors, 224
Flow systems, 137, 140
Fluorescein, label, 155
Fluorescence, 2, 12, 19, 22
  acceptors, 24
  excitation transfer, 155
  lighting, 225
  polarisation, 155
  quantum yield 19, 23
Fluorodinitrobenzene, 82, 165, 170
Folic acid, assay, 203, 204
Formaldehyde, 34
Forskål, 89, 90
Free energy, 12
Free radicals, 32
Friday Habor, 100
Fusogens, 109, 110

β-Galactosidase
  conjugates, 158
  fluorometric assay, 158
Gas chromatograph, 35
Gentamicin, 198, 201
  serum, 57, 58, 156, 199, 200, 202
Giant amoeba, 107
Giant axon, squid, 106
Giant muscle fiber
  barnacle, 106
  crab, 106
Gibbs-Duhem equation, 12
Glass beads, enzyme immobilization, 183, 186
Glass rods, enzyme immobilization, 82, 186
Globulin, 143
Glomerular disease, 209
Glowworm, 3
Glucose
  assay, 6, 63, 81, 83, 84, 142, 180, 185
    csf, 142
    interferences, 14
    plasma, 142-143
    serum, 140, 142
    urine, 140, 142
Glucose oxidase, 144
  immobilized, 140
  label, 167, 168
Glucose-6-phosphate, assay, 83, 84, 185
Glucose-6-phosphate dehydrogenase
  assay, 81, 84, 184
  immobilized, 84
  label, 170
Glutaraldehyde, 166
Glycerol, assay, 63
L-Glycerol-3-phosphate, assay, 81, 83
Glycogen, 97
Glycolysis, 205
Glycolytic pathway, 210
Gram negative organism, 199
Granulocytes, 209
Granulomateous disease, chronic, 209
Green fluorescent protein, 100, 101
Grignard reagents, 31-32
  alkyl, 32
aryl, 31, 32
Ground state, 12, 13
Growth index, 57
Guanosine 5'-diphosphate, assay, 62
Guanosine 5'-monophosphate, assay, 62
Guanosine 5'-triphosphate, assay, 62
Gymnoblastic hydroids, 92

Halistaurin, 93
Harvey, 3, 5
HeLa cells, 54, 56
Hemagglutination inhibition, 155
Hematin, 147
  detection limit, 139, 142
Heme, 139

[Heme]
labels, 169
Hemin, 30, 139, 160
Hemoglobin, 139, 142
Hemolytic disease, acquired, 202
Hepatitis B surface antigen, 156
Hepatocyte, membrane, 112
Hexokinase
assay, 64, 84, 185
deficiency disease, 206
immobilization, 84
Hippopodius, 101
Horse-chestnut, 4
Horseradish peroxidase (*see* Peroxidase)
Humus, 4
Hydrazine, 31
Hydrogen-air flames, 35, 36
Hydrogen peroxide, 136
detection limit, 6, 140, 142, 145, 146
enzymatic production, 140, 143
Hydrogen sulfide, 35
Hydroids, 91, 94
Hydroperoxides, 34
intermediates, 24
3-Hydroxybutyrate, assay, 81, 83
D-3-hydroxybutyrate dehydrogenase
assay, 81, 84
in pancreatic islets, 82
Hydroxyl radical, 32
Hydroxysteroid dehydrogenase, 81, 186
Hydrozoan medusa, 94
Hypersurface, 13
Hypochlorite, 30, 32, 34
Hypoxanthine, 142, 143

Image intensification, 119
Imidazole acetate, 52
Imidazoles, 24
Immobilization procedures
chemical, 181
physical, 180
Immobilized enzymes, 82, 84
[Immobilized enzymes]
luciferase, 183, 184, 185
oxidoreductase, 183, 184, 185
Immune complexes, 209
Immunoassay
heterogeneous, 154, 166
homogeneous, 100, 155, 156, 159, 166
labels, 155
luminescent
advantages/disadvantages, 157-159
alpha-fetoprotein, 161
anti-human serum albumin, 162
bioton, 160
cortisol, 162
digoxin, 160
2,4-DNB, 172
2,4-DNB-β-alanine, 172
IgG, 160, 161, 162
insulin, 160, 162
methotrexate, 161
parathyroid hormone, 161
progesterone, 161
sensitivity, 157, 159
Serratia macescans, 207
staphylococcal enterotoxin B, 162
Sindbis virus, 162
T_4, 160
testosterone, 161
Immunocytochemical assay, 166
Immunoelectrode, 180
Immunoglobulin G, 160, 161, 168
cell surface, 161, 168
Infusoria cells, 54, 56
Inhibition assays, 136-139
Inhibition zone, 190
reading system 203
Inosine, 205
Insecticides, 139
Instrumentation, 79-80
Insulin, 160, 163, 166, 168
Internal standard, 80, 233, 234
Intracellular free calcium, 97
method comparison, 123
quantification, 116-119

Iodine, 30, 154
Ion-exchange chromatography, 137
Iron (II), 32
Islet cell membrane, rat, 112
Isoluminol, 31
  derivatives, 162
  label, 159, 160, 161, 163, 167
  quantum efficiency, 31

Jellyfish, 90
Jobin-Yvon Photometre Pico-ATP, 221, 223
Jones reductor column, 139
Klebsiella, 196, 200

Labels
  ATP, 170, 171-173
  bacterial luciferase, 161, 166
  cyclic AMP, 167
  cytochromes, 169
  erythrocyte, 155
  ferrocene, 155
  firefly luciferase, 161, 166
  fluorescent, 155
  glucose oxidase, 167, 168
  glucose-6-phosphate dehydrogenase, 169
  heme, 169
  isoluminol, 160, 161, 162, 163, 167
  lucigenin, 163
  luminol, 160, 161, 162, 163
  NAD, 82, 170, 171, 172
  organometallic, 155
  peroxidase, 146, 155, 167, 168
  pyruvate kinase, 168
  radioactive, 146, 155
  spin, 155
  T_2, 155
β-Lactamase, 198
Lactate dehydrogenase, assay, 81, 83, 140, 142, 184
Lactobacilli, 196
Lactobacillus casei, 302
Lactoperoxidase, 147, 160
Laminaria, 100, 102
Lanthanum (III), Ca channel blocker, 95
Lasers, chemical, 34, 35
Lauryl aldehyde, 161
Leukartiana octona, 92
Leukocytes, 58
LIETS, 186
Life detection, 44, 54
Light
  emitting diodes, 238
  flash height, 235
  integration, 235-236
Lipase, assay, 84
Liposomes, 108, 109
Liquid scintillation counter, 220, 231
  light measurement, 80
Lithium
  diphenyl phosphide, 32
  organophosphides, 32, 37
LKB Luminometer 1250, 217, 220, 240
Loligo pealii, 103
Lophine, 4, 24
  oxidation, 25
  structure, 5
Luciferase, bacterial, 76, 161
  assay, 76
  contaminants, 78
  immobilization, 84, 183
  inhibitors, 77
  label, 161, 166
  purification, 79
  reaction
    sequence, 76
    time course, 76
  stability, 79
Luciferase, firefly, 3, 4, 161
  analytical interference, 53-54
  applications, 54-65
  buffers, 51-52
  competitive inhibitors, 45, 52
  coupled enzyme systems, 48
  effect of anions, 51
  effect of cations, 49
effect of ionic strength, 51

[Luciferase, firefly]
immobilized, 183
inhibitors, 44
kinetics, 46
label, 161, 166, 170, 171-173
pH and, 49
pH optimum, 51
product complexes, 45
product inhibition, 45
purification, 50
quantum yield, 45
SDS gel electrophoresis, 50
temperature, 49, 50
Luciferin, firefly, activation, 44
Luciferin, *Pholas*, peroxidase and, 147, 169
Luciférine, 3
Luciferin sulfokinase, 99
Luciferyl sufate, 99
Lucigenin, 4, 5, 136
NADH assay, 25
oxidation, 25, 26
peroxidase and, 144
quantum yield, 25
structure, 5, 26
Lumac
Biocounter 2000, 225, 245
Biocounter 2010, 224, 245
Celltester 1030, 225, 247
Celltester 1060, 225
Lumacounter 2080, 224, 244
Lumatec 1010, 220, 225, 247
Luminometer 1070, 220, 225, 246
Luminescence
arylmagnesium halides, 31-32
asparagine, 4
classification, 2
definition, 1-2
energetics, 12-19
enhancement, 167
historic aspects, 2-5
humus, 4
quantum efficiency, 6
reagents, 50-51
stability, 7
toxicity, 7
[Luminescence]
sensitised, 167
sensitivity, 6
terminology, 1-2
time course, 7
uric acid, 4
urine, 4
Luminescent assays
adenylate kinase, 64
ADP, 61, 62
AENADH-biotin, 83
alcohol (*see* Ethanol)
alcohol dehydrogenase, 81, 83, 184
AMP, 61, 62
androsterone, 81, 83
apyrase, 64
ATP, 6, 184, 185
ATPase, 185
B_{12}, 139, 142
calcium, 98
calibration curves, 80
cholesterol, 7
chromium (III), 137, 138, 142
coenzyme A, 63
cortisol, 168
cost, 7
creatine kinase, 50, 61, 64
creatine phosphate, 63
cyclic AMP, 61, 62, 167
cyclic AMP phosphodiesterase, 64
cyclic GMP, 63
cyclic GMP phosphodiesterase, 64
N(2,4-dinitrophenyl)-β-alanine, 172, 173
1,3-diphosphoglycerate, 63
ethanol, 6, 83, 84, 185
ferritin, 139, 142
FMN, 81, 83
$FMNH_2$, 184
folic acid, 200
glucose, 6, 63, 81, 83, 84, 142, 180, 185
glucose-6-phosphate, 83, 84, 185
glucose-6-phosphate dehydrogenase, 81, 84, 184
glycerol, 63

[Luminescent assays]
L-glycerol-3-phosphate, 81, 83
guanosine 5'-diphosphate, 62
guanosine 5'-monophosphate, 62
guanosine 5'-triphosphate, 62
hematin, 142
hemoglobin, 142
hexokinase, 64, 84, 185
hydrogen peroxide 6, 142, 145, 146
3-hydroxybutyrate, 81, 83
D-3-hydroxybutyrate dehydrogenase, 81, 84
hypoxanthine, 142, 143
internal standards, 80
lactate dehydrogenase, 81, 83, 140, 142, 184
life, 44, 54
lipase, 84
luciferase, bacterial, 76
malate, 81, 83
malate dehydrogenase, 81, 184
methotrexate, 161
mixing efficiency, 7
myoglobin, 139, 142
NAD, 81, 83
NADH, 6, 80, 83, 141, 142, 184
NADPH, 80, 83, 184
nucleotide phosphokinase, 64
oxaloacetate, 81, 83
oxygen, 83
permanganate, 25
peroxidase, 143
3',5'-phosphoadenosine phosphase, 97, 98, 99
3',5'-phosphoadenosine phosphosulfate, 97, 98. 99
phosphoenol pyruvate, 62
progesterone, 161, 167
protease, 82
pyruvate, 81
pyruvate kinase, 64, 185
specificity, 7
speed, 7
testosterone, 81, 83, 186
thyroxine, 163
TNT, 81, 83

[Luminescent assays]
triglycerides, 63
uric acid, 143
uridine 5'-triphosphate, 62
vitamins, 7, 56, 58, 200-201
Luminescent cofactor immunoassay, 156-157
Luminescent enzyme immunoassay, 156
Luminescent enzyme multiplied immunoassay technique, 156, 169-170
Luminescent reagents
stability, 7
toxicity, 7
Luminol, 4, 5, 30, 136
aprotic solvent and, 30
catalysts, 30
conjugates, 30, 160, 161, 162, 163
coupling, 30
derivatives, 162
interferences, 7
oligomers, 31
oxidants, 30
polymers, 31
quantum yield, 17, 30
reaction mechanism, 29
structure, 5
substituents, 162
Luminometer
automation 239-241
battery operated, 220
data processing, 139-241
detector, 229-231
microprocessor control, 239-241
mixing, 221
portable, 220, 232
reaction vessel, 221, 229
sample injector, 233-234
sensitivity, 238
temperature control, 232-233
Lumisome, 109, 110
Lymphocytes
ATP levels, 204
mouse, 166
transformation, 207-209

Macaire, 3
Macrophage, 56, 209
Magnesium ion, firefly luciferase and, 4, 44
Malate, assay, 81, 83
Malate dehydrogenase
  assay, 81, 184
  label, 159
Marwell International
  Model 302, 226, 252
  Model 340/50, 226, 249
Medaka, 107, 119
Medusa aequorea, 90
Medusae, 90, 91
Membrane vesicles, 111
  photoprotein containing, 112
  sonicated, 109
Mercury (II) ion, 49
Methotrexate
  conjugate, 166
  immunossay, 161
N-Methylacridone, 25
Methylene blue
  lactate dehydrogenase assay, 140
  NADH assay, 141
Michaelis-Menten equation, 46, 52
Microcomputer
  Apple, 235, 236, 237, 240
  PET, 235, 236, 237, 240
  HP97S, 236, 237, 240, 244
Microenvironment, 182, 183
Microinjection, 103
  pressure, 105, 108
Microperoxidase, 147, 160
Microprocessor, 235
Minimum inhibitory concentration, 198
Mitochondria, 65
Mitoplasts, 65
Mitosis, 207
Mixing efficiencey, 221
Mnemiopsin, 93
Mnemiopsis spp., 101
Molecular
  beams, 34
  orbital
    antibonding, 14, 17
[Molecular]
    bonding, 14
    η, 17
    σ, 17
    π, 17
  vibrations, 15
Monocytes, 209
Murexide, 97
Muscle cells
  amphibian, 106
  cardiac, 107, 108
  human, 107
  rat, 107
  smooth, 105, 107
Myocardial infarction, 61
Myoglobin, 139, 142
Myristic acid, 83

NAD
  assay, 81, 83
  label, 170
NADH
  assay, 6, 80, 83, 141, 142, 184
    detection limit, 6, 77
    linear range, 77
    pH optima, 77
    temperature optimum, 77
NADH:FMN oxidoreductase, 76, 169
  commercial, 78
  immobilization, 84
  purification, 79
  storage, 79
  yield, 79
NADPH, assay, 80, 83, 184
NADPH:FMN oxidoreductase
  commercial, 78
  immobilization, 84
  purification, 79
  storage, 79
  yield, 79
NASA, 221
Nerve gas, 139
Neurospora crassa, 59
Neurotransmitters, 97
New Brunswick
  fermenter, 78

[New Brunswick]
Lumitran L2000, 221, 226
Lumitran L3000, 221, 226
Nicotinamide 6-(2-aminoethyl) purine dinucleotide, 82, 172
Nitric acid, 36
Nitriles, 32
Nitrofurantoin, 198
Nitrogen
afterglow, 36
oxides, 136
trifluoride, 35
Non-radiative transition (*see* Radiationless transition)
Nucleotide phosphokinase, assay, 64
Nucleotide-releasing agent, 195

Obelia, 94, 109
Obelia geniculata, 100, 101, 102
Obelin, 93, 94, 98, 111
extraction, 102-103
purification, 102-103
Organic free radicals, 155
Organometallics, 31-32, 155
Oxalic acid esters, 27
chemiluminescence, 27
lighting systems, 30
Oxaloacetate, assay, 81, 83
Oxidase enzymes, 136
Oxidative phosphorylation, 61, 65
Oxygen, 4
assay, 83
firefly luciferase substrate, 45
singlet, 23
Oxyluciferin, firefly, 44
Ozone, 34, 35, 136

Packard, Pico-Lite, 221, 226, 250
Pancreatic islets, 59
Parathyroid hormone, 161
Pelagia, 94, 95
Pelagia noctiluca, 101
Pelagin, 93
Peltier device, 229
Peptococcus prevotii, 59
Periodate oxidation, 163, 165
Periphylla, 101
Permanagate, 25, 30, 36
Peroxidase, 143, 144, 145, 207
assay, 143
detection limit, 158
labels, 146, 167, 168
Peroxylactones (*see* 1,2-Dioxetanones)
Peroxyl radicals, 32
Peroxyoxalate (*see* 1,2-Dioxetane-diones)
Perylene, 140
PET computer, 235, 236, 237, 240
Phagocytosis, 205, 230
Phenanthrene quinone, 32
Phenytoin, 156, 170
Pholas dactylus, 147, 158
Phosphate, plasma, 205
Phosphatidyl choline, 109
Phosphatidyl serine, 109
3',5'-Phosphoadenosine
phosphate, assay, 97, 98, 99
phosphosulphate, assay, 97, 98, 99
sulfatase, 99
Phosphoenol pyruvate, assay, 62
Phosphorescence, 2, 22, 23, 229
Phosphors, 36
Phosphorus, 34, 35
Photobacterium fisheri, growth, 78
Photocathode, 229, 230
Photochemical reactions, 16
Photodiodes, 229, 234
Photoluminescence, 2
Photographic detection systems, 137
Photomultiplier photometer, discrete sampling, 137
Photomultiplier tube, 137
bialkali, 230
dynamic range, 229
photocathode, 229
bialkali, 230
S4, 230
S11, 230
S20, 230
photosensitive area, 230

[Photomultiplier tube]
sensitivity, 230
spectral response, 49
Photon, energies, 12
Photophosphorylation, 49, 61, 65
Photoproteins, 92
calcium-activated, 92
calcium affinity, 122
introduction into cytoplasm, 109
preparation, 100-103
reactions, 94, 95
release, 121
toxicity, 122
Phytohemagglutinin, 207, 208
Plasmodium, slime mold, 107
Platelets, 59
adenosine nucleotide reservoirs, 202
aggregation, 58
ATP levels, 202
ATP release, 58
viability, 206-207
Plymouth Sound, 100
Polyacrylamide gel, 145
Polystyrene tube test, 192-193
Polyacrylic hydrazide, 184
Polyethlene glycol 6000, 109, 110
Polymorphonuclear granulocytes, 56
Porphyrins, 207
Potassium ion, 94
Potential energy hypersurface, 14
Primidone, 156
Pronase, 95
Progesterone
assay, 161, 167
conjugate, 161, 167
Prostaglandins, 7
Protease, assay, 82
Proteus, 196
Proteus mirabilis, 56
Protozoa, 93
Providentia, 196
Pseudomonas, 196
Pyrogallol, 34, 169, 211
chemiluminesence, 4
structure, 5
Pyrophosphatase, inorganic, 45
Pyrophosphate, 44, 62
Pyruvate, assay, 81
Pyruvate kinase
assay, 64, 185
deficiency disease, 206
label, 163

Quantum
efficiency, 6
yield, 17, 22

Radiationless transition, 16, 17, 19, 22
Radioisotopes, 154
Radiolaria, 92, 93
Radioluminescent photon standard, 231
Radiomatic Instruments and Chemical Co., 229, 256
Reaction vessel, 221, 229
Red blood cells, 56
Redox dyes, 190
Renilla, 92
Renilla kollikeri, 101
Renilla mulleri, 101
Renilla reniformis, 101
Renillin, 98, 99, 100
Respiratory burst, 209
Rhodamine, 155
Rifampicin, 201, 202
Rubrene, 27
anion radical, 33
cation radical, 33
electron-transfer chemiluminescence, 33

SAI ATP Photometer
Model 100, 227, 251
Model 2000, 221, 227, 252
Model 3000, 221, 227
Salivary gland cells, 119
Sarcoplasmic reticulum, 112
Schiff base, 163
Scyphozoa, 91, 101

Scyphozoan jellyfish, 94
Sea combs, 91
Sea jellies, 91
Sea pansies, 91
Sea pens, 91
Sea urchin eggs, 107, 119
Selenium, 35
Selenomonas ruminantium, 59
Semipermeable membrane, 180
Sendai virus, 109, 110, 114
Sensitized chemilumnescence, 140, 141
Sepharose, immobilized enzymes, 183, 185
Serratia marcescens, 211
Signal-to-noise ratio, 49
Silicon photodiode, 229, 234
Siloxene
  hydroxy, 37
  oxidation, 36
Sindbis virus, 168
Singlet oxygen, 23
Siphonophores, 91, 95, 101
Skan Bioluminescence Analyzer, 221, 227, 253
Sodium
  atoms, 35
  borohydride, 163
  chloride cool flame, 35
Solid supports, 182
Sonoluminescence, 2
Spacer groups, 158, 182
Specific protein binding reactions, 82, 154-174
Spectrophotometric assay, ATP, 191
Spermatazoa, 58, 59, 206
Splash zone, 90
Squid axon, 106, 115, 117, 122
Staphylococcal enterotoxin B, 168
Staphylococcus aureus, 56
Starfish oocyte, 107
Stomatoca, 92
Storage pool disease, 206, 207
Streptococcus, 196
Stylatula elonga, 101
Sulfite, 32
Sulfer, 32, 35, 36
Sulfur dioxide, 35
Synaptic membrane, 112

T_2, 155
Tes buffer, 115
Testosterone
  assay, 81, 83, 186
  conjugate, 165
Tetra-alkyldioxetanes, 24
Tetra-aminoethylenes, 27
Tetradecanal, 77
Tetra- kis-(dimethylamino)ethylene, 27
Tetramethyl-1,2-dioxetane, 24
Tetramethyloxamide, 27
Tetramethylpiperidinoxyl, 155
Tetramethylurea, 27
Thalassicola, 93, 94
Thalassicolin, 93
Thermal enzyme probes, 180
Thermoluminescence, 2
Thymidine, 207
Thymocytes, 207
Thyroxine, 156, 160, 163
Tissue typing, 209
TNT
  assay, 81, 83
  reductase, 81
Tobramycin, 201, 202
Trace metal analysis, 137
Triatomic molecule, 13
  potential energy surface, 13, 15
Triboluminescence, 2
Trichloracetic acid, 53
Triglycerides, assay, 63
Trinitrotoluene (*see* TNT)
2,4,5-Triphenylimidazole (*see* Lophine)
Triphenyl phosphite, 34
Triplet acceptors, 24
Triton X-100, 58, 113
  lysis of nonbacterial cells, 55, 195
Trypan blue exclusion test, 207

Turbidimetric assay, microbial growth, 190, 198, 199, 203
Turner ATP Photometer, Model 20, 221, 227, 254
Two-step reactions, 23

Urea, 202
Uremia, 202
Urethane, 122
Uric acid, 4, 7, 142, 143, 180
Uridine 5'-triphosphate, assay, 62
Urinary tract infections, 196
Urine, 4, 55
  bacteria in, 55, 195-197

Videotape, 119
Viral membranes, 109
Viscosity, 221, 233
Vision, 97
Vitamins
  assay, 7, 56, 58, 203-205
  B_{12}, 139, 142
Vitatect luminometer
  Model II, 227
  Model IIS, 228, 255
  Model III, 228

Weak chemiluminescence, 32
White blood cells, 56
Woodward's reagent K, 166

Xanthine oxidase, 147

Zinc ion, 49
Zooanthids, 91